HANDBUCH DER EXPERIMENTELLEN PHARMAKOLOGIE

BEGRÜNDET VON A. HEFFTER
FORTGEFÜHRT VON W. HEUBNER

ERGÄNZUNGSWERK

HERAUSGEGEBEN VON

O. EICHLER UND A. FARAH

PROFESSOR DER PHARMAKOLOGIE AN DER UNIVERSITÄT HEIDELBERG

PROFESSOR DER PHARMAKOLOGIE AN DER STATE UNIVERSITY OF NEW YORK

VIERZEHNTER BAND · TEIL 2

THE ADRENOCORTICAL HORMONES

THEIR ORIGIN, CHEMISTRY, PHYSIOLOGY, AND PHARMACOLOGY

PART 2

SUB-EDITORS

HELEN WENDLER DEANE

AND

BETTY L. RUBIN

SPRINGER-VERLAG
BERLIN HEIDELBERG GMBH 1964

THE ADRENOCORTICAL HORMONES
THEIR ORIGIN · CHEMISTRY
PHYSIOLOGY AND PHARMACOLOGY

PART 2

CONTRIBUTORS

M. S. GOLDSTEIN · F. G. HOFMANN · R. LEVINE · E. H. SOBEL

SUB-EDITORS

HELEN WENDLER DEANE
AND
BETTY L. RUBIN

SPRINGER-VERLAG
BERLIN HEIDELBERG GMBH 1964

ISBN 978-3-662-13134-3 ISBN 978-3-662-13132-9 (eBook)
DOI 10.1007/978-3-662-13132-9

Originally published by Springer-Verlag oHG. Berlin · Göttingen · Heidelberg 1964
Softcover reprint of the hardcover 1st edition 1964

Library of Congress Catalog Card Number Agr 25-699

Titel-Nr. 5708

Contents

Authors and addresses

MAURICE S. GOLDSTEIN, M. D., Department of Metabolic and Endocrine Research, Michael Reese Hospital, Chicago 16, Illinois, U. S. A.

FREDERICK G. HOFMANN, Ph. D., Department of Pharmacology, Columbia University, College of Physicians and Surgeons, New York 32, N. Y., U. S. A.

RACHMIEL LEVINE, M. D., Department of Medicine, New York Medical College, New York 29, N. Y., U. S. A.

EDNA H. SOBEL, M. D., Department of Pediatrics, Albert Einstein College of Medicine, Bronx 61, N. Y., U. S. A.

THE ADRENOCORTICAL HORMONES
THEIR ORIGIN · CHEMISTRY
PHYSIOLOGY AND PHARMACOLOGY

Chapter 7

Glucocorticoid action of adrenocortical hormones*

By

RACHMIEL LEVINE and MAURICE S. GOLDSTEIN

Introduction

Classification is by its very nature restrictive, although it may often serve as a useful didactic device. The term "glucocorticoid" gained general acceptance about a decade ago, since it referred to the potent activity exerted by certain of the adrenal steroids on blood sugar level, glycogen storage and carbohydrate turnover. True as these observations were, they have had the unfortunate consequence of suggesting that the myriad effects of this group of compounds, recorded in the literature, are somehow the results of a primary effect on intermediary metabolism, specifically that of carbohydrates.

As we shall see, the "glucocorticoids" exert a very considerable influence, under many circumstances, on the rates of exchange of minerals and water; in other words, they are also "mineralocorticoids." Their effects upon life-saving adaptive responses and upon a host of inflammatory, allergic and hematopoietic disturbances are still in the descriptive stage and have not been related to any known metabolic action.

For purposes of this review *the term "glucocorticoids" will apply to those steroids,* occurring in nature and produced in the laboratory, *which in their range and probable mode of activity resemble cortisol* (hydrocortisone).

In view of the vastness of the literature on the subject, it would be foolhardy indeed to try for a complete or comprehensive approach. Instead the attempt will be made to assess our present state of knowledge of steroid action in many areas, and especially to indicate the gaps in understanding.

In addition to the selected bibliography, the following group of reviews and monographs of the past ten years will aid the reader in finding detailed information on particular aspects. Each reference is briefly characterized as to its content and views. These comments are for the convenience of the reader, and they attempt to fill out the limited information presented in this chapter.

Ciba Colloquia on Endocrinology. These published reports of symposia conducted in London in the 1950s are published in the U.S.A. by Little, Brown and Company, Boston, Massachusetts, and in Great Britain by J. and A. Churchill, London. They constitute a wide variety of presentations by current workers in the field.

Vol. 1: *Steroid Hormones in Tumor Growth and Steroid Hormones and Enzymes* (1952). The effects of adrenocortical steroids on the course of neoplasia and the status of adrenocortical function in the presence of neoplasia were reported. The influence of glucocorticoid administration on a variety of enzyme systems was discussed.

* This chapter was submitted in August 1960; a few selected references were added in July 1962.

Vol. 3: *Hormones, Psychology and Behavior* (1952). A range of interests was covered in respect to the influence of glucocorticoids on behavioral responses, as well as to the glucocorticoid status in various behavioral states and disorders.

Vol. 4: *Anterior Pituitary Secretion and Hormonal Influence in Water Metabolism* (1952). Book II of this volume contains considerable detail about the various relationships of the glucocorticoids to other hormones with respect to their influence on water and electrolyte excretion.

Vol. 5: *Bioassay of Anterior Pituitary and Adrenal Cortical Hormones* (1953). An entire section was devoted to chemical measurement and bioassay of glucocorticoids in blood and urine, with a comparison of these methods.

Vol. 7: *Synthesis and Metabolism of Adrenal Cortical Steroids* (1954). Additional data on the intermediary metabolism and biosynthesis of the glucocorticoids are available in this volume and supplement the review by HECHTER and PINCUS listed below (cf. DORFMAN, Chapter 3, Part 1 of this Handbuch volume).

Vol. 8: *The Human Adrenal Cortex* (1955). A very wide range of articles was presented which extended from studies of the adrenal cortex itself to studies of adrenal function in a variety of human somatic and psychological stressful situations, and clinical conditions.

DEBODO, R. C., and N. ALTSZULER: *Insulin Hypersensitivity and Physiological Insulin Antagonists.* Physiol. Rev. **38:** 389—445 (1958). The subject of this review went beyond glucocorticoids, but the influence and role of these steroids in relation to insulin, other hormones and carbohydrate metabolism was thoroughly handled.

DIRSCHERL, W.: *Über die Einwirkung von Steroidhormonen auf Gewebestoffwechsel und Fermente.* Ergeb. Physiol. **48**: 112—229 (1955). This review was an extensive coverage of the *in vitro* experiences with glucocorticoids.

HECHTER, O., and G. PINCUS: *Genesis of the Adrenocortical Secretion.* Physiol. Rev. **34:** 459—496 (1954). This lucid account of intermediary metabolism and biosynthesis of adrenal steroids and the influence of ACTH on these processes is complemented by Vol. 7 of the Ciba Colloquia referred to above. It reviewed results obtained with isolated perfused adrenals in considerable detail.

Hydrocortisone, Its Newer Analogs and Aldosterone as Therapeutic Agents. Ann. N. Y. Acad. Sci. **61** (Art. 2): 281—636 (1955). Many authors contributed extensive as well as preliminary trials in the clinical application of these steroid agents. Taken with the symposium edited by SOFFER and ORR listed below, these constitute a fairly complete clinical evaluation of therapeutic responses to the steroids and their analogues.

RAMEY, E. R., and M. S. GOLDSTEIN: *The Adrenal Cortex and the Sympathetic Nervous System.* Physiol. Rev. **37:** 155—195 (1957). This review interprets data in the older literature that suggested a major functional interaction between the glucocorticoids and the sympathetic neurohumors.

SAYERS, G.: *Adrenal Cortex in Homeostasis.* Physiol. Rev. **30:** 241—320 (1950). This is one of the most complete reviews available of the experimental data obtained in animal experiments on virtually all aspects of adrenocortical steroid function.

SOFFER, L. J., and R. H. ORR (Eds.): *Symposium on the Biological and Clinical Investigation of Newer Hydrocortisone Analogs.* Metabolism (Baltimore) **7:** 383—573 (1958). A series of articles was collected that extended to recent years information and experience with the newer glucocorticoid analogues obtained since those reported in the Symposium of the New York Academy of Sciences listed above.

THORN, G. W., D. JENKINS, J. C. LAIDLAW, F. C. GOETZ, J. F. DINGMAN, W. L. ARONS, D. H. F. STREETEN and B. H. MCCRACKEN: *Pharmacological Aspects of Adrenocortical Steroids and ACTH in Man.* New Engl. J. Med. **248:** 232—245,

284—294, 323—337, 369—378, 414—423, 588—601, 632—646 (1953). This extensive review parallels and complements the review by SAYERS, listed above, with the major emphasis being on experiences in man.

WOODBURY, D. M.: *Relation Between Adrenal Cortex and the Central Nervous System.* Pharmacol. Rev. **10:** 275—357 (1958). This review covers virtually all aspects, both clinical and experimental, of adrenal steroid interaction with the nervous system.

The many effects of this group of steroids may be classified under three major headings:

(1) Metabolic — actions concerned with the transformations of the foodstuffs and their intermediates.

(2) Anti-stress — all those physiological phenomena that are involved in the adaptive response to major stress stimuli.

(3) Anti-inflammatory — actions influencing the local and general responses to "irritants," whether chemical, microbiological, immunological or other.

This grouping has the advantage not only of convenience for discussion, but is an expression of the belief (already strongly supported by data) that one basic action underlies all the effects within each grouping. Ultimately, of course, it may well be shown that all three groups of activities derive from one specific chemical event on the molecular level, but that is for the future. Indeed the clinician's hope is that the anti-inflammatory actions may be separated from the other effects, making long-term therapeutic application less hazardous. We shall return to this problem again in the discussion of the relation between chemical structure and action (p. 18).

Table 1 lists the major natural and synthetic glucocorticoids, their common names and synonyms, and their approximate potencies in metabolic, anti-stress and anti-inflammatory activities, compared to cortisol as a reference standard. The sodium-retaining activities, assayed against deoxycorticosterone, are also included.

The metabolic effects of glucocorticoids

In 1908, BIERRY and MALLOIZEL and, in 1909, PORGES observed hypoglycemia in Addisonian patients and in adrenalectomized dogs. At the same time it was also concluded (ATHANASIU and GRADINESCO 1909) that it is the function of the adrenal gland to maintain the vascular and other fluid compartments of the body in a normal state, perhaps by regulating the integrity of the capillaries. In the literature of the following 20 years, one finds that *adrenocortical dysfunction* was identified with either (1) *low blood sugar and body glycogen* or (2) *changes in water and electrolyte balance.* BRITTON and SILVETTE (BRITTON 1930, BRITTON and SILVETTE 1934, 1937a, BRITTON et al. 1938) emphasized particularly the disturbances in carbohydrate levels, since under their experimental conditions and in the mammalian species they studied (cat, marmot, opossum), the blood sugar and glycogen changes were more apparent, by far, than the shifts in water, sodium and potassium distribution.

Data gathered from man and the dog with adrenal insufficiency favored, for some time, the view that electrolyte regulation was the primary and prepotent concern of the adrenal cortex (LUCAS 1926, ROGOFF and STEWART 1926, HASTINGS and COMPERE 1931, LOEB 1932, HARROP et al. 1933a). The lowering of blood sugar and glycogen levels was ascribed to the poor food intake due to anorexia. The administration of adequate amounts of sodium chloride did indeed improve food intake, blood sugar values and glycogen content. The general impression prevailed for

Table 1. *Relative potencies of natural and synthetic glucocorticoids, based on data given in reviews cited above*

Group	Individual compounds	Metabolic + anti-stress activities[1]	Anti-inflammatory activity[1]	Na^+ retention[2]
Naturally occurring steroids	Cortisol (Hydrocortisone)	1	1	1
	Cortisone	0.8	0.8	1
	Corticosterone	0.3	—[3]	1.5
	11-Dehydrocorticosterone	0.2	—	2.0
	11-Deoxycortisol	0	0	0
1-Dehydro steroids	1-Dehydrocortisone (Prednisone)	4—5	4	0
	1-Dehydrocortisol (Prednisolone)	5	5	0
9α-Halogenated steroids	9α-Fluorocortisone	5	10	—
	9α-Fluorocortisol	10—20	10—15	600
9α-Halogenated, 1-dehydro steroids	9α-Fluoroprednisone	30	20	600
	9α-Fluoroprednisolone	50	30	600
2α-Methyl derivatives	2α-Methylcortisol	5	5	120
	2α-Methyl, 9α-fluorocortisol	10—20	10	1500
6α-Methyl derivatives	6α-Methylprednisolone	10—15	6	Na^+diuresis
	6α-Methyl, 9α-fluoroprednisolone	150	25	0.5
16α-Hydroxy derivatives	16α-Hydroxy, 9α-fluoroprednisolone	12	5	Na^+diuresis
	16α-Hydroxy, 2α-methyl, 9α-fluorocortisol	2	—	30

[1] All values equated to 1.0 for cortisol.
[2] All values equated to 1.0 for deoxycorticosterone (= approx. 1.0 for cortisol).
[3] No data available.

some time that the carbohydrate disturbances of adrenalectomy were secondary to the electrolyte shifts (Loeb 1933, Harrop et al. 1935, Allers and Kendall 1937). A truer picture emerged when (1) it was shown that adrenal extirpation modifies experimental diabetes profoundly (Hartman and Brownell 1934, Long and Lukens 1936), and (2) it became evident that the adrenal cortex elaborates not one but a group of hormones with different physiological effects (Mason et al. 1936, DeFremery et al. 1937, Steiger and Reichstein 1937).

The adrenalectomized animal suffers from several different deviations that result in hypoglycemia and low carbohydrate stores.

A. Food intake and sugar absorption

The *anorexia* that develops in adrenal insufficiency bears a relation to hyponatremia and dehydration. When these deviations are corrected by administering saline, with or without the addition of mineralocorticoids, appetite returns. Such recovery occurs in many species, especially the rat (Anderson et al. 1940a, Anderson 1943). In the dog and in man, however, while appetite improves greatly under these circumstances, it is difficult to obtain a completely normal food intake. Large doses of the glucocorticoids can lead to increases in appetite, an effect due in all probability to the action of these steroids upon the central nervous system (euphoria, etc.) and upon the rate of insulin production and release (see below).

The untreated adrenalectomized individual also suffers from a *diminution in the rate of absorption of sugars* by the intestinal tract. Restoration of normal electrolyte levels seems to restore the absorptive capacity to normal in all species studied (Deuel et al. 1937, Althausen et al. 1939, Anderson 1940). At one time

VERZÁR advocated the view that adrenal hormones promote the rate of sugar phosphorylation, thereby increasing the rate of intestinal absorption (VERZÁR 1939). This view was based upon indirect evidence and upon the assumption that phlorhizin was a specific inhibitor of phosphorylation. These assumptions are not valid (FERREBEE 1940). Indeed, the most recent work in this field indicates that sugars like glucose are transported across cell barriers in an active fashion by the aid of an as yet undefined transport system, but that this process precedes the phosphorylation of the sugar (ROTHSTEIN 1954, CRANE and KRANE 1956, CRANE 1960).

B. Maintenance of body stores of carbohydrates

The saline- or mineralocorticoid-treated adrenalectomized animal maintains normal levels of sugar in the blood, and of glycogen in liver and muscles, when fed *ad libitum* (ANDERSON et al. 1940a, LONG et al. 1940, ANDERSON 1943). However, when food is withdrawn, liver glycogen diminishes rapidly, muscle glycogen declines modestly and the blood sugar level falls (LONG et al. 1940, LONG 1942). When the normal animal is fasted, the initial decline in liver glycogen is followed by reappearance of moderate amounts in the hepatic cells. The blood sugar of the intact animal is maintained at normal or near normal values during long periods of fasting.

The total body stores of carbohydrate existing at the beginning of a fast could not maintain the normal rate of obligatory carbohydrate utilization for more than 10 to 12 hours. In the normal animal, fat and protein are then "mobilized." The fatty acids are either used directly, in many tissues, or transported to the liver and transformed, in part, to "ketone bodies," which are then readily utilized for the production of energy. The amino acids, resulting from protein mobilization, are deaminated in the liver and their keto-residues are, for the most part, transformed to glucose-6-phosphate, the immediate precursor of blood glucose. The maintenance of nearly normal blood sugar values during prolonged fasting in the intact animal is due to this ability to manufacture glucose from non-carbohydrate sources, primarily from body proteins. *The falling blood sugar of the fasting adrenalectomized animal indicates an impaired ability to set in motion an adequate rate of sugar production by the liver.*

The administration of *glucocorticoids* permits the fasting, adrenalectomized animal to maintain body carbohydrate stores in the face of continued utilization, i. e., these steroids *promote gluconeogenesis* (LONG et al. 1940, INGLE 1942, LONG 1942, SWINGLE and REMINGTON 1944). This conclusion is strengthened by the fact that the administration of glucocorticoids during fasting leads to an increased nitrogen excretion in the urine (LONG 1942, INGLE 1944, 1949).

The influence of the glucocorticoids on gluconeogenesis *becomes* strikingly *evident in* experimental and clinical *diabetes*. Removal of the adrenal glands from the depancreatized, phlorhizinized or alloxanized animal sharply diminishes (1) the degree of glycosuria, (2) the level of the blood sugar and (3) the extent of nitrogen excretion during fasting. Ketosis is decreased or abolished. Death in hypoglycemia may even occur, if the fasting period is prolonged (HARTMAN and BROWNELL 1934, LONG and LUKENS 1936, BLOOMFIELD 1939, LONG et al. 1940, WELLS and KENDALL 1940). However, if a normal amount of mixed food is ingested and the electrolyte balance is maintained, the manifestations of diabetes return almost completely (WELLS and CHAPMAN 1940, WELLS 1940, WELLS and KENDALL 1940, 1941, SWINGLE and REMINGTON 1944, INGLE 1944). The glucocorticoids render a diabetic, adrenalectomized animal severely glycosuric even though it is fasting; and they aggravate the degree of glycosuria and hyperglycemia of fed diabetic or diabetic, adrenalectomized preparations. In all such instances nitrogen excretion rises (WELLS and KENDALL 1940, WELLS 1940, INGLE 1945).

The diabetic state (SOSKIN and LEVINE 1952), induced by pancreatectomy, alloxan or phlorhizin, *consists of:*

(1) a lowered rate of glucose utilization at normal or increased blood sugar levels, and especially the rate of its transformation to fat (STETTEN and BOXER 1944);

(2) an increased rate of fat mobilization, and its breakdown to CO_2 and ketone bodies;

(3) an increased rate of hepatic gluconeogenesis from protein and release of the newly-formed sugar.

It is evident therefore that the glucocorticoids may aggravate the degree of diabetes by affecting one or more of the characteristics just listed.

I. Protein mobilization and gluconeogenesis

The best-documented action of these hormones is that of increasing gluconeogenesis from protein. When an adrenalectomized, diabetic animal is fed protein, the resulting increases in glucose and nitrogen excretion are not significantly different from those exhibited by the diabetic animal with adrenal glands (WELLS and KENDALL 1940, INGLE 1944, 1949). Since fed protein is hydrolyzed in the gastrointestinal tract and thus reaches the liver in the form of amino acids, it can be deduced that the liver cell does not require glucocorticoids for the reactions of deamination, the urea cycle, or for the transformation of α-keto-acids to glucose-6-phosphate and, eventually, free sugar. When the animal is forced to utilize endogenous protein for sugar formation, the *glucocorticoids become necessary for that set of processes by which cellular proteins are mobilized and appear finally as circulating amino acids in the blood.* In the eviscerated, nephrectomized animal, the body urea level remains constant because of the loss of the liver, and the rate of protein catabolism can now be judged by the rise in amino-acid level of the blood. This rate is slower in the previously adrenalectomized animal and can be increased to normal and above by the glucocorticoids (ENGEL 1950, 1951). Unfortunately, we do not understand the manner in which proteins are "mobilized," nor do we at present know the behavior of the enzyme systems active in endocellular proteolysis.

There seem to be no depot or storage proteins as such, corresponding to depot fat or glycogen. However, there is an enormous variety of protein molecular species, each thought to have a characteristic life span, ranging from 10 to 260 days (SPRINSON and RITTENBERG 1949, TARVER 1954). Because of the comparatively rapid action of the glucocorticoids, it would be reasonable to suppose that their effects are exerted or become evident primarily on those protein species that have a short half-life (SOBEL et al. 1958). As yet *it cannot be stated categorically whether the glucocorticoids enhance mobilization and proteolysis, or exert their influence by retarding protein formation from amino acids* – the so-called anti-anabolic effect. The result to the body as a whole would be identical – net protein decrease (CLARK 1950, ENGEL 1950, 1951, HOBERMAN 1950, SAYERS 1950).

The effect is not exerted uniformly on all tissues of the body. In the animal or man given large doses of glucocorticoids, the major losses of proteins and cellular elements occur in the lymphatic nodes, the thymus, the spleen, the subcutaneous, interstitial and bony tissues, then later in the skeletal muscle fibers (ALBRIGHT et al. 1941, ALBRIGHT 1942–1943, DOUGHERTY and WHITE 1943, DOUGHERTY et al. 1944, BAKER 1951). The epidermis, the liver, kidneys, heart and brain are not affected measurably (BAKER 1950, 1951, BAKER and INGLE 1951).

It can be surmised that *glucocorticoids*, by their described effects on protein metabolism, would *hinder total body growth.* This they do, as demonstrated during their chronic adminstration to animals and man. This action is sufficiently powerful to overcome the anabolic action of pituitary growth hormone (BECKS et al. 1944, BAKER 1950).

Are the glucocorticoids the agents that act directly to produce a negative nitrogen balance, or are these steroids substances which by their presence in tissues enable more direct effector materials to operate ? This question arises out of the experiments of INGLE and others (INGLE 1944, 1949, 1952, ENGEL 1950) upon which *the theory of the "permissive" nature of steroid action* is founded. It was known for some time that severe injury elicits hypertrophy and hyperfunction of the adrenal cortex. Among other metabolic changes following such a stressful stimulus, nitrogen excretion increases sharply. It was logical to suppose that the increased nitrogen excretion was due to the increase in glucocorticoid output of the adrenal. This hypothesis was supported by the fact that such a stress did not increase the urinary nitrogen excretion of adrenalectomized animals in good electrolyte balance. However, INGLE showed that one could obtain a typical negative nitrogen balance in the stressed, adrenalectomized animal when maintaining the animal on a steady, *small* amount of glucocorticoid. This amount of steroid, in the absence of injury, did not itself change the nitrogen balance (INGLE et al. 1947b). We may suppose, then, that the presence of a basal concentration of glucocorticoid in tissues enables or "permits" them to respond to a variety of non-hormonal catabolic stimuli. On the other hand we know that the administration of large amounts of glucocorticoids alone, in the absence of any evident stimuli, does give rise to an extensive negative nitrogen balance (INGLE 1949, ENGEL 1950). It may be possible to suggest some working hypotheses to reconcile the data, as follows:

(1) *The glucocorticoids may somehow act directly to impede protein formation from amino acids.* Many situations such as those elicited by injury or other stresses cause *protein breakdown*, which *proceeds by enzymatic routes different from those of the anabolic path.* In the adrenalectomized animal it might be difficult to show net catabolism because of the presence of uninhibited anabolism. The administration of a small amount of a glucocorticoid would slow anabolism sufficiently to permit the catabolic effect to become evident.

(2) *Protein catabolic rates may depend upon the presence of both a tissue factor "X" released by stress and of glucocorticoids.* The effect obtained would depend on the product of the concentrations of these two factors. Thus, "weak" stressful stimuli, present at all times, require large amounts of the glucocorticoids to effect protein loss, whereas a "strong" stimulus evokes sufficient "X" to produce a negative nitrogen balance in the presence of only small amounts of the corticoids.

This increase in supply of amino acids derived from endogenous protein accounts for the actions of glucocorticoids in raising blood sugar level, and in increasing liver glycogen and urinary urea. This *gluconeogenetic action* is quite intense. It was shown by WELT et al. (1952) that, in the normal rat, cortisone will increase hepatic sugar production and output by a factor of 7 or 8. If one were to assume continuance of normal rates of extrahepatic utilization, the blood sugar should rise steadily and promptly. This does not generally occur in the normal animal except after enormous doses of the glucocorticoids given for a long period of time (CONN et al. 1948, INGLE 1948). In depancreatized animals, steroids immediately and profoundly raise blood and urinary sugar (INGLE 1948, 1953). In the normal animal the glucocorticoids either directly or indirectly affect the β cells of the pancreatic islets. The islet cells hyperfunction, secrete more insulin and become hypertrophic. The extra insulin helps in the disposal of the newly produced sugar (HAUSBERGER and

Ramsay 1953, 1959, Kinash and Haist 1954). Only when the β cell is no longer able to respond adequately does the animal become significantly hyperglycemic. During the last decade, many nondiabetic individuals have been given large doses of various glucocorticoids in the therapy of some chronic "collagen" or allergic disorders. All developed increased protein catabolism, as evidenced by some degree of increased nitrogen excretion, thinning of the subcutaneous tissue, loss of protein matrix of bone, and muscular weakness and atrophy, if kept on chronic therapy. Only a small percentage (2 to 6%) developed overt diabetes (Thorn et al. 1953). This group probably represents those possessing an inherited inadequacy of pancreatic islet response. Those not developing diabetes showed a tendency to obesity, which is the result of two factors – increased food intake and the availability of sufficient insulin to enhance lipogenesis (Millstein and Hausberger 1956).

II. Release of hepatic glucose

According to our present concepts, the liver forms sugar by a reversal of the glycolytic scheme. Glucose-6-phosphate is formed and the free sugar released by the organ is derived from it by the action of a specific phosphatase. It has been shown that glucocorticoid administration leads to an increased activity of liver glucose-6-phosphatase; the effect, however, seems to be secondary to the enhanced sugar production (Ashmore et al. 1956, Weber et al. 1956, Kvam and Parks 1960).

Whether the regulation of glucose-6-phosphatase activity is also responsible for another observed effect of glucocorticoids on release of hepatic sugar is not known. This is the finding that epinephrine is less effective in eliciting hyperglycemia in the saline-treated adrenalectomized animal than in the normal. This lack of response to epinephrine occurs despite the presence of adequate glycogen stores in the liver. A small, maintenance amount of a glucocorticoid makes hepatic glycogen again normally responsive to breakdown by epinephrine (DeBodo et al. 1952, DeBodo and Sinkoff 1953, DeBodo and Altszuler 1958).

The actions of *glucocorticoids in speeding up the rate of formation of sugar by the liver and in making the liver glycogen sensitive to the effect of epinephrine* explain in some measure the counteraction to insulin sensitivity provided by these steroids. The liver of the adrenalectomized animal is "hypoglycemia unresponsive." It cannot respond rapidly to hypoglycemia with increased glucose secretion since glycogenolysis is depressed and the liver's ready supply of glucose is diminished. The chronic administration of large doses of glucocorticoids may produce insulin resistance of marked degree. In part, this may be due to the already enhanced level of endogenous insulin, so that a comparatively small exogenous dose may seem ineffective (DeBodo and Altszuler 1958). In addition, insulin resistance may be the result of the glucocorticoids indirectly opposing such insulin effects as the rapid uptake of glucose for lipogenesis. When resistance develops to very large doses of insulin, the glucocorticoids might also be evoking the appearance of serum and tissue proteins capable of binding insulin. Whether glucocorticoids increase the rate of breakdown of insulin is unknown (DeBodo and Altszuler 1958).

III. Peripheral metabolism of glucose

In the preceding section, we have considered the effects of the glucocorticoids on the hepatic production of glucose, its storage as glycogen and its release as free blood sugar. What effect do these steroids exert on the extrahepatic utilization of the blood glucose? In order to be exposed to the intracellular enzyme systems, extracellular sugar has to be transported across cell barriers. It can be shown by the use of non-utilizable sugars that are transported by the glucose transfer system, that the removal of the source of *glucocorticoids* or their administration in moderatly large doses *exert no significant effect on the rate of sugar transport* (LEVINE and GOLDSTEIN 1953). In the eviscerated, previously adrenalectomized animal, the blood sugar falls more rapidly than in the control animal, but this increased utilization is brought back to a more normal rate if the electrolyte levels are carefully adjusted and if shock is avoided (RUSSELL 1943, COHN et al. 1952). The diaphragm of the adrenalectomized rat takes up glucose at a faster than normal rate (KRAHL and CORI 1947). However, if the animal is force-fed to avoid the effects of anorexia, glucose consumption by the diaphragm *in vitro* is reduced to a normal rate (COHN and KOLINSKY 1952).

Similarly, with high levels of glucocorticoids, the rate of glucose utilization is not affected. In rats in which cortisone increased gluconeogenesis 7 to 8 fold, CO_2 production from glucose was not decreased (WELT et al. 1952). In the adrenalectomized dog, cortisone did not decrease glucose utilization. It was rather increased, but this was probably due to secondarily evoked insulin secretion (DEBODO and ALTSZULER 1958).

On the other hand, under conditions of carbohydrate loading, data from INGLE's laboratory led to the conclusion that the glucocorticoids in some manner interfere with rapid disposal of carbohydrate. Thus, in eviscerated rats given glucose plus adrenal extracts containing cortisol, the blood sugar level attained was significantly higher than in the animals given glucose but not receiving the glucocorticoid. Despite the difference in blood sugar levels, the total amounts of sugar utilized during the experimental period were not significantly different (INGLE et al. 1947b). Also, intact normal rats given a very high carbohydrate diet and large doses of glucocorticoids become severely diabetic and insulin-resistant (INGLE 1948). It appears then that, directly or indirectly, *in the presence of a carbohydrate load, the glucocorticoids may interfere with its rapid disposal.* The site of this effect should probably be looked for in the most important system for utilizing large amounts of glucose – the lipogenetic system of the fat cell (WELT and WILHELMI 1950, BRADY et al. 1951).

IV. Fat mobilization

Fasting and pancreatectomy lead to suppression of lipogenesis and to an enhancement of fat mobilization from the depots. These effects seem to depend upon the presence of at least normal pituitary and adrenocortical activity (MACKAY and BARNES 1936, VERZÁR and LASZT 1936, LUKENS 1953). The administration of glucocorticoids may on occasion lead to actual lipemia in rabbits but not in man (KOBERNICK and MORE 1950, BLOOM and PIERCE 1952, PIERCE and BLOOM 1952).

In the intact organism, evidences of *fat-mobilizing activity* are seen best following overwhelming doses *of the glucocorticoids;* otherwise the increased insulin secretion probably overcomes and obviates these effects. Whether the fat-mobilizing action is direct or acts by enabling such factors as pituitary growth hormone or "lipemia-producing" peptides to operate has not been clearly established (SEIFTER and BAEDER 1956, OLSON and VESTER 1960). In human beings suffering from Cushing's disease, there occurs premature and rather severe atherosclerosis. Cortisone intensifies cholesterol-induced atherogenesis in experimental animals (STAMLER et al. 1951). Such phenomena are additional pieces of evidence that the glucocorticoids can intensify fat mobilization. Recent work on the mechanism of such effects suggests that the glucocorticoids exert an enabling or permissive influence on fat mobilization. The direct agents are probably the catecholamines (WOOL et al. 1954, RESHEF and SHAPIRO 1960).

C. Concluding remarks

It is at the moment impossible to single out any particular well-documented metabolic effect of the glucocorticoids as being primary. In the first place the final effect is the same, no matter what mechanism one considers to be operating. In the second place one can choose experimental conditions such that the action of the glucocorticoids seems merely "permissive" to more direct effectors. In our opinion, *it seems most likely that the protein anti-anabolic or catabolic effect is* the most likely contender for being *the primary effect of these steroids*. At least, such a concept permits uniting the gluconeogenetic, lympholytic, osteoporotic and anti-proliferative (anti-inflammatory) activities.

Anti-stress and adaptive activities of glucocorticoids

The role of the glucocorticoids in bodily adaptation to stressful stimuli is emphasized by two sets of observations. (1) The one group relates to the lowered adaptive capacity of the Addisonian patient or the adrenalectomized animal, even when the electrolyte disturbances are controlled. (2) The other set of data refers to the changes seen in adrenal structure and function during a stress episode (TEPPERMAN and ENGLE 1942, SELYE 1946, 1951, SAYERS 1950, INGLE 1952).

A. Vascular collapse: End result of stress in the adrenalectomized organism

The extreme sensitivity of the Addisonian patient or of the adrenalectomized animal to sudden or severe degrees of environmental change or to injurious stimuli is a universally established phenomenon. It is nonspecific in the sense that it refers to almost any stimulus or change, whether it be a rise or fall of the ambient temperature, a "toxic" chemical substance, severe muscular exertion, mechanical injury, or other severe stress. Such stressful disturbances generally require two types of adjustment on the part of the organism. One adaptive response is related to the type of stimulus, calling forth a specific set of physiological adjustments required to maintain the *milieu interieur* and the integrity of the bodily mechanisms. Specific physiological mechanisms are called into play and their rates adjusted to meet specific displacements. Thus, cold requires adjustment in the heat-loss mechanisms and a rise in heat production. Mechanical injury, such as a bone fracture, requires those local and general adjustments that help in repair, i. e., the inflammatory response preparatory to healing, and supply of materials to the site of repair. Muscular exercise induces a rise in the rates of oxygen intake, blood flow, and cardiac activity, all needed to meet the increased metabolic turnover of work.

When the stimulus is not intense, the adrenalectomized, salt-treated animal can and does adjust his bodily responses. As a matter of fact, if such animals are "trained," i. e., exposed to particular stresses slowly and gradually, then they adjust without lasting detriment, even to degrees of disturbance that would prove lethal without "training" (NOBLE and COLLIP 1942). When the degree of "stress" is more intense, however, the adrenalectomized animal not only fails to adjust, it succumbs (SWINGLE et al. 1933, 1938). The mode of death is independent of the type of stress involved. It always involves *a generalized, peripheral vascular collapse*, usually unresponsive to nonspecific shock therapy such as fluid or vasoconstrictor drugs. *This type of "shock," however, does respond specifically to large doses of glucocorticoids.*

It would seem that in addition to specific responses, moderate to intense stresses call forth intense activity in the peripheral circulatory system, especially of the smallest blood vessels. In the normal organism this leads only rarely to a shock-like state, while in the absence of glucocorticoids, peripheral vascular collapse is the rule. This collapse interferes with the specific adaptations that the adrenalectomized animal could otherwise make.

The above discussion may be illustrated by detailed description of the response to *one kind of stress,* namely *muscular exercise.* One of the well known characteristics of the adrenalectomized animal is its susceptibility to muscular fatigue, even if its general condition has been well maintained by good intake of food and the administration of a mineralocorticoid, such as deoxycorticosterone (EVERSE and DEFREMERY 1932, INGLE 1936, INGLE et al. 1952). The initial contractile or work response to a stimulus is within the normal range. However, continuation of the work load is soon followed by weakening contractions and finally by complete fatigue. Pretreatment with any of the glucocorticoids in adequate dosage normalizes work performance. This action of the glucocorticoids parallels their effect on liver glycogen deposition and has been used as one assay technique for this group of steroids. It could be argued therefore that the failure in work performance of the adrenalectomized animal is due to a breakdown in some aspect of the metabolism of carbohydrates in muscle.

However, it can be shown that *muscular fatigue is always preceded by a fall in blood pressure* to less than 70 mm of Hg, and that continuation of the work stimulus beyond this to the point of fatigue leads to irreversible shock. On the other hand, as long as blood pressure is even temporarily maintained by nonsteroidal vasoactive materials or by transfusions, the muscular performance continues in good fashion. Furthermore, muscle removed from adrenalectomized animals dying in shock after complete fatigue *in vivo* is able to exhibit adequate normal work capacity when stimulated *in vitro* (RAMEY et al. 1950, 1951).

In the case of muscular work it is immediately evident that adaptation of the small blood vessels is useful in increasing the blood flow to the working parts of the body. The *glucocorticoids provide* the body with *the* ready *ability to open and close vascular areas* for judicious redistribution of blood flow. We shall defer the discussion of the possible mechanism of action of the glucocorticoids in this regard until we consider its relation to the vasoactive functions of the autonomic nervous system (p. 14ff.).

B. Increased glucocorticoid turnover: Result of stress in organisms possessing adrenal glands

The same stress stimuli that evoke irreversible shock in the adrenalectomized animal unless glucocorticoids are provided evoke, in the normal animal, a set of responses that depend upon the presence of the adrenal glands. There is evidence that such stimuli lead to temporary hypertrophy and hyperfunction of the adrenal cortex associated with hypersecretion of the glucocorticoids (TEPPERMAN and ENGEL 1942, TEPPERMAN et al. 1943, SELYE 1946, 1951). *During* this phase of *stress,* one *can observe shrinkage of lymphatic organs* and the thymus, *lymphopenia, eosinopenia, granulocytosis, hyperglycemia, glucose tolerance disturbances, a negative nitrogen balance,* and *loss of body potassium.* These effects are in essence the same as can be obtained by the administration of large amounts of exogenous glucocorticoids. It would be logical enough to conclude that these phenomena are solely the result of the adrenal hyperfunction evoked by the stressful stimulus. However, *many* if not all *of these effects can be obtained* in the absence of the adrenals *when only a "maintenance" amount of a glucocorticoid is provided* – an amount which by itself, in the absence of the stressful stimulus, does not produce such effects. The *concept of the "permissive" nature of glucocorticoid* action derives from such observations. According to this concept the stressful stimulus, in the presence of "normal" amounts of glucocorticoids, leads to the metabolic changes seen in stress. The additional secretion of more glucocorticoid material is neither critical nor necessary for their initial appearance (HASTINGS and COMPERE 1931, SAYERS 1950, INGLE 1952).

There is, however, *indirect evidence that during a stressful period administered glucocorticoids are utilized at a greater than normal rate.* Thus, replacement doses of 3 to 5 ml of aqueous extract of beef adrenal, 5 mg of cortisone acetate, or 3 to 4 mg of cortisol daily will sustain the adrenalectomized adult rat through a wide variety of moderate adjustments. In the absence of severe stresses, this daily steroid replacement will allow for normal growth and nitrogen retention; regulation of blood pressure in the normal range; maintenance of blood sugar through periods of fasting; regulated retention of water or diuresis, as well as appropriate renal loss of sodium, chloride and potassium in the maintenance of body fluid composition. By contrast, the supervention of a severe and sustained stress such as forced, sustained muscular work or severe hemorrhage makes necessary a two to five fold increase of the dose in order to maintain the regulated norms of blood pressure, glucose, water, and other features mentional above. The *large doses of corticoids* required for sustaining optimal resistance to stress in the adrenalectomized animal *are destructive when administered to the non-stressed animal,* leading to *hypercorticalism.* Hypertension, edema, sodium retention and potassium loss, severe nitrogen loss and hyperglycemia appear and lead to death if continued. The severely stressed animal certainly "requires," can "handle," "dipose of" or "utilize" large amounts of corticoids, whereas the non-stressed animal is overwhelmed by the same dose levels of steroids (SAYERS 1950, RAMEY et al. 1951, INGLE 1952). These data could just as well be interpreted as demonstrating interference with action at effector sites as increased "utilization."

Whatever may be the mechanism whereby the glucocorticoids protect the organism against peripheral vascular collapse resulting from a severe stress stimulus this action is clearly life-saving. Are the metabolic changes seen during stress directly associated with this life-saving property of the glucocorticoids. For example, what positive adaptive value has the negative nitrogen balance that follows injury ? In what way does the dissolution of the lymphatic tissues aid in adaptation, defense or repair ? At present, no definitive answers are at all apparent to such questions.

Effects of glucocorticoids on nervous function

The varied effects of glucocorticoids on the nervous system have been reviewed *in extenso* recently (WOODBURY 1958). Reports of corticoid effects on metabolism and function of nervous tissue are legion and variable. All manner of preparations, from the central nervous system of the intact animal to isolated nerves, have been studied. Techniques of examination have varied from behavioral measurements to neuro-electrical recording, from studies of reflex responses to determination of biochemical constitution. Clear and unequivocal demonstration of patterns of relationship between glucocorticoids and various aspects of nervous function are lacking. A simple catalogue of reported effects would yield conflicting statements about the presence or absence of steroidal influence and the direction of influence upon many parameters of nervous activity. The inference from the more cautious work is that shifts in the sodium and potassium *milieu* of the nerve cell account for most of the reported over-all nervous-system changes in relation to adrenocortical status.

A. Influence in unstressed organisms

Adrenalectomized mammals in which electrolyte and water patterns have been corrected by diet or deoxycorticosterone often exhibit well organized and sustained volitional and involuntary nervous behavior at rest. In the nearly ideal physical environment of the laboratory, or hospital, the exquisite sensitivity to stress of the salt-maintained adrenalectomized animal or man is not obvious. Without steroid-hormone replacement, the central nervous system can regulate muscular activity, temperature control, visceral function and appetite within moderate ranges. Consciousness, learning, memory, instinctive patterns of behavior, and emotional stability with appropriate range of response persist without glucocorticoids. Against this background, the changes in neuronal and synaptic performance revealed by neurophysiological techniques after adrenalectomy must be considered as largely non-critical for central nervous system activity unless there are severe adaptive demands.

In similar fashion, many of the effects of glucocorticoid overdosage in man and animals can be obviated or markedly reduced by electrolyte and water correction. Increased intake of potassium salts, to offset the potassium loss and sodium retention that accompany the administration of excess glucocorticoids, corrects many of the disturbances in nervous function while not affecting the disturbed functions of other organs or systems (RAGAN et al. 1950, GREENMAN et al. 1951, LIDDLE et al. 1952, RAGAN 1953, THORN et al. 1953). Mental aberrations during therapy with glucocorticoids correlate best with electrolyte disturbances or prior history of mental disease. Emotional patterns both in hypercorticalism and in adrenal insufficiency are neither consistent in occurrence nor characteristic in appearance (BONNYCASTLE and BRADLEY 1956). (See also Chapter 6, Part 1 of this Volume.)

Over and above changes in behavior and nerve function attributable to changes in electrolyte milieu, there remains a body of data indicating an influence of steroids on neuronal and motor endplate function. Unfortunately, the changes are not consistent. Depending on dose, manner of recording, or the exact nature of the experimental preparation, either an augmentation or inhibition of a given nerve function can be demonstrated with either excessive or deficient levels of hormones (WOODBURY 1958). The only definite statement that can be made is that nervous tissue depends on adrenocortical hormones for normal function, as do renal, cardiac, hepatic and other tissues. The inability to define the steroid dependence of nervous tissue underlines how far we are from understanding the primary action of steroids at the molecular level.

Possibly some insight into steroid action is revealed in one specific area of nervous function, namely, the response of some target organs of the sympathetic nervous system. *For* one major area — *tissue response to the epinephrines — dependence on glucocorticoids can be described in some detail.* The small blood vessels of the body exhibit a sensitive glucocorticoid dependence in stressful situations that parallels or mirrors the total organism's steroid dependence in stress. When a change in the animal's status evokes vasoconstriction by the sympathetic neurohumors, the presence of glucocorticoids becomes crucial for two aspects of the response of the small blood vessels. *Steroidal support is necessary for adequate and sustained vasoconstriction.* And, further, *steroidal support becomes essential for maintenance of the anatomical integrity of the small blood vessels.* Recent reviews describing these phenomena and discussing their implications are available (GOLDSTEIN and RAMEY 1957, RAMEY and GOLDSTEIN 1957).

B. Influence in stressed organisms

A progressive failure of circulation can be shown to underlie and precede defects typical of adrenalectomy such as easy fatigability on muscular exertion or sensitivity to trauma, hemorrhage, and noxious agents, e. g., formalin (ZWEIFACH et al. 1947, RAMEY et al. 1950, 1951, FRITZ and LEVINE 1951). Arterial and venous blood pressures and cardiac rate, when studied during such "stresses," suggest that a *functional failure at the small blood vessel site*, rather than at the heart, *leads to the profound vascular collapse seen in adrenal insufficiency.* Direct microscopic visualization of vascular beds at various sites, but especially in the splanchnic beds, reveals vasoconstriction in response to bodily insults. In the intact animal, this is sustained for long periods before ultimate recovery. In the absence of glucocorticoids, the initial constrictive response is poorly sustained and gives way to progressive dilation of the capillaries. Stasis of blood in engorged capillaries is soon associated with aneurysmal dilations extending from the venules progressively back through to the arterioles. Fluid escapes the confines of the endothelium and, in time, escape of formed elements gives additional evidence of loss of endothelial integrity. Diapedesis spots the vascular bed with multiple sites of extravascular accumulations of red blood cells. As this progressive disruption of the circulation in small blood vessels occurs, the general state of the animal deteriorates. The poor vascular response to the epinephrines and the anatomical disruption appear to be two different aspects of vascular collapse.

Three points in the description of this phenomenon deserve emphasis. (1) The vasculature of the adrenalectomized animal is well organized in the absence of stress. (2) The initial vasoconstrictive response to the sympathetics is adequate and virtually normal. (3) But once deterioration of small blood vessels has begun, it is progressive and self-sustaining and no longer requires the continued application of the external insult. Adrenocortical steroids of the glucocorticoid variety will arrest and reverse this small vessel destruction and produce return of capillary tone and intravascular flow of blood. Steroids effect this repair generally when given systemically, or effect a local repair when applied topically. Sustained vasoconstriction, with retention of anatomical integrity, at high levels of sympathetic neurohumor activity is dependent upon a direct action of glucocorticoids on small blood vessels.

Administration of epinephrine or norepinephrine produces the same picture of vascular collapse in the adrenalectomized animal. The effects of systemic administration of the epinephrines can be imitated locally by topical application of the neurohumors to an exposed vascular bed (FRITZ and LEVINE 1951, REIS 1960). Disruption of small blood vessels by the epinephrines is a local phenomenon, as is vascular protection by glucocorticoids. *The intimate circulation itself is a major site of epinephrine-steroid interaction.*

From these findings it can be proposed that *a major defect exhibited by the adrenalectomized animal* is the inability of his tissues to sustain the adaptive responses necessary to combat stress. *The physically damaging agent is* not the external insult, as a rule, but rather *the increased levels of elicited epinephrines acting on glucocorticoid-deficient vasculature.* One can further attribute the survival of the salt-maintained adrenalectomized animal, at rest, to an environment that minimizes external stimuli eliciting endogenous sympathetic nervous responses (HOSKINS 1922, GOLDSTEIN and RAMEY 1957).

In support of this view is the striking protection yielded a stressed adrenalectomized animal by any agent that interrupts the flow of sympathetic nervous impulses to tissues. *Adrenolytic agents active at effector sites, ganglionic blockades, spinal anesthetics, and central nervous system depressants all share the property of protecting adrenal-insufficient animals from progressive and fatal vascular collapse, resulting from various stresses* (FREEDMAN and KABAT 1940, KLEINBERG et al. 1942, SCHACHTER 1945, REMINGTON et al. 1948, WIGGERS et al. 1948, GOLDSTEIN et al. 1952). This strongly suggests that adrenal insufficiency as a syndrome is in large part due to the absence of adrenal steroids in the presence of an active sympathetic nervous system. Such a dual consideration is not unique in endocrinology. The diabetes of pancreatectomy, in similar fashion, is a syndrome arising from the absence of insulin in the presence of anterior pituitary secretions; diabetes insipidus results from the absence of anti-diuretic hormone in the presence of adrenocortical hormones, and so on.

Seeking the mechanism of action of steroids in the phenomena associated with adaptations to stress, forces additional consideration of the epinephrines. It is possible to expose an animal to ever increasing amounts of these neurohumors and build up a curious and dramatic resistance (ESSEX 1952). By training, one can achieve epinephrine resistance to many times the lethal dose for an untrained animal. Dogs can be adapted to survive as much as 3000 μg of epinephrine per kg of body weight, given intravenously, with only brief and minimal discomfort. Despite this tolerance, the response of various systems (including the vascular) to minimal amounts of 1 to 5 μg is unchanged. Resistance to toxicity has been developed, not reduced capacity to respond physiologically. Such resistance is independent of the adrenal cortex and persists after adrenalectomy. Further, such adapted adrenalectomized animals are exceedingly resistant to stress and can survive surgery, hemorrhage, and severe exercise (GOLDSTEIN and RAMEY 1956).

Adaptation to the epinephrines frees the organism from its profound dependence on adrenal steroids. In all likelihood, this explains the adaptation to chronic exposure to stress. With time, repeated exposure to a bodily insult often leads to a progressively diminishing need for steroids. The hypertrophied adrenal cortex regresses in size and activity towards the resting state (HEROUX and HART 1954, NICHOLLS et al. 1956).

These two classes of agent, cortical steroids and epinephrines, are intimately related. Situations that elicit increased secretion of one also elicit more of the other (GOLDSTEIN and RAMEY 1957). They both act locally on small blood vessels. The influences of both the sympathetic neurohumors and the glucocorticoids are ubiquitous. Removal of, or adaptation to, excessive sympathetic influence in tissues leads to a diminished need for adrenal steroids. While pointing to no specific mechanism for the action of either agent, this close local relationship of the two offers an investigative avenue of approach.

C. Relation to pathogenesis of mental disorders*

The effects of corticoids on mental behavior and status have been widely studied and extensively reviewed (BROWNE 1952a, b, CLEGHORN 1952, THORN et al. 1953, BONNYCASTLE and BRADLEY 1956, WOODBURY 1958). Since the initial recognition of Addison's disease and Cushing's syndrome, mental aberrations of various kinds have been observed in patients with either hypofunction or hyperfunction of the adrenal cortex. With the use of large therapeutic doses of ACTH and glucocorticoids, an extensive variety of euphoric states, psychological disturbances and frank psychoses have been precipitated. Many of these mental abnormalities can be obviated by adjustment of electrolyte balance, but a considerable incidence of behavioral disorders must be attributed to direct action of the corti-

* This topic is considered further in Part 1, Chapter 6 (Sub-Eds.).

coids. It is not possible to predict in any precise manner what behavioral disorders will be induced by glucocorticoids in man or experimental animals. Responses are highly individual. Some predisposition, underlying instability or prior history of mental illness seems responsible for the occurrence and also for the characteristics of the resulting disorder. Despite a high incidence of behavioral disorders, patients with either hypoadrenalism, even in Addisonian crisis, or with hypercorticalism, even in the florid state of Cushing's syndrome, may retain well organized and fully rational mental faculties.

The conscious aspects of central nervous system function, like the neurophysiological and electroencephalographic parameters, exhibit a highly variable, complicated, and poorly defined functional relationship to the glucocorticoids. It is not surprising, therefore, that extensive studies of psychotic patients with respect to adrenal cortical function have yielded no general agreement. Increased, decreased, and even simultaneously variable functions have been claimed. Despite the generally controversial nature of the relation of glucocorticoids to psychoses, therapeutic adrenalectomies and therapeutic trials of ACTH and glucocorticoids have been undertaken. As an isolated example, a recent 6-year follow-up of schizophrenic patients after bilateral adrenalectomy concluded that no beneficial effect was obtained (APTER 1958). The opposite position, therapeutic treatment with excess steroids, has resulted in similarly negative results.

It must be concluded, therefore, that the effects on mental behavior of excesses or deficiencies of glucocorticoids are unique to these situations and do not apply in mental disease generally.

Anti-inflammatory effects of glucocorticoids

The effectiveness of glucocorticoids in reducing the tissue phenomena of inflammation was not originally nor can it now be predicted on the basis of knowledge of their metabolic or antistress activities. HENCH, KENDALL and their associates first used cortisone in active rheumatoid arthritis because of the known remissions of the acute disorder during pregnancy and in certain liver diseases with jaundice. HARNIG had used other steroid hormones previously without success. Since it was noted that in addition to the sex steroids, adrenal glucocorticoids were also produced in larger than normal amounts during pregnancy, cortisone was tried as a therapeutic agent (HENCH et al. 1949a, b). A presumed clinical relationship between rheumatoid arthritis and rheumatic fever led to its trial also in the latter disorder (HENCH et al. 1949b). For some time a group of diseases of unknown etiology, characterized by pathological disturbances of the connective tissue "ground substance" and believed to arise from hyperergic reactions to microbiological products or drug allergens, have been classified as belonging to the vague but useful clinical-pathologic compartment of "collagen diseases." Rheumatoid arthritis can be viewed as a member of these "collagen diseases," which include the polyangiitides, lupus erythematosus, dermatomyositis and drug sensitivity. The therapeutic successes achieved with cortisone, cortisol and ACTH in this group of diseases (BAEHR and SOFFER 1950, CAREY et al. 1950, DUBOIS et al. 1952, SOFFER and BADER 1952) led at first to the viewpoint that *the glucocorticoids may interfere with the production of tissue antibodies*. Support for this view was seen in the known action of glucocorticoids on lymphatic tissues and the presumed relation of these tissues to antibody production (DOUGHERTY and WHITE 1944, DOUGHERTY et al. 1944, 1945, EISEN et al. 1947, BAKER 1951, BAKER et al. 1951, DOUGHERTY 1952). Although such an action may bear some relationship to the observed effect on the inflammatory response, it cannot be the primary or chief factor. It can be shown that *response to the induced antibodies is suppressed rather than the production of the antibodies.* On withdrawing the protective administration of glucocorticoids, the presence of the antibodies is manifested by inflammatory reaction (FISCHEL 1950). Further, the anti-inflammatory action of these steroids is not limited to conditions involving antibody production. It was soon realized that the glucocorticoids exert a damping effect on inflammations induced experimentally by local instillation of irritant materials, such as turpentine, or even on those caused by pellets of cotton (TAUBENHAUS and AMROMIN 1950, MICHAEL and WHORTON 1951). It was also shown that a local anti-inflammatory action can be achieved by local injection of the steroids into the site of the introduction of the irritant (TAUBENHAUS and AMROMIN 1950, HOLLANDER et al. 1951).

Despite the fact that the phenomena of inflammation were well and lucidly described many years ago and that the train of histopathological events seems clearly defined, we still do not understand the biochemical mechanisms operative in the cellular and extracellular changes of the inflammatory process. It is not surprising therefore that *those actions of the glucocorticoids collectively called anti-inflammatory can only be summarized descriptively* rather than placed on the basis of chemical mechanism.

Early in the inflammatory response the small vessels dilate, the endothelial lining becomes changed, as evidenced by sticking of the leukocytes, and the capillaries become more permeable. Gradually the injury progresses to bulging of the capillary walls, diapedesis and disruption. This process can be followed, for example, in the rabbit ear chamber or with other techniques of *in vivo* visualization. The prior administration of glucocorticoids minimizes all the steps in this process, as if the presence of a certain concentration of glucocorticoids in and around the capillary bed shielded the endothelial cells from the initial chemical event set in motion by the agent of injury (MENKIN 1951, 1953, EBERT and BARCLAY 1952). This protective effect is very similar to the prevention of the irreversible capillary injury following a major stress or the administration of a toxic dose of norepinephrine to the adrenalectomized animal. (See page 14.)

While this action may of itself account for a major portion of the anti-inflammatory effect, since it is a suppression of the initial event in a chain reaction, it probably is not the only site or mode of action of the glucocorticoids. Multiplication of fibroblasts around the area of injury is inhibited and the proliferation of blood vessels is sharply reduced, as is the amount of granulation tissue (GORDON and KATSH 1949, RAGAN et al. 1950, SPAIN et al. 1950, TAUBENHAUS and AMROMIN 1950, SACERDOTE DE LUSTIG and MANCINI 1951). Abscesses, if formed, have extremely thin fibrous tissue walls. Epithelial proliferation seems not to be greatly affected. *The cellular elements affected by the glucocorticoids in inflammation are the same as those that are "inhibited" by these steroids in the absence of inflammation* (BAKER 1950, 1951, BAKER et al. 1951). In patients with Cushing's syndrome, secreting excessive amounts of glucocorticoids, the components of the supporting or connective tissue suffer more severely than do other cell groups. The subcutaneous tissue becomes thin, losing both cells and extracellular matrix; osteoid cells and their protein products are disrupted (SOBEL et al. 1958). It would seem therefore that *there is some connection between the* "antigrowth" or *anti-anabolic actions of the glucocorticoids and their anti-inflammatory activity.*

Accumulation of lymphocytes in inflammatory areas is of course universal, and, in prolonged or chronic inflammation, this may be the predominant cell type. That glucocorticoids lead to atrophy of lymphatic tissues has already been noted. The function of lymphocytes in inflammation is still unknown: RICH (1936) remarked that "... they have the appearance of phlegmatic spectators passively watching the turbulent activities of the phagocytes." It is possible that at least a portion of the blood globulins is manufactured by the lymphatic tissue. It may well be that the suppressive effect of the glucocorticoids upon this tissue explains the reduction of the blood globulin concentration to the normal range by the administration of steroids in various inflammatory states (FISCHEL 1950, BUNIM et al. 1952).

While there is certainly no unanimity of opinion or in experimental data, the preponderance of the evidence favors the view that *glucocorticoids* do not interfere directly with antibody production but *inhibit the extent of tissue reaction following the binding of certain antibody complexes in cells* (FISCHEL 1950). Thus in the experimental form of nephritis produced by the immunological technique of administering anti-kidney serum, cortisone treatment prevents the inflammatory reaction. When steroid administration is stopped abruptly, the expected reaction then appears, many days after the anti-kidney serum was given. Steroid actions appear to affect some primary event common in cell reactions to a great variety of stressful agents, rather than directly to neutralize the agents themselves (WEDGEWOOD et al. 1951, THORN et al. 1953).

Since the glucocorticoids act on the reactive process irrespective of the nature of the inciting agent, they interfere with the protective functions of inflammation as well as with the crippling aspects. Walling off a bacterial infection by means of the "pyogenic" membrane is thus done away with. Hence, there is a tendency for *viral and bacterial infections to spread* during the time when the glucocorticoids are used to inhibit an "inflammatory" disorder — of the collagen disease type or of the hyperergic group of syndromes (SHWARTZMAN 1953, KASS and FINLAND 1948).

When a quiescent peptic ulcer is present in a patient given prolonged glucocorticoid therapy, the continuous struggle between the breakdown and the healing process by fibrosis is tipped against healing. Gastrointestinal bleeding from a peptic ulceration is one of the serious complications of chronic steroid therapy (GRAY et al. 1951, KIRSNER et al. 1952, ROBERT and NEZAMIS 1958).

Undesirable pathological events during glucocorticoid therapy are sometimes referred to as side-reactions. This is incorrect. They are the actions to be expected from large doses of the steroids. Collectively they form an artificially produced Cushing's syndrome: (1) plethoric facies and erythrocytosis, (2) retention of salt and water, (3) thin skin with purple striae, (4) trunk obesity, (5) osteoporosis and possible vertebral collapse, (6) hyperglycemia and glycosuria, (7) psychotic episodes usually of the hypomanic variety.

A study of Table 1 (p. 4) demonstrates that the steroid nucleus requires particular substituents or certain bond modifications to bring out or intensify the "metabolic" activities. Although it is possible to modify the degree of metabolic activity, thus far it has proved impossible, by structural modifications, to influence separately to any significant degree the "anti-inflammatory" and the "metabolic" activities. Indeed, our previous discussion has led tentatively to *the supposition that the inhibition of proliferation of certain cell types is another expression of an intense anti-anabolic action of the glucocorticoids.* Steroids with strong mineralocorticoid activity (such as deoxycorticosterone), instead of being anti-inflammatory, may actually enhance the inflammatory response. The failure to dissociate anti-inflammatory and metabolic effects of glucocorticoids in a convincing manner strengthens the hypothesis that glucocorticoid action, when it is finally described on a molecular level, will be found to affect a single reaction. Such a hypothetical chemical reaction between a glucocorticoid and a specific cell constituent may underlie all the phenomena described as "effects."

Effects of glucocorticoids on seemingly unrelated systems

Despite the enormous number of investigations dealing directly or indirectly with glucocorticoid action, we are far from an understanding of the molecular happenings that initiate the steroid effects on tissues. The attempts to demonstrate effects *in vitro* on excised tissues, homogenates and enzyme extracts have yielded either negative, uncertain or uninterpretable results (UMBREIT 1951, DIRSCHERL 1955). Nevertheless it is possible to develop a hypothesis or hypotheses to correlate many of the effects of glucocorticoids. *A review of the "metabolic" activities leads to the suspicion that the basic effect in this sphere is the anti-anabolic* or protein-catabolic *effect*, although nothing is known as to its mechanism, or even whether this reflects lysis of certain cell systems or reflects an inhibition of development and growth of cells. *The anti-inflammatory* and some of the adaptive *responses can be traced to* more primary, although unknown, *actions affecting the state of the capillary endothelium, in addition to inhibiting cell reactivity in the connective tissue systems.*

However, *even these vague generalizations do not begin to satisfy the question of mode of action of the glucocorticoids on certain aspects of bodily function.* We shall consider two such areas in which the mode of hormone action is completely unknown: (1) the effects of these steroids upon blood cells and the clotting mechanism; and (2) their actions upon water diuresis.

A. Effects on blood cells, platelets, and clotting time

The glucocorticoids have been found to exert a varying set of effects on the formed elements of the blood and on the hematopoietic system, which it is difficult or impossible to group jointly, at this time, with the other known activities of these steroids. Some of these data are derived from experimental animals and from normal humans, others from observations on human beings with disease, during glucocorticoid administration.

In adrenal insufficiency there is found an *absolute lymphocytosis*. Glucocorticoids will severely depress the lymphocyte count in normal as well as in adrenalectomized animals and humans (DOUGHERTY and WHITE 1944). Although there are some studies tending to show that this may be due to rapid cell destruction (HEILMAN 1945, FELDMAN 1950), the evidence is by no means unequivocal (BALDRIDGE et al. 1951). At least this action on the circulating lymphocyte is consistent with the regression caused by the glucocorticoids in the lymphatic tissues. This effect may be the expression of an intense anti-anabolic action on cells and tissues which "turn over" very rapidly.

The eosinophil leukocyte is similarly affected by the glucocorticoids. In normal animals the diurnal variation in eosinophil count is inversely related to the diurnal variation in glucocorticoid secretion by the adrenals. The fall in eosinophils produced by ACTH administration is a fairly good index of adrenal cortical reactivity, although it has been shown that substances like epinephrine will cause a drop in eosinophil cell count in an adrenalectomized animal or human being providing a maintenance amount of glucocorticoids had been previously given (HENRY et al. 1953, THORN et al. 1953).

The glucocorticoids generally tend to raise the total leukocyte count by increasing *the number of granulocytes.* This effect is not sustained on chronic administration (HILLS et al. 1948).

As has been mentioned above, *in* naturally occurring *Cushing's syndrome* as well as in the drug-induced variety, *the red blood cell count tends to be high* (14 to 17 g hemoglobin per 100 ml). There is usually no increase in the blood volume, so that one speaks of an *erythrocytosis* rather than a polycythemia. Because of this observation, clinicians have given glucocorticoids for hypoplasia or aplasia of the erythroid marrow with good, albeit temporary, success in stimulating erythropoiesis (BOUSSER and BOUILLER 1958). This action is not understood and seems in contradiction to the picture of inhibition of cell proliferation and of protein anabolism that is in general the hallmark of the glucocorticoids. However, it is possible that this action of glucocorticoids in stimulating red blood cell production is secondary. In other words, this may be still another example of steroid counteraction of the reaction to an injury – which here would be suppression of hematopoiesis.

Somewhat more typical are the effects of the glucocorticoids upon certain aspects of blood clotting disorders. Thrombocytopenic purpura is a disease in which two factors seem to be involved. The number of platelets in the circulating blood is very low. By itself, this would not lead to spontaneous bleeding. The capillary endothelium seems also to be affected because of an auto-antibody reaction. The glucocorticoids may be used to control the bleeding tendency without affecting the number of circulating platelets. This action of glucocorticoids seems consistent with their tendency to make tissues non-reactive to injurious agents, and especially to protect, in some manner, the capillary endothelium (WHITELAW 1951).

B. Effects on water diuresis

Many of the glucocorticoids have an influence on the excretion and absorption of sodium, potassium and therefore water similar to that of deoxycorticosterone. Such activity is of course the predominant one of the mineralocorticoids. As is evident from Table 1, changes in chemical configuration at certain carbon atoms of the steroid nucleus may either enhance or almost completely abolish the mineralocorticoid activity. On the other hand, there is one action that is independent of the mineralocorticoid activity but is definitely related to the degree of glucocorticoid effect – and that is the reaction to a water load.

It was observed some years ago that *in adrenal insufficiency there is a lag in the excretion of a water load* (LEVY et al. 1946, LOTSPEICH 1949). The lag is evident when the water is administered either orally or intravenously, and therefore it is not due to an absorptive defect in the gastrointestinal tract (THORN et al. 1953). The administration of saline, with or without mineralocorticoids, does not repair this defect even when glomerular filtration rates have been restored to normal levels (SOFFER and GABRILOVE 1952, GARROD et al. 1955). The ability to excrete a water load quickly is restored specifically by the glucocorticoids, in good accord with the degree of their other metabolic activities. This effect is exerted either on the kidney tubule, perhaps by opposing an antidiuretic factor (GAUNT et al. 1949, ROEMMETT et al. 1952, SOFFER et al. 1957), or on the tissues in general, by preventing overhydration of cells (THORN et al. 1953).

Whatever may be the true site of this action, it is difficult to see any particular relation of this effect to the major groupings of activities described above.

Even though the glucocorticoids are not "omnipotent," their effects are so widespread that the molecular mode of action will have to be looked for in a property which is fundamental for almost all cellular systems. The present state of knowledge allows only a descriptive and classifying approach to this problem.

Literature

ALBRIGHT, F.: Cushing's syndrome. Harvey Lect. **38**, 123—186 (1942/43).

— W. PARSON and E. BLOOMBERG: Cushing's syndrome interpreted as hyperadrenocorticism leading to hypergluconeogenesis. J. clin. Endocr. **1**, 375—384 (1941).

ALLERS, W. D., and E. C. KENDALL: Maintenance of adrenalectomized dogs without cortin through control of the mineral constituents of the diet. Amer. J. Physiol. **118**, 87—94 (1937).

ALTHAUSEN, T. L., E. ANDERSON and M. STOCKHOLM: Effect of adrenalectomy and of NaCl on intestinal absorption of dextrose. Proc. Soc. exp. Biol. (N. Y.) **40**, 343—344 (1939).

ANDERSON, E.: The physiology of the salt-treated adrenalectomized animal. In: Essays in Biology, pp. 33—49. Berkeley: University of California Press 1943.

— V. HERRING and M. JOSEPH: Salt after adrenalectomy; carbohydrate stores in adrenalectomized rats given various levels of sodium chloride. Proc. Soc. exp. Biol. (N. Y.) **45**, 488—493 (1940a).

— M. JOSEPH and V. HERRING: Salt after adrenalectomy; growth and survival of adrenalectomized rats given various levels of NaCl. Proc. Soc. exp. Biol. (N. Y.) **44**, 477—481 (1940b).

APTER, N. S.: Bilateral adrenalectomy in chronic schizophrenic patients: 6 years later. Amer. J. Psychiat. **115**, 55—59 (1958).

ASHMORE, J., A. B. HASTINGS, F. B. NESBETT and A. E. RENOLD: Studies on carbohydrate metabolism in rat liver slices. VI. Hormonal factors influencing glucose-6-phosphatase. J. biol. Chem. **218**, 77—88 (1956).

ATHANASIU, J., et A. GRADINESCO: Les capsules surrénales et les échanges entre le sang et les tissus. C. R. Acad. Sci. (Paris) **149**, 413 (1909).

BAEHR, G., and L. J. SOFFER: Treatment of disseminated lupus erythematosus with cortisone and adrenocorticotropin. Bull. N. Y. Acad. Med. **26**, 229—234 (1950).

BAKER, B. L.: Modification of body structure by adrenocortical secretions with special reference to regulation of growth. In: R. C. CHRISTMAN, Ed., Pituitary-Adrenal Function, pp. 88—95. Washington: Amer. Assoc. Advanc. Sci. 1950.

— Comparison of histological changes induced by experimental hyperadrenocorticalism and inanition. Recent Progr. Hormone Res. **7**, 331—373 (1951).

— D. J. INGLE and C. H. LI: Histology of lymphoid organs of rats treated with adrenocorticotropin. Amer. J. Anat. **88**, 313—350 (1951).

BALDRIDGE, G. D., A. M. KLIGMAN, M. J. LIPNIK and D. W. PILLSBURY: *In vitro* effects of cortisone on mesodermal tissue. Arch. Path. **51**, 595—596 (1951).

BECKS, H., M. E. SIMPSON, W. MARX, C. H. LI and H. M. EVANS: Effects of adrenocorticotropic hormone (ACTH) on osseous system in normal rats. Endocrinology **34**, 305—310 (1944).

BIERRY, H., et L. MALLOIZEL: Hypoglycémie après descapsulation, effets de l'infection d'adrénaline sur les animaux descapsulés. C. R. Soc. Biol. (Paris) **65**, 232—234 (1908).

BLOOM, B., and F. T. PIERCE, JR.: Relationship of ACTH and cortisone to serum lipoproteins and atherosclerosis in humans. Metabolism **1**, 155—162 (1952).

BLOOMFIELD, A. L.: Coincidence of diabetes mellitus and Addison's disease; effect of cortical extract on glycemia and glycosuria. Bull. Johns Hopk. Hosp. **65**, 456—465 (1939).

BONNYCASTLE, D. D., and A. J. BRADLEY: Blocking action of dilantin upon the pituitary-adrenal system. Fed. Proc. **15**, 403 (1956).

BOUSSER, J., et J. M. BOUILLER: Le traitement des pancytopénies idiopathiques sans splénomégalé. Sang **29**, 150—157 (1958).

BRADY, R. O., F. D. W. LUKENS and S. GURIN: Synthesis of radioactive fatty acids *in vitro* and its hormonal control. J. biol. Chem. **193**, 459—464 (1951).

BRITTON, S. W.: Adrenal insufficiency and related considerations. Physiol. Rev. **10**, 617—682 (1930).

—, and H. SILVETTE: On the function of the adrenal cortex—general carbohydrate and circulatory theories. Amer. J. Physiol. **107**, 190—206 (1934).

— — A comparison of sodium chloride and carbohydrate changes in adrenal insufficiency and other experimental conditions. Amer. J. Physiol. **118**, 595—599 (1937a).

— — The adrenal cortex and carbohydrate metabolism. Cold Spr. Harb. Symp. quant. Biol. **5**, 357 (1937b).

— — and R. F. KLINE: Adrenal insufficiency in American monkeys. Amer. J. Physiol. **123**, 705—711 (1938).

BROWNE, J. S. L.: The interplay between endocrine disturbances and psychological aberrations. Ciba Found. Colloquia Endocr. **3**, 112—118 (1952a).

— Effects of ACTH and cortisone on behavior. Ciba Found. Colloquia Endocr. **3**, 197—204 (1952b).

BUNIM, J. J., A. G. KUTTNER, J. S. BALDWIN and C. MCEWEN: Cortisone and corticotropin in rheumatic fever and juvenile rheumatoid arthritis. J. Amer. med. Ass. **150**, 1273—1278 (1952).

CAREY, R. A., A. M. HARVEY, J. E. HOWARD and P. F. WAGLEY: Effect of adrenocorticotropic hormone (ACTH) and cortisone on drug hypersensitivity reactions. Bull. Johns Hopkins Hosp. **87**, 354—386 (1950).

CLARK, I.: Effect of cortisone on protein metabolism in the rat as studied with isotopic glycine. Fed. Proc. **9**, 161 (1950).

CLEGHORN, R. A.: Endocrine influence on personality and behavior. In: Biology of Mental Health and Disease, pp. 265—273. New York: Hoeber 1952.

— Steroid hormones in relation to neuropsychiatric disorders. In: H. HOAGLAND, Ed., Hormones, Brain Function, and Behavior, pp. 3—25. New York: Academic Press 1957.

COHN, C., and M. KOLINSKY: Influence of previous diet on carbohydrate utilization by diaphragms of adrenalectomized rats. Metabolism **2**, 146—152 (1952).

— B. KATZ, H. HUDDLESTUN, M. KOLINSKY and R. LEVINE. Utilization of carbohydrate by extrahepatic tissues of the adrenalectomized dog. Amer. J. Physiol. **170**, 87—93 (1952).

CONN, J. W., L. H. LOUIS and C. E. WHEELER: Production of temporary diabetes mellitus in man with pituitary adrenocorticotropic hormone. J. Lab. clin. Med. **33**, 651—666 (1948).

CRANE, R. K.: Intestinal absorption of sugars. Physiol. Rev. **40**, 789—825 (1960).

—, and S. M. KRANE: On the mechanism of the intestinal absorption of sugars. Biochim. biophys. Acta **20**, 568—569 (1956).

— — Studies on the mechanism of the intestinal active transport of sugars. Biochim. biophys. Acta **31**, 397—401 (1959).

DEBODO, R. C., and N. ALTSZULER: Insulin hypersensitivity and physiological insulin antagonists. Physiol. Rev. **38**, 389—445 (1958).

—, and M. W. SINKOFF: Anterior pituitary and adrenal hormones in the regulation of carbohydrate metabolism. Recent Progr. Hormone Res. **8**, 511—570 (1953).

— — S. P. KIANG and H. DEN: Prevention of growth hormonal-induced diabetes in hypophysectomized dogs by adrenocortical steroids. Proc. Soc. exp. Biol. (N. Y.) **81**, 425—430 (1952).

DEUEL, H. J., L. F. HALLMAN, S. MURRAY and L. T. SAMUELS: Sexual variation in carbohydrate metabolism; rate of absorption of glucose and of glycogen formation in normal and adrenalectomized rats. J. biol. Chem. **119**, 607—615 (1937).

DIRSCHERL, W.: Über die Einwirkung von Steroidhormonen auf Gewebestoffwechsel und Fermente. Ergebn. Physiol. **48**, 112—229 (1955).

DOUGHERTY, T. F.: Studies of antiphlogistic and antibody suppressing function of pituitary adrenocortical secretions. Recent Prog. Hormone Res. **7**, 307—330 (1952).

—, and A. WHITE: Effect of pituitary adrenotropic hormone on lymphoid tissue. Proc. Soc. exp. Biol. (N. Y.) **53**, 132—133 (1943).

— — Influence of hormones on lymphoid tissue structure and function. Role of pituitary adrenotrophic hormone in regulation of lymphocytes and other cellular elements of blood. Endocrinology **35**, 1—14 (1944).

— — and J. H. CHASE: Relationship of effects of adrenal cortical secretion on lymphoid tissue and on antibody titer. Proc. Soc. exp. Biol. (N. Y.) **56**, 28—31 (1944).

— J. H. CHASE and A. WHITE: Pituitary-adrenal cortical control of antibody release from lymphocytes: explanation of anamnestic response. Proc. Soc. exp. Biol. (N. Y.) **58**,135—140 (1945).

DUBOIS, E. L., R. R. COMMONS, P. STARR, C. S. STEIN, JR., and R. MORRISON.: Corticotropin and cortisone treatment for systemic lupus erythematosus. J. Amer. med. Assoc. **149**, 995—1002 (1952).

EBERT, R. H., and W. R. BARCLAY: Changes in connective tissue reaction induced by cortisone. Ann. intern. Med. **37**, 506—518 (1952).

EISEN, H. N., M. M. MAYER, D. H. MOORE, R. R. TARR and H. C. STOERK: Failure of adrenal cortical activity to influence circulating antibodies and gamma globulin. Proc. Soc. exp. Biol. (N. Y.) **65**, 301—306 (1947).

ENGEL, F. L.: Studies on the site and mode of action of the adrenal cortex in protein metabolism. In: R. C. CHRISTMAN, Ed., Pituitary-Adrenal System, pp. 62—78. Washington: Amer. Assoc. Advanc. Sci. 1950.

— Roles of the adrenal cortex and stress in the regulation of protein metabolism. Recent Progr. Hormone Res. **6**, 277—313 (1951).

ESSEX, H. E.: Further observations of certain responses of tolerant and control animals to massive doses of epinephrine. Amer. J. Physiol. **171**, 78—86 (1952).

EVERSE, J. W. R., and P. DEFREMERY: A method measuring fatigue in rats and its application for testing suprarenal cortical hormone (Cortin). Acta brev. neerl. Physiol. **2**, 152 (1932).

FELDMAN, J. D.: *In vitro* reaction of cells to adrenal cortical steroids with special reference to lymphocytes. Endocrinology **46**, 552—562 (1950).

FERREBEE, J. W.: The effect of adrenalectomy on the phosphorylation of vitamins B_1 and B_2. J. biol. Chem. **136**, 719—728 (1940).

FISCHEL, E. E.: Relationship of adrenal cortical activity to immune responses. Bull. N. Y. Acad. Med. **26**, 225—260 (1950).

FREEDMAN, A. M., and H. KABAT: The pressor response to adrenalin in the course of traumatic shock. Amer. J. Physiol. **130**, 620—626 (1940).
DEFREMERY, P., E. LAQUEUR, T. REICHSTEIN, R. W. SPANHOFF and I. E. UYLDERT: Corticosteron. Nature (Lond.) **139**, 26 (1937).
FRITZ, I., and R. LEVINE: Action of adrenal cortical steroids and norepinephrine on vascular responses of stress in adrenalectomized rats. Amer. J. Physiol. **165**, 456—465 (1951).
GARROD, O., S. A. DAVIES and G. F. CAHILL: The action of cortisone and deoxycorticosterone acetate on glomerular filtration rate, sodium and water exchange of the adrenalectomized dog. J. clin. Invest. **34**, 761—776 (1955).
GAUNT, R., J. H. BIRNIE and W. J. EVERSOLE: Adrenal cortex and water metabolism. Physiol. Rev. **29**, 281—310 (1949).
GOLDSTEIN, M. S., and E. R. RAMEY: Relation of adrenal cortex to epinephrine conditioning. Fed. Proc. **15**, 78 (1956).
— — Non-endocrine aspects of stress. Perspect. Biol. Med. **1**, 33—47 (1957).
— — I. FRITZ and R. LEVINE: Reversal of effects of stress in adrenalectomized animals by autonomic blocking agents. Amer. J. Physiol. **171**, 92—99 (1952).
GORDON, A. S., and G. F. KATSH: Relation of adrenal cortex to structure and phagocytic activity of macrophagic system. Ann. N. Y. Acad. Sci. **52**, 1—30 (1949).
GRAY, S. J., J. A. BENSON, JR., H. M. SPIRO and R. W. REIFENSTEIN: Effects of ACTH and cortisone upon stomach. Gastroenterology **19**, 658—673 (1951).
GREENMAN, L., T. S. DANOWSKI, R. TARAIL, J. H. PETERS, F. WEIGAND and F. MATEER: Studies of prevention of some of toxic effects of prolonged cortisone and ACTH therapy. J. clin. Invest. **30**, 644 (1951).
HARROP, G. A., L. J. SOFFER, R. ELLSWORTH and J. H. TRESCHER: Studies on the suprarenal cortex. J. exp. Med. **58**, 17—38 (1933a).
— A. WEINSTEIN, L. J. SOFFER and J. H. TRESCHER: Addison's disease. J. Amer. med. Assoc. **100**, 1850—1855 (1933b).
— L. J. SOFFER, W. M. NICHOLSON and M. STRAUSS: Studies on the suprarenal cortex. J. exp. Med. **61**, 839—860 (1935).
HARTMAN, F. A., and K. A. BROWNELL: Relation of adrenals to diabetes. Proc. Soc. exp. Biol. (N. Y.) **31**, 834—835 (1934).
HASTINGS, A. B., and E. L. COMPERE: Effects of bilateral suprarenalectomy on constituents of blood of dogs. Proc. Soc. exp. Biol. (N. Y.) **28**, 376—378 (1931).
HAUSBERGER, F. X., and J. A. RAMSAY: Steroid diabetes in guinea pigs; effects of cortisone administration on blood and urinary glucose, nitrogen excretion, fat deposition, and the islets of Langerhans. Endocrinology **53**, 423—435 (1953).
— — Islet hypertrophy in obesity of mice bearing ACTH-secreting tumors. Endocrinology **65**, 165—171 (1959).
HEILMAN, D. H.: Effect of 11-dehydro-17-hydroxycorticosterone and 11-dehydrocorticosterone on migration of macrophages in tissue culture. Proc. Mayo Clin. **20**, 318—320 (1945).
HENCH, P. S., E. C. KENDALL, C. H. SLOCUMB and H. F. POLLEY: Effect of a hormone of adrenal cortex (17-hydroxy-11-dehydrocorticosterone: compound E) and of pituitary adrenocorticotropic hormone on rheumatoid arthritis. Proc. Mayo Clin. **24**, 181—197 (1949a).
— C. H. SLOCUMB, A. R. BARNES, H. L. SMITH, H. F. POLLEY and E. C. KENDALL: Effects of the adrenal cortical hormone 17-hydroxy-11-dehydrocorticosterone (compound E) on acute phase of rheumatic fever: preliminary report. Proc. Mayo Clin. **24**, 277—297 (1949b).
HENRY, W. L., JR., L. OLINER and E. R. RAMEY: Relationship between actions of adrenocortical steroids and adrenomedullary hormones in the production of eosinopenia. Amer. J. Physiol. **174**, 455—458 (1953).
HEROUX, O., and J. S. HART: Adrenal cortical hormone requirement of warm and cold acclimated rats after adrenalectomy. Amer. J. Physiol. **178**, 449—452 (1954).
HILLS, A. G., P. H. FORSHAM and C. A. FINCH: Changes in circulating leukocytes induced by administration of pituitary adrenocorticotrophic hormone (ACTH) in man. Blood **3**, 755—768 (1948).
HOBERMAN, H. D.: Endocrine regulation of amino acid and protein metabolism during fasting. Yale J. Biol. Med. **22**, 341—367 (1950).
HOLLANDER, J. L., E. M. BROWN, JR., and R. A. JESSAR: Hydrocortisone and cortisone injected into arthritic joints: comparative effects of and use of hydrocortisone as local antiarthritic agent. J. Amer. med. Ass. **147**, 1629—1635 (1951).
HOSKINS, R. G.: The relation of the adrenals to the circulation. Physiol. Rev. **2**, 343—358 (1922).
INGLE, D. J.: Work capacity of adrenalectomized rat treated with cortin. Amer. J. Physiol. **116**, 622—625 (1936).
— Problems relating to the adrenal cortex. Endocrinology **31**, 419—438 (1942)

INGLE, D. J.: The physiological action of the adrenal hormones. In: F. R. MOULTON, Ed., Chemistry and Physiology of Hormones, pp. 83—103. Washington: Amer. Assoc. Advanc. Sci. 1944.
— Experimental glycosuria in the rat. Recent Progr. Hormone Res. **2**, 229—254 (1948).
— Some studies on the role of the adrenal cortex in organic metabolism. Ann. N. Y. Acad. Sci. **50**, 576—595 (1949).
— The role of the adrenal cortex in homeostasis. J. Endocrinol. **8**, xxiii—xxxvii (1952).
— Some studies on experimental diabetes. Lancet **73**, 470—478 (1953).
—, and R. C. MEEKS: Comparison of some metabolic and morphologic effects of cortisone and hydrocortisone given by continuous injection to rats. Amer. J. Physiol. **170**, 77—80 (1952).
— E. O. WARD and M. H. KUIZENGA: The relationship of the adrenal glands to changes in urinary non-protein nitrogen following multiple fractures in the force-fed rat. Amer. J. Physiol. **149**, 510—515 (1947a).
— M. C. PRESTRUD, J. E. NEZAMIS and M. H. KUIZENGA: Effect of adrenal cortex extract upon the tolerance of the eviscerated rat for intravenously administered glucose. Amer. J. Physiol. **150**, 423—427 (1947b).
— J. E. NEZAMIS and E. H. MORLEY: Comparative value of cortisone, 17-hydroxycorticosterone and adrenal cortex extract by continuous intravenous injection in sustaining ability of adrenalectomized rat to work. Endocrinology **50**, 1—4 (1952).
KASS, E. H., and M. FINLAND: Corticosteroids and infections. Advanc. intern. Med. **9**, 45—80 (1958).
KINASH, B., and R. E. HAIST: Effect of ACTH and of cortisone on the islets of Langerhans and the pancreas in intact and hypophysectomized rats. Amer. J. Physiol. **178**, 441—444 (1954).
KIRSNER, J., A. P. KLOTZ and W. L. PALMER: Unfavorable course of gastric ulcer during administration of ACTH and cortisone. Gastroenterology **20**, 27—29 (1952).
KLEINBERG, W., J. W. REMINGTON, V. A. DRILL and W. W. SWINGLE: The nervous factor in the circulatory failure induced in adrenalectomized dogs, by intestinal stripping and a single stage bilateral adrenalectomy. Amer. J. Physiol. **137**, 362—370 (1942).
KOBERNICK, S. D., and R. H. MORE: Diabetic state with lipaemia and hydropic changes in pancreas produced in rabbits by cortisone. Proc. Soc. exp. Biol. (N. Y.) **74**, 65—67 (1950).
KRAHL, M. D., and C. F. CORI: The uptake of glucose by the isolated diaphragm of normal, diabetic, and adrenalectomized rats. J. biol. Chem. **170**, 607—618 (1947).
KVAM, D. C., and R. E. PARKS: Hydrocortisone-induced changes in hepatic glucose-6-phosphatase activity. Amer. J. Physiol. **198**, 21—24 (1960).
LEVINE, R., and M. S. GOLDSTEIN: Effect of insulin on rate of transfer of sugars across cell barriers. Brookhaven Symp. Biol. **5**, 77—84 (1952).
LEVY, M. S., M. H. POWER and E. J. KEPLER: The specificity of the "water test" as a diagnostic procedure in Addison's disease. J. clin. Endocr. **6**, 607—632 (1946).
LIDDLE, G. W., P. H. FORSHAM and L. L. BENNETT: Quantitative study of prevention of ACTH-induced sodium retention by potassium salts. J. clin. Endocr. **12**, 926 (1952).
LOEB, R. F.: Chemical changes in blood in Addison's disease. Science **76**, 420—421 (1932).
— Effects of sodium chloride in patient with Addison's disease. Proc. Soc. exp. Biol. (N. Y.) **30**, 808—812 (1933).
LONG, C. N. H.: A discussion of the mechanism of action of adrenal cortical hormones on carbohydrate and protein metabolism. Endocrinology **30**, 870—883 (1942).
—, and F. D. W. LUKENS: The effects of adrenalectomy and hypophysectomy upon experimental diabetes in the cat. J. exp. Med. **63**, 465—490 (1936).
— B. KATZIN and E. G. FRY: The adrenal cortex and carbohydrate metabolism. Endocrinology **26**, 309—344 (1940).
LOTSPEICH, W. D.: Effect of adrenalectomy on the renal tubular reabsorption of water in the rat. Endocrinology **44**, 314—316 (1949).
LUCAS, G. W. H.: Blood and urine findings in desuprarenalectomized dogs. Amer. J. Physiol. **77**, 114—125 (1926).
LUKENS, F. D. W.: Hormonal influences in the synthesis of fat from carbohydrate. Ciba Colloquia Endocr. **6**, 55—69 (1953).
MACKAY, E. M., and R. H. BARNES: Fasting ketosis in the pregnant rat as influenced by adrenalectomy. Proc. Soc. exp. Biol. (N. Y.) **34**, 682—683 (1936).
MASON, H. L., C. S. MYERS and E. C. KENDALL: Clinical studies of the suprarenal cortex. J. biol. Chem. **116**, 267—276 (1936).
MENKIN, V.: Cortisone and mechanism of increased capillary permeability in inflammation. Fed. Proc. **10**, 91 (1951).
— Significance of accumulation of cortisone in inflamed area. Brit. J. exp. Path. **34**, 412—419 (1953).
MICHAEL, M., JR., and C. M. WHORTON: Delay of early inflammatory response by cortisone. Proc. Soc. exp. Biol. (N. Y.) **76**, 754—756 (1951).

MILLSTEIN, S. W., and F. X. HAUSBERGER: Liver glucose-6-phosphatase activity and intracellular distribution after cortisone adminstration. Diabetes **5**, 89—92 (1956).
MUEHRCKE, R. C., J. L. LEWIS and R. M. KARK: Effects of cortisone and heparin on eosinophils in defibrinated human blood. Science **115**, 377—378 (1952).
NICHOLLS, D., F. C. HEAGY and R. J. ROSSITER: Phosphorus metabolism of adrenal gland of rat; effect of exposure to cold environment for eight days. Canad. J. Biochem. **34**, 543—553 (1956).
NOBLE, R. L., and J. B. COLLIP: Adrenal and other factors affecting experimental traumatic shock in rats. Quart. J. exp. Physiol. **31**, 201—210 (1942).
OLSON, R. E., and J. W. VESTER: Nutrition-endocrine interrelationships in the control of fat transport in man. Physiol. Rev. **40**, 677—733 (1960).
PADAWER, J., and A. S. GORDON: Hormonal influences upon the distribution and fate of the eosinophilic leukocyte. J. clin. Endocr. **12**, 922 (1952).
PIERCE, F. T., JR., and B. BLOOM: Relationship of ACTH and cortisone to serum lipoproteins of rabbit. Metabolism **1**, 163—171 (1952).
PORGES, O.: Hypoglykämie bei Morbus Addisonii sowie bei nebennierenlosen Hunden. Z. klin. Med. **69**, 341 (1909).
RAGAN, C.: Corticotropin, cortisone and related steroids in clinical medicine: practical considerations. Bull. N. Y. Acad. Med. **29**, 355—376 (1953).
— E. L. HOWES, C. M. PLOTZ, K. MEYER, J. W. BLUNT and R. LATTES: Effect of ACTH and cortisone on connective tissue. Bull. N. Y. Acad. Med. **26**, 251—254 (1950).
RAMEY, E. R., and M. S. GOLDSTEIN: The adrenal cortex and the sympathetic nervous system. Physiol. Rev. **37**, 155—195 (1957).
— — and R. LEVINE: Mechanism of muscular fatigue in adrenalectomized animals. Amer. J. Physiol. **162**, 10—16 (1950).
— — — Action of norepinephrine and adrenal cortical steroids on blood pressure and work performance of adrenalectomized dogs. Amer. J. Physiol. **165**, 450—455 (1951).
REIS, D. J.: Potentiation of the vasoconstrictor action of topical norepinephrine after topical application of certain adrenocortical hormones. J. clin. Endocr. **20**, 446—456 (1960).
REMINGTON, J. W., N. C. WHEELER, G. H. BOYD, JR., and H. M. CADDELL: Protective action of dibenamine after hemorrhage and after muscle trauma. Proc. Soc. exp. Biol. (N. Y.) **69**, 150—151 (1948).
RESHEF, L., and B. SHAPIRO: Effect of epinephrine, cortisone and growth hormone on release of unesterified fatty acids by adipose tissue *in vitro*. Metabolism **9**, 551—555 (1960).
RICH, A. R.: Inflammation in resistance to infection. Arch. Path. **22**, 228—254 (1936).
ROBERT, A., and J. E. NEZAMIS: Effect of cortisol on gastric ulcers of the Shay rat. Proc. Soc. exp. Biol. (N. Y.) **98**, 9—12 (1958).
ROEMMELT, J. C., O. W. SARTORIUS and R. F. PITTS: Excretion and reabsorption of sodium and water in the adrenalectomized dog. Amer. J. Physiol. **159**, 124—136 (1949).
ROGOFF, J. M., and G. N. STEWART: Adrenal insufficiency in dogs. Amer. J. Physiol. **78**, 711—729 (1926).
ROTHSTEIN, A.: The enzymology of the cell surface. In: L. V. HEILBRUNN, Ed., Protoplasmatologia Vol. 11/E/4. Vienna: Springer 1954.
RUSSELL, J. A.: The adrenals and hypophysis in the carbohydrate metabolism of the eviscerated rat. Amer. J. Physiol. **140**, 98—106 (1943).
SACERDOTE DE LUSTIG, E., y R. E. MANCINI: Acción de la deoxicorticosterona y cortisona sobra la fibrilogenesis en cultivos de fibroblastos. Rev. Soc. argent. Biol. **27**, 324—330 (1951).
SAYERS, G.: The adrenal cortex and homeostasis. Physiol. Rev. **30**, 241—320 (1950).
SCHACHTER, R. J.: The use of cholinesterase in shock. Amer. J. Physiol. **143**, 552—557 (1945).
SEIFTER, J., and D. H. BAEDER: Occurrence in plasma of an extractable lipid mobilizer (L. M.). Proc. Soc. exp. Biol. (N. Y.) **91**, 42—45 (1956).
SELYE, H.: The general adaptation syndrome and the diseases of adaptation. J. clin. Endocr. **6**, 117—199 (1946).
— First Annual Report on Stress. Montreal: Acta 1951.
SHWARTZMAN, G. (Ed.): The effect of ACTH and cortisone upon infection and resistance. N. Y. Acad. Med., Section on Microbiology, Symposium No. 6, 1953.
SOBEL, H., S. GABAY and C. JOHNSON: Effect of cortisone on connective tissues of the rat. Proc. Soc. exp. Biol. (N. Y.) **99**, 296—299 (1958).
SOFFER, L. J., and R. BADER: Corticotropin and cortisone in acute disseminated lupus erythematosus: results of long-term use. J. Amer. med. Ass. **149**, 1002—1008 (1952).
—, and J. L. GABRILOVE: Simplified water loading test for the diagnosis of Addison's disease. Metabolism **1**, 504—510 (1952).
— A. GUTMAN, J. GELLER and J. L. GABRILOVE: The role of adrenal steroids on renal function and electrolyte metabolism. Bull. N. Y. Acad. Med. **33**, 665—680 (1957).

SOSKIN, S., and R. LEVINE: Carbohydrate metabolism, 3 rd ed. Chicago: University of Chicago Press 1952.
SPAIN, D. M., N. MOLOMUT and A. HABER: Biological studies on cortisone in mice. Science **112**, 335—337 (1950).
SPRINSON, D. B., and D. RITTENBERG: Rate of interaction of the amino acids of the diet with tissue proteins. J. biol. Chem. **180**, 715—726 (1949).
STAMLER, J., R. PICK and L. N. KATZ: Intensification of cholesterol-induced atherogenesis by cortisone in chick. Circulation **4**, 461—462 (1951).
STEIGER, M., and T. REICHSTEIN: Deoxycorticosterone. Helv. chim. Acta **29**, 1164 (1937).
STETTEN, DE W., JR., and G. E. BOXER: Studies in carbohydrate metabolism. III. Metabolic defects in alloxan diabetes. J. biol. Chem. **156**, 271—278 (1944).
SWINGLE, W. W., and J. W. REMINGTON: The role of the adrenal cortex in physiological processes. Physiol. Rev. **24**, 89—127 (1944).
— W. M. PARKINS, A. R. TAYLOR and H. W. HAYS: A study of the circulatory failure and shock following trauma to the healthy, vigorous adrenalectomized dog. Amer. J. Physiol. **124**, 22—29 (1938).
— J. J. PFIFFNER, H. M. VARS, P. A. BOTT and W. M. PARKINS: Function of adrenal cortical hormone and cause of death from adrenal insufficiency. Science **77**, 58—64 (1933).
TARVER, H.: Peptide and protein synthesis. Protein turnover: In: H. NEURATH and K. BAILEY, Eds., The proteins, Vol. II/B, p. 1259ff. New York: Academic Press 1954.
TAUBENHAUS, M., and G. D. AMROMIN: Effects of hypophysis, thyroid, sex steroids, and adrenal cortex upon granulation tissue. J. Lab. clin. Med. **36**, 7—18 (1950).
TEPPERMAN, J., and F. L. ENGEL: Metabolic determinants of adrenal size and function. Josiah Macy, jr., Foundation Conference Report 1942.
— — and C. N. H. LONG: A review of adrenocortical hypertrophy. Endocrinology **32**, 373—402 (1943).
THORN, G. W., D. JENKINS, J. C. LAIDLAW, F. C. GOETZ, J. F. DINGMAN, W. L. ARONS, D. H. P. STREETEN and B. H. McCRACKEN: Pharmacological aspects of adrenocortical steroids and ACTH in man. New Engl. J. Med. **248**, 232—245, 284—294, 323—337, 369—378, 414—423, 588—601, 632—646 (1953).
UMBREIT, W. W.: Influence of adrenalectomy and cortisone treatment on enzymatic reactions in rat tissues. Ann. N. Y. Acad. Sci. **54**, 569—574 (1951).
VERZÁR, F.: Die Funktion der Nebennierenrinde. Basel 1939.
—, u. L. LASZT: Nebennierenrinde und Fettwanderung. Biochem. Z. **288**, 356—358 (1936).
WEBER, G., C. ALLARD, G. DE LAMIRANDE and A. CANTERO: Liver glucose-6-phosphatase activity and intracellular distribution after cortisone administration. Endocrinology **58**, 40—50 (1956).
WEDGWOOD, R. J., C. VAN A. HAWN and C. A. JANEWAY: Mechanisms of action of ACTH in experimental nephritis due to foreign protein. In: J. R. MOTE, Ed., Proceedings of the Second Clinical ACTH Conference, Vol. 1, 108—114. Philadelphia: Blakiston 1951.
WELLS, B. B.: The influence of crystalline compounds separated from the adrenal cortex on glucoconeogenesis. Proc. Mayo Clin. **15**, 294—297 (1940).
—, and A. CHAPMAN: Influence of pituitary, adrenal and thyroid on gluconeogenesis in phlorhizin diabetes. Proc. Mayo Clin. **15**, 503—505 (1940).
—, and E. C. KENDALL: Influence of adrenal cortex in phlorhizin diabetes. Proc. Mayo Clin. **15**, 565—573 (1940).
— — Influence of hormones of adrenal cortex on ketonuria in rats treated with phlorhizin. Proc. Mayo Clin. **16**, 113—116 (1941).
WELT, I. D., and A. E. WILHELMI: Effect of adrenalectomy and of adrenocorticotrophic and growth hormones on synthesis of fatty acids. Yale J. Biol. Med. **23**, 99—111 (1950).
— D. STETTEN, D. J. INGLE and E. H. MORELY: Effect of cortisone upon rates of glucose production and oxidation in the rat. J. biol. Chem. **197**, 57—66 (1952).
WHITELAW, M. J.: A case of Cooley's anemia treated with ACTH. In: J. R. MOTE, Ed., Proceedings of Second Clinical ACTH Conference, Vol. 2, 195—198. Philadelphia: Blakiston 1951.
WIGGERS, H. C., R. C. INGRAHAM, F. ROEMHILD and H. GOLDBERG: Vasoconstriction and the development of irreversible hemorrhagic shock. Amer. J. Physiol. **153**, 511—520 (1948).
WOODBURY, D. M.: Relation between the adrenal cortex and the central nervous system. Pharmacol. Rev. **10**, 275—357 (1958).
WOOL, I. G., M. S. GOLDSTEIN, E. R. RAMEY and R. LEVINE: Role of epinephrine in the physiology of fat mobilization. Amer. J. Physiol. **178**, 427—432 (1954).
ZWEIFACH, B. W., E. SHORR, S. BAEZ and S. J. ROSENFELD: Hepatorenal factors in circulatory hemostasis: relation of adrenals to formation of a renal vaso-excitor principle. J. clin. Endocr. **7**, 460—461 (1947).

Chapter 8

Adrenocortical insufficiency*

By

FREDERICK G. HOFMANN and EDNA H. SOBEL

Introduction

The numerous and profound biochemical and physiological disturbances that characterize adrenocortical insufficiency bear striking witness to the biological importance of adrenocortical hormones. A vast number of investigations of various aspects of adrenocortical insufficiency have been undertaken in the past 30 years; our capacity to *describe* this condition has grown accordingly. There has been no parallel growth, however, in our ability to *comprehend* the fundamental nature of adrenocortical insufficiency. The mechanism(s) of action of adrenocortical hormones remains but poorly understood; as a result, one rarely knows which are primary and which are secondary consequences of adrenocortical insufficiency.

It will be evident at many points in this chapter that extraneous blocks to our understanding of adrenocortical function also exist. They derive, in the main, from the fact that experimental conditions differ from one laboratory to another. After adrenalectomy, some investigators provide physiological saline as a drinking fluid; others administer adrenocortical hormones for a few days postoperatively or throughout the entire period of observation; still others provide no maintenance therapy whatsoever. These variations seem to be a matter of no great consequence in certain studies, for many upsets occasioned by adrenalectomy are not benefited by the administration of saline or hormones such as deoxycorticosterone. In many studies, however, the supportive measures employed may have partially remedied the state of complete adrenocortical insufficiency.

The timing of observations must also be considered. Many of the disturbances of adrenocortical insufficiency grow more intense with the passage of time: what appears inconsequential 48 hours after adrenalectomy may be dramatically evident in the preterminal period. Findings at these two periods of time cannot be compared, and yet disputes have arisen from just such comparisons.

If adrenalectomy is incomplete or if accessory adrenocortical tissue is present in the organism, erroneous findings result. In some reports, it is impossible to determine if attempts have been made to assess these factors. All of the exogenous and endogenous factors that can influence survival in states of adrenocortical insufficiency (see pp. 64—68) have been sources of obfuscation at one time or another.

For a variety of reasons, the discussion of adrenocortical insufficiency that follows is by no means complete. To avoid duplication of other chapters in the three parts of this volume, certain topics, such as the derangements in the intermediary

* This chapter was submitted in the summer of 1961; a few modifications were introduced in 1962.

metabolism of carbohydrates, proteins and lipides that occur in adrenocortical insufficiency, have been omitted. After the historical introduction, the principal emphasis has been given to reports published in the period 1950 through 1960. The discussion of the diagnosis and treatment of Addison's disease is not intended to supplant more detailed medical guides, to which reference is made, but rather to consider these subjects in relationship to the entire topic of adrenocortical insufficiency.

The most valuable general reviews on adrenocortical insufficiency are those of HARTMAN and BROWNELL (1949) and NOBLE (1950, 1955). Since 1945, selected aspects of this topic have been treated by PARKES (1945), KENDALL (1948), SORKIN (1949), GAUNT et al. (1949), MORGAN (1951), THORN (1951), NABARRO (1956), RAMEY and GOLDSTEIN (1957) and WOODBURY (1958).

Few investigations have been made of adrenocortical insufficiency in the lower vertebrates; the presently available data have been ably reviewed by CHESTER JONES (1957). (See also Chapters 1 and 4, Part 1 of this Volume.)

Historical review of adrenocortical insufficiency

A. Addison's reports

During the 18th and 19th centuries, great progress was made in accumulating knowledge about the structure and comparative anatomy of the adrenal glands (Chapter 1; SORKIN 1957). However, no advance was made in understanding their role in physiology until 1849, when THOMAS ADDISON suggested a possible connection between postmortem evidence of disease of the "supra-renal capsules" and the signs and symptoms presented by the patient. The earlier speculations about the possible meaning of these structures were reviewed by BIEDL (1913), ROWNTREE and SNELL (1931) and BOURNE (1949).

Reference is made most frequently to ADDISON's monograph (1855), which was published 6 years later than his presentation at a meeting of the South London Medical Society (ADDISON 1849). That this earlier description was received with considerable skepticism is evident from the discussion following the paper, which provoked the chairman of the meeting, JOHN HILTON, to rise to ADDISON's defense (discussion of ADDISON 1849). The critics said the adrenals could not be important since they were smaller in adults than in newborn infants, but HILTON stated he had found them "a good size even in the aged."

ADDISON's modest paper in 1849 included 5 patients who had insidious onset of *languor*, restlessness, pallor, *decreased muscle strength*, weakness of heart action, faintness and slight edema. Two of the patients recovered, but postmortem examination of three showed disease of the adrenals. By 1855 ADDISON's description of changes in skin pigmentation included vitiligo. The details of findings at postmortem examination of 11 patients (including one described by BRIGHT in 1831) make it evident that the clinical picture was not always due to adrenal insufficiency, since in 4 cases only one adrenal was affected by tuberculosis or metastatic carcinoma. Of the other 7 patients, 5 had bilateral tuberculosis of the adrenals, one had metastases to both adrenals, and in one both adrenals were atrophic (the combined weight was given as 49 grains, or 3.3 g). ADDISON was careful to include a differential diagnosis in his monograph, emphasizing malignant disease of the viscera, miasmic poisoning, and anemic disease, in which abnormalities of the spleen and lymphatic organs should aid in the clinical differentiation. He clearly regarded the *increased pigmentation of the skin* as the finding that was always present in the newly described disease.

The impact of ADDISON's reports and the subsequent efforts to unravel the role of the adrenal glands in body processes have been reviewed by several authors, whose individual perspectives provide fascinating reading (GREENHOW 1875, BIEDL 1913, VINCENT 1922, SHARPEY-SCHAFER 1924, TRENDELENBURG 1929, BRITTON 1930, ROWNTREE and SNELL 1931, KEPLER 1935, GROLLMAN 1936, SWINGLE 1937, SOFFER and ENGEL 1942, GOLDZIEHER 1944, BOURNE 1949, GAUNT and EVERSOLE 1949, SORKIN 1956, 1957). A summary of the work leading to the eventual acceptance of ADDISON's hypothesis follows.

B. Period of exploration

I. Early extirpation experiments

Attempts to establish the vital importance of the adrenal glands were rendered difficult by the results of early extirpation experiments in animals, which failed to reproduce the signs seen in patients.

BROWN-SÉQUARD (1856a, b) convinced himself that the adrenals were essential to life, since the animals he used (dogs, cats, rabbits and guinea pigs) died more rapidly after their removal than after extirpation of the kidneys. Evidently he was not disturbed by the fact that animals lived at best a few hours longer after unilateral than after bilateral adrenalectomy. He was able to prolong life for some hours in either type of experiment by the injection of blood from a healthy animal, and so suggested that a toxic substance accumulated in the blood of operated animals. GRATIOLET (1856) then felt encouraged to publish the results of his experiments with guinea pigs, reportedly done in 1853 and 1854. He showed that removal of the right adrenal or of both adrenals always led to death within 24 hours, although animals could survive removal of the left adrenal for months. (Apparently this discrepancy can be attributed to the greater incidence of hemorrhage when the right adrenal is removed, and suggests that death following bilateral adrenalectomy might also have been from hemorrhage.) TROUSSEAU (1856) also supported BROWN-SÉQUARD's contentions, since rabbits affected with a natural disease in which inflammation of the adrenals occurred had signs similar to those observed in animals following experimental extirpation; furthermore healthy animals developed the same symptoms when injected with blood from diseased rabbits.

PHILIPEAUX (1856) concluded that the adrenals were not essential to life, but related death following bilateral extirpation to the operation itself, since he found that rats survived when the left adrenal was removed two weeks after the right. In a rebuttal to this denial of his work BROWN-SÉQUARD (1857) suggested that if white rats survive adrenalectomy it must be because they do not accumulate pigment, which he supposed to be the cause of death. However, SCHIFF (1863) observed that non-albino rats were also able to survive adrenalectomy, provided time was allowed for them to become accustomed to laboratory conditions before surgery was undertaken. The impression at the time was that the adrenal glands are not essential for life. (See BIEDL 1913.)

II. Early clinical observations, hypothesis of nervous etiology

The conclusion that animals can survive adrenalectomy caused considerable difficulty for those clinicians seeking to convince the skeptics that Addison's disease is an entity. The hypothesis was proposed that the disease in man is due to destruction of ganglia of the sympathetic nervous system adjacent to the adrenals.

In 1859, WILKS, ADDISON's colleague, expressed the opinion that some of the doubts were based on erroneous views, which he attempted to correct by pointing out (1) that such conditions as jaundice or pityriasis must not be confused with the skin changes described by ADDISON, and (2) that a gradual increase in pigmentation of the skin would not have time to develop in a person who died early in the course of the disease. He also described the prominent Peyer's patches of the disease for the first time. WILKS had more difficulty in dealing with the criticism that the adrenals could be removed from lower animals with impunity, and suggested that these reports, if true, did not mean the disease would not eventually be fatal. He suggested that loss of function of the semilunar ganglia and solar plexus might be responsible for the symptoms of the disease, although he did not doubt that the lesion of the adrenal itself was related to tuberculosis.

However, only two years after ADDISON's death, WILKS (1862) decided that it was not established that the disease was connected with destruction of the adrenals, but that involvement of the sympathetic nerves was most important in explaining the signs and symptoms. He believed that of the patients described by ADDISON those with vitiligo should not be considered as having *Morbus Addisonii;* he also did not accept the patients who had carcinoma, and in this he may have been correct. Among the 25 patients WILKS reported as truly afflicted with this disease was one who habitually ate highly seasoned dishes and butter; another died after having received 2 small doses of morphine. These statements were perhaps the first indications that adrenal insufficiency may lead to a craving for salt and to an increased sensitivity to morphine.

A good deal of the difficulty experienced by Addison's contemporaries seems to have been related to the emphasis they placed on pigment changes in the skin, which was described in great detail in all case reports. In the Croonian Lectures for 1875, GREENHOW reviewed the 330 cases that had been reported as instances of Addison's disease in the preceding 20 years. He devoted much space to descriptions of the gross and microscopic appearance of the skin as well as to details of the clinical course and the pathological findings. In his efforts to differentiate between true and spurious reports, he started from the premise that in all genuine instances the disease must be due to a common lesion, which he considered to be an inflammatory involvement of the sympathetic chains adjacent to the adrenals. He attributed the misconceptions which had arisen to ADDISON; he criticized the inclusion of a patient with atrophy of the adrenals, and the use of the term "bronzed skin disease," which, he said, had identified the "disease itself with its most conspicuous but least important feature." Nevertheless, GREENHOW's review includes much valuable material. He emphasized the occurrence of gastric ulcers and enlargement of lymphatic structures of the intestinal tract in patients with Addison's disease. In 1877, WILKS returned to some extent to ADDISON's opinions about the role of the adrenal glands.

The theory that the signs and symptoms of Addison's disease are due to involvement of the sympathetic nerves was slow to die; it was mentioned as a possibility as late as 1924 (G. STEWART).

C. Period during which endocrine nature of adrenal cortex was established

I. Compensatory hypertrophy and accessory adrenal tissue

STILLING (1889) demonstrated in rabbits that compensatory hypertrophy of the remaining adrenal gland followed unilateral adrenalectomy. He observed that the average weight of an individual adrenal in rabbits was 10 mg/100 g of body weight, whereas, some months after the left adrenal had been destroyed, he found regularly that the remaining adrenal weighed 43 mg/100 g. STILLING also reported the enlargment of accessory adrenals after partial adrenalectomy and that these nodules were composed entirely of cortical tissue. That remaining cortical tissue hypertrophied after partial adrenalectomy was confirmed by STREHL and WEISS (1901), B. MOORE and PURINTON (1901), T. ELLIOTT and TUCKETT (1906), CROWE and WISLOCKI (1914), and T. ELLIOTT (1914/15). The discovery that in many animals, especially rats, accessory adrenocortical tissue hypertrophied if the animals survived adrenalectomy for a significant time explained some of the earlier conflict about the vital role of the adrenal glands (BIEDL 1913, G. STEWART 1924).

As experimental techniques improved, it became clear that bilateral adrenalectomy was always fatal to animals lacking accessory glands [ABELOUS and LANGLOIS 1891 (frogs), 1892a (guinea pigs), LANGLOIS 1893a, b (dogs), SODDU 1898 (dogs), STREHL and WEISS 1901 (a variety of mammalian species), BIEDL 1913 (dogs, cats and rabbits)].

II. Relative importance of cortex and medulla

Clinical reports relating symptoms of adrenal insufficiency to atrophy of the cortex and experimental observations that animals could survive adrenalectomy in the presence of accessory cortical tissue could have been interpreted to mean that it was the cortex and not the medulla which was essential to life. PHILLIPS (1899) found the medulla in better condition than the cortex in 17 patients with simple atrophy of the adrenal glands. SIMMONDS (1903) reported 2 patients in whom atrophy of the adrenal cortex was found at postmortem examination; distinct medullary tissue was found in both instances.

However, the relative insignificance of the medulla was not accepted readily. Even the clear-cut work done by BIEDL in 1899 (BIEDL 1913, pp. 150—152) did not succeed in settling the issue. He *removed all the interrenal (cortical) tissue from elasmobranchs (sharks), in which cortical and medullary components are anatomically separate. Those animals which survived the surgical procedure soon became weak and died.* This finding was confirmed for dogs by CROWE and WISLOCKI (1914), who removed the cortex in 4 stages from a dog that remained in good health until the last fragment of hypertrophied cortex had been removed. The medulla was well-preserved. They noted that only cortical tissue underwent compensatory hypertrophy. Subsequently WHEELER and VINCENT (1917) destroyed the medulla of one gland by cauterization and found that animals from which the other gland had been entirely removed survived without symptoms, providing some cortical tissue remained; no details of the experiments were given. HOUSSAY and LEWIS (1923) *removed one medulla and a whole gland from dogs, which remained in good condition for months but died subsequent to removal of the cortical remnant.* ZWEMER (1926/27a) repeated these experiments in cats with similar results; remnants of the medulla were found in only 2 of 10 "cortical glands" created by removal of medullary tissue and zona reticularis under a dissecting microscope. As late as 1935 KEPLER, however, stated the view probably still held by some physicians (as GOLDZIEHER 1944) that the symptoms in chronic adrenal insufficiency may be partly due to destruction of the medulla.

III. Theories of adrenal function

With the successful demonstration of the necessity of the adrenal glands, speculation concerning their function was renewed. There were two opposing theories: (1) that the organs serve to detoxify noxious substances, and (2) that they are endocrine glands.

1. Theories of anti-toxic function

The theory that toxic substances accumulate in the blood of adrenalectomized animals, suggested by BROWN-SÉQUARD in 1856, was revived. ABELOUS and LANGLOIS (1891, 1892a, b) and BROWN-SÉQUARD (1893) reported that the blood of adrenalectomized animals was toxic to intact ones, and that life of deficient animals could be prolonged for a short time by giving blood, saline, or an extract of adrenal glands. ALBANESE (1892a) suggested that the adrenals serve to neutralize the products of muscle fatigue, while CHARRIN and LANGLOIS (1894) concluded, from *in vitro* experiments, that adrenal tissue was able to fix toxic substances, or to transform them chemically. ALBANESE (1892b) and MARINO-ZUCCO and DUTTO (1894) thought that the detoxifying function of the adrenals was the neutralization of "neurine" (apparently choline), which was found in large quantities in the urine of Addisonians, and to which adrenalectomized animals were sensitive. BOINET (1896) considered neurine of little specific importance, since adrenalectomized rats were also sensitive to atropine. Others suggested that exogenous toxins were neutralized by the adrenals (BERNARD and BIGART 1902, OPPENHEIM and LOEPER 1903a, b).

SODDU (1898) performed experiments in dogs adrenalectomized in 2 stages in an attempt to determine whether the animals died because of damage to the sympathetic nervous system or because the adrenals were essential organs. Since he believed that death was due to alterations in the blood, he removed 20 to 25% of the total blood volume and replaced it with what he considered an innocuous substance – physiological saline solution. By this maneuver he succeeded in prolonging the lives of his experimental animals for some hours; he concluded that the saline probably had helped by diluting a toxic substance, although he raised the possibility that it might have served to maintain a normal physiological property of the blood. This appears to have been the first demonstration that physiological saline benefits the adrenalectomized mammal.

Even after clear evidence of the potency of extracts from the adrenal cortex had been acquired, the hypothesis persisted for some time that death in adrenal insufficiency is due to accumulation of a toxic substance (SILVETTE and BRITTON 1932, ROGOFF 1944).

2. Theories of secretory function

Strehl and Weiss (1901) favored the idea that the adrenal secretes a substance into the blood stream. The data on which they based this concept were not especially sound (see Biedl 1913, p. 148), yet they were inspired to try the effects of adrenal venous blood or of the adrenal gland itself on survival after adrenalectomy. Intravenous administration of either of these extracts to adrenalectomized rabbits led to death from emboli. Extracts prepared by other investigators seeking to determine whether the adrenals were in fact ductless glands with a secretory product were also toxic to adrenalectomized animals.

Bernard et al. (1903) proposed that the high lecithin content of the adrenal indicates that the gland produces it in an active secretory process.

IV. Extracts of adrenal glands

Success in isolation of an active material from the adrenal glands, at first limited to the medulla, gave great strength to the hypothesis of secretory function.

1. Medullary extracts

Finally an active extract was prepared from the adrenal glands. However, it was found to be present in much larger amounts in the medulla than in the cortex, and it could not be extracted from the caseated medulla and sclerosed cortex of patients with Addison's disease (Oliver and Schäfer 1895). This exciting discovery of adrenaline, or epinephrine, retarded the progress of research on a secretory product from the adrenal cortex for many years, although it clearly established the existence of an internal secretion from the gland.

Treatment of Addisonian patients or adrenalectomized animals with medullary extracts was not successful. Boinet (1903) commented on severe symptoms following administration of epinephrine to patients, and warned against its use. Biedl (1913, pp. 166—167) specifically denied that adrenaline could be regarded as substitution therapy since it did not prolong the lives of adrenalectomized animals.

2. Cortical extracts

Since choline (neurine) could be extracted in larger amounts from the adrenal cortex than the medulla, some investigators suggested that this substance might be antagonistic to epinephrine and might in fact play a role in adrenal physiology. The false trail of neurine production by the adrenal gland is reviewed interestingly by E. Schäfer (1908) and Biedl (1913, pp. 274—279). Biedl himself paid much attention to the lipides present in the adrenal cortex and attempted to characterize them after extraction and separation with organic solvents. However, he neither administered any of his fractions to adrenalectomized animals nor regarded them as secretory products (Biedl 1913, p. 292).

The rich lipide content of the adrenal cortex led to theories that the adrenals might serve as a source of lipides for the body or the central nervous system (Sharpey-Schäfer 1924) or that, by virtue of the high lipide content, the gland was able to absorb and neutralize toxic products of metabolism (Bonnamour 1902, quoted by Biedl 1913).

Efforts to obtain active extracts from the adrenal cortex continued to be generally unsuccessful until 1927, when Hartman et al. and Rogoff and Stewart simultaneously reported the preparation of extracts that were effective in prolonging the lives of adrenalectomized animals. Also in 1927 Kühl described an improvement in the muscle function of adrenalectomized guinea pigs following injection of a lipide-soluble extract of the adrenals. Goldzieher (1928) prepared an extract, which he called "interrenin", by repeated extraction and precipitation from lipide solvents, after initial extraction of ground adrenal glands with dilute hydrochloric acid. Unfortunately the activity of his material was not demonstrable beyond doubt.

Rogoff and Stewart (1928b) were able to prolong life in adrenalectomized dogs for about 10 days by daily intravenous administration of 0.5 to 1.0 ml of extracts of dog or sheep adrenals. The immediate results were less dramatic than those obtained with Ringer's solution. Since they offered no details as to the means of preparation or the amount of gland represented by each ml of extract, the potency of the material cannot be evaluated. Hartman et al. (1928a, b) stated that one ml of "cortin" represented 0.2 g of beef adrenal cortex; this probably explains its ineffectiveness in animals with advanced symptoms of insufficiency and its inability to protect against the stress of surgery or infection. In 1930 HARTMAN et al. improved the potency of their material by using diethyl ether as the solvent; with this extract adrenalectomized cats survived 80 to 207 days (as compared to 2- to 35-day survival times for untreated animals). The dose used was equivalent to 10 to 20 g of fresh cortex daily, but more was needed in the presence of infection.

A lipide extract, each ml of which represented 30 g of fresh cortex, was prepared by PFIFFNER and SWINGLE (1929). This material caused considerable local irritation (SWINGLE and PFIFFNER 1930a), but an aqueous fraction made from the lipide extract was able to maintain adrenalectomized cats in good condition for 100 days (SWINGLE and PFIFFNER 1930b) and was potent enough to revive moribund animals (SWINGLE and PFIFFNER 1930c).

Further purification led to extracts giving even better results in animals. Preparations which could be given intravenously (5 to 7 ml, 3 times daily) were able to restore cats in coma to normal activity, and as much as 32 ml could be given intravenously to a 2-kg cat without evoking evidence of toxicity (SWINGLE and PFIFFNER 1931a, b, PFIFFNER and SWINGLE 1931). Oral administration was found effective in cats, although the dose required was substantially greater than that effective by the parenteral route (BRITTON and SILVETTE 1931b).

ROGOFF and STEWART (1929) were the first to treat patients having Addison's disease with extract, which they gave in capsules by mouth although they had found the oral route ineffective in dogs and cats (G. STEWART and ROGOFF 1929/30). ROWNTREE and GREENE (1930) *were able to carry a patient through* 3 *crises with the Swingle and Pfiffner extract;* preparations which had been freed of epinephrine could be given intravenously in 20 to 50 ml doses. Further reports of experience in treating patients with adrenal insufficiency suggested that extracts had limited usefulness, as many symptoms persisted and patients often died despite administration of what were considered large amounts (ROWNTREE et al. 1931a, b, HARROP et al. 1931a, HARROP and WEINSTEIN 1932). HARTMAN et al. (1932b) gave the equivalent of 1200 g of cortex daily to a patient in severe crisis and emphasized that the dose required in infection was much greater than that adequate for maintenance.

Some of the disappointing results of treatment with extract and skepticism about its effectiveness were probably due to the rather large amounts needed in comparison to the relatively small weight of the adrenal glands. BRITTON (1932) was vehement on this point, since cats, whose adrenals weighed only 100 mg per kg, needed the extract from 40 g to relieve symptoms of insufficiency. However, the endocrine nature of the adrenal cortex was now firmly established.

D. Adrenocortical dependence on the anterior pituitary

The earliest suggestion of an association between the adrenals and the pituitary was the observation by BENDER and LERI (1903) that in anencephalic fetuses the adrenals were very small; they thought the finding indicated dependence of the adrenal on the brain. MEYER (1912) also thought that adrenal atrophy in anencephaly was a result of malformation of the brain. KOHN (1922), however, concluded that malformation of the anterior pituitary was the cause of poor development of the adrenals.

Experimental work on pituitary: adrenal relationships was initiated by ASCOLI and LEGNANI (1912), who observed atrophy of the inner zones of the adrenal cortex in hypophysectomized dogs. P. SMITH and SMITH (1922, 1923) found atrophic adrenals in hypophysectomized tadpoles, and were able to restore the glands to normal size by injections of anterior pituitary extracts. P. SMITH (1927) made similar observations in the rat. ATWELL (1937) summarized some of this early work. (See also Chapter 1, Part 1 of this Volume).

The effectiveness of "cortin" in repairing a defect resulting from hypophysectomy was noted by BAIRD et al. (1933) in an investigation of decreased tolerance for cold in hypophysectomized rats. In 1932 HOUSSAY et al. demonstrated that hypophysectomy abolished the glycosuria of partially pancreatectomized dogs. However, HOUSSAY (1936) was not sure that this effect of hypophysectomy could be ascribed to adrenal insufficiency, since adrenalectomy was less effective than hypophysectomy in modifying diabetes. LONG and LUKENS (1936) found that adrenalectomy did ameliorate experimental diabetes in cats, and suggested that hypophysectomy might be effective because of the consequent adrenal atrophy.

LONG (1937) demonstrated that anterior pituitary extract displayed no diabetogenic properties when given to adrenalectomized animals. COREY and BRITTON (1937), who used adrenalectomized-hypophysectomized rats, suggested that the adrenals are primarily concerned in the change in carbohydrate metabolism following hypophysectomy.

The recognition of pituitary insufficiency in man was delayed, since pituitary lesions lead to such a variety of manifestations (BIEDL 1913). A few years later atrophy of the thyroid, gonads and adrenals was described as associated with insufficiency of the anterior pituitary (SIMMONDS 1918). BAUER (1935) *clearly stated that atrophy of the adrenal cortex in Simmonds' disease was a result of pituitary disease.*

In 1937 INGLE and KENDALL induced atrophy of the adrenals by giving cortical extract to intact rats. The chain of evidence linking these 2 glands was completed the following year, when INGLE et al. (1938) proposed the working hypothesis that anterior pituitary secretion is regulated by an excess or deficiency of cortical hormone and thus is able to meet the physiological requirements of the organism.

In 1940, SWANN reviewed the progress made in defining the relationship between pituitary and adrenals.

E. Signs of complete and partial adrenal insufficiency

Clear demonstration that removal of the adrenal glands is incompatible with survival in experimental animals still did not dispose of the difficulty that the signs produced by adrenalectomy in animals do not bear much relationship to those seen in human beings afflicted with Addison's disease. Sometimes the differences were attributed to the fact that the duration of the disease was much shorter in experimental animals, and NOTHNAGEL, TISSONI and others made unsuccessful attempts to induce chronic lesions in animals (see BIEDL 1913, p. 138). OPPENHEIM and LOEPER tried unsuccessfully to reproduce the disease as seen in man by injecting cultures of tubercle bacilli (1903a) and extracts of tissues from people with tuberculosis (1903b) into the adrenal glands of rabbits. The efforts of LACASSAGNE and SAMSSONOW (1923) to induce partial insufficiency in rabbits by local irradiation demonstrated once again that destruction of the medulla is quite harmless; only those animals in which destruction of the cortex was achieved died. In 1932 ROGOFF (1932a, 1944) undertook the production of chronic adrenal insufficiency in dogs and cats by the application of ligatures to the adrenal blood vessels. Sixty per cent of the animals developed anorexia, gastrointestinal upsets, neuromuscular and circulatory collapse.

The ultimate establishment of the basic similarity between experimental and spontaneous adrenal insufficiency had to await the evolution of methods permitting detailed study of the alterations in function which follow adrenal insufficiency, regardless of its etiology, experimental or natural.

As operative techniques improved so that animals survived longer after adrenalectomy, observations concerning its effects became more meaningful (ROGOFF and STEWART 1921, HARTMAN et al. 1922). GIBELLI (1908/09) reported the presence of stomach ulcers in 2 of 13 rabbits and 3 of 11 dogs after bilateral adrenalectomy; he never observed ulcers in animals after unilateral adrenalectomy. These observations were confirmed by FINZI in 1913. T. ELLIOTT (1914/15) noted ulcers in cats that had survived 2-stage bilateral adrenalectomy for more than 6 days. He described the condition of these long-survivors as including weakness, decreased blood pressure, decreased body temperature unless artificial heat was supplied, and softness of the eye balls.

The biochemical findings of many investigators provided isolated bits of information. Decreased blood sugar concentration (PORGES 1909/10), decreased liver glycogen content (PORGES 1910, SCHWARZ 1910) and the amelioration of diabetic glycosuria (FROUIN 1908) were all noted after adrenalectomy. An increased concentration of red blood cells was found and ascribed to dehydration (GRADINESCU 1913). T. ELLIOTT (1914/15) found the blood pressure of adrenalectomized cats decreased. Other observations included elevated blood urea nitrogen (PORAK and CHABANIER 1914), acidosis (ROWNTREE 1925, YONKMAN 1926), elevated plasma phosphate (LUCAS 1926) and sulfate concentrations (SWINGLE and WENNER 1927), decreased electrical conductivity of the blood (ROGOFF and STEWART 1926b), and elevated potassium and decreased sodium concentrations (BAUMANN and KURLAND 1927).

Efforts to unravel the interrelationships of these abnormalities led to much speculation; the period from 1920 to the mid-1930s was exciting indeed.

The disturbances of function which were under discussion at this time will be considered in separate categories of dehydration, renal function, acidosis, distribution of electrolytes and water, and carbohydrate metabolism.

I. Dehydration

Dehydration and hemoconcentration were observed before modern biochemical tools were available. T. ELLIOTT (1914/15) described decreased blood pressure and softness of the eyeballs in adrenalectomized cats. Both GRADINESCU (1913) and LUCAS (1926) interpreted an increased number of red blood cells as a reflection of decreased plasma volume, possibly consequent to increased capillary permeability. The same suggestion was made by DONÁTH (1914) as an explanation for increased blood concentration in adrenalectomized cats; he attributed their lowered blood pressure to loss of fluid from the blood.

BRITTON (1930) reasoned that fluid could not be lost because of increased capillary permeability, as no edema developed; he suggested that the anorexia of adrenalectomized animals might be the basis for their dehydration.

The dehydration of adrenal insufficiency, with its accompanying hypotension and hemoconcentration, was recognized as being very similar to that occurring in vascular shock. SWINGLE et al. (1933) therefore attempted to account for the entire chain of events on the basis of loss of fluid from the circulation by transudation through the capillary walls, with the increased osmotic pressure of the plasma, assumed to occur since plasma proteins were increased, being insufficient to draw fluid back into the circulation. The only difference they recognized between adrenal insufficiency and traumatic shock was the low blood sugar of adrenalectomized animals. This claim led to rather violent attacks by other investigators (BRITTON and SILVETTE 1933, FREEMAN 1933, CATTELL 1933). SWINGLE (1937) pointed out that the idea that dehydration was a consequence of altered renal function was untenable, since signs could develop and be relieved in anuric animals.

An understanding of the genesis of dehydration in adrenal insufficiency was not achieved until electrolyte changes in the blood were understood (p. 37ff.).

II. Renal function

The first substance found to be increased in the concentrated blood of adrenalectomized animals was urea nitrogen (PORAK and CHABANIER 1914). This observation was confirmed in 1916 by E. MARSHALL and DAVIS; in addition they observed diminished ability to excrete a test dose of creatinine or urea. Since adrenalectomized cats died before severe uremia had developed, MARSHALL and DAVIS suggested that renal failure might be secondary to low blood pressure or to the absence of a substance secreted by the adrenals.

G. STEWART (1924), considering possible causes for the abrupt development of signs of deficiency in animals that had remained well for some days after adrenalectomy, proposed that a disruption of renal function might be responsible, with the kidney becoming unable to eliminate toxic substances. He regarded low blood pressure as purely a terminal phenomenon with no relationship to adrenal insufficiency. Elevation of blood urea nitrogen (BUN) and its decrease after the intravenous administration of fluids were described in patients by ROWNTREE (1925), who suggested that this finding was secondary to low blood pressure. LUCAS (1926) emphasized that the rises in blood urea nitrogen and non-protein nitrogen were less than those observed after ligation of the ureters and probably were not the cause of death. Other observations taken as indications of impaired renal function were: decreased nitrogen excretion (LUCAS 1926), relatively poor excretion of dye (YONKMAN 1928) and impaired excretion of a water load (ROWNTREE 1925).

Somewhat more detailed studies of renal function were made by SILVETTE and BRITTON (1932), who tested the ability of adrenalectomized rats to excrete phenolsulfonphthalein. The difference between its excretion by intact animals (71% of the dose) and adrenalectomized ones (67%) may be regarded as negligible; however there was a marked difference in the output of the administered water (83% in intact as compared to 35% in operated animals). GERSH and GROLLMAN (1939) examined the renal function of adrenalectomized rats, cats and dogs by administering ferrocyanide intravenously and examining the kidney for areas in which dye had been localized as an indication of the site and intensity of water reabsorption. Their findings indicated that an increased amount of water was reabsorbed in the descending limb of the loop of Henle in animals with moderate signs of adrenal insufficiency, while no fluid filtered through the glomerulus when the animal had been prostrate at the time of sacrifice. They expressed the opinion that any electrolyte changes in adrenalectomized animals were secondary to renal dysfunction, although no pathological alterations were observed in the kidneys. HARRISON and DARROW (1939) made detailed studies of renal function in 2 dogs before and after adrenalectomy. Their investigation, carried out by determination of urea and creatinine clearances and measurements of rates of electrolyte excretion, established clearly that, after adrenalectomy, Na^+ reabsorption was decreased and K^+ and PO_4^{3-} reabsorption increased in relation to the concentrations of these substances in the plasma. Since the glomerular filtration rate (GFR) of intact dogs decreased when their Na^+ intake was restricted, it was suggested that the abnormally low GFR after adrenalectomy was secondary in part to Na^+ depletion. Treatment with sodium salts resulted in only partial restoration of renal function, while the defect could be repaired entirely by the administration of adrenocortical extract.

TALBOTT et al. (1942) were unable to restore the reduced GFR of patients with adrenal insufficiency to normal by treatment with adrenal extract; it seems probable that the doses they used were too small. They found that the GFR was impaired much more than were other parameters of renal function (such as renal blood flow, ability to concentrate the urine and to excrete phenolsulfonphthalein). Although there was considerable improvement in renal function when the patients were treated with deoxycorticosterone acetate, these authors suggested that multiple factors were probably involved, including decreased tone of efferent arterioles and structural changes in the kidneys.

III. Acidosis

The results of investigations failed for some time to explain the relative acidosis characteristic of adrenocortical insufficiency, since its origin was at first sought in an accumulation of acid substances. ROWNTREE (1925) noted acidosis in patients with adrenal insufficiency, and administered sodium bicarbonate to combat it. In 1926, YONKMAN found that the CO_2 content of the blood of adrenalectomized dogs was decreased, as was the pH. SWINGLE (1926/27) concluded that the high PO_4^{3-} concentration in the blood of adrenalectomized cats was an important factor in the development of uncompensated metabolic acidosis and that the rise was greater than could be accounted for by the degree of hemoconcentration. Unfortunately, little attention was paid to the slight decrease in Na^+ concentration found in the blood of 7 of 20 animals; few determinations of Na^+ were made "since the level of this substance is not known to vary greatly in any pathological condition."

In further studies of the acidosis of adrenalectomized cats (SWINGLE and EISENMAN 1926/27), total acid was found diminished, essentially because of decreased CO_2 (which could not be attributed to overventilation). The metabolic acidosis was considered to be due to an increase in unidentified organic acids since it could not be accounted for entirely by the increase in PO_4^{3-} and ketone bodies. The erroneous interpretation came about because an increase in total base cation was expected to occur as a compensation for acidosis. A decrease in total base was found in 3 animals, but this was thought to be a consequence of dehydration. Studies in dogs, from which larger blood samples could be obtained, demonstrated a marked increase in SO_4^{2-} (to 12 mg%) as coma developed; however even the combined SO_4^{2-} and PO_4^{3-} was regarded as probably inadequate to account for the degree of acidosis (SWINGLE and WENNER 1927). Although they carried out careful studies of plasma cation concentration in adrenalectomized cats and rabbits, BAUMANN and KURLAND (1927), in considering acid-base balance, commented only that they found no evidence of alkalosis.

The explanation of the relative acidosis in adrenal insufficiency awaited more detailed study of sodium and chloride metabolism.

IV. Sodium and chloride metabolism

In 1926 ROGOFF and STEWART (1926b) found that the electrical conductivity of the blood of adrenalectomized animals was decreased, and commented that chloride must be lost in excess of water. Decreased blood Cl^- concentration in adrenalectomized dogs was reported by BANTING and GAIRNS (1926). LUCAS (1926) suggested that the decrease in chloride concentration could be a consequence of PO_4^{3-} elevation.

The acidosis of adrenal insufficiency, considered a sign of "intoxication," remained unexplained until studies of sodium balance demonstrated that there was loss of base rather than accumulation of acid. In 1927 BAUMANN and KURLAND found a definite decrease in Na^+ concentration in the plasma of adrenalectomized cats and rabbits. They found that intravenous administration of sodium salts prolonged life. BRITTON (1930) also mentioned decreased Na^+ and Cl^- concentrations in the blood, but did not connect these alterations with hemoconcentration and dehydration.

That the decrease of total base was entirely the result of decreased Na^+ was established by LOEB (1932) in studies of 3 patients with Addison's disease. He was impressed by the improvement in a patient given 12–15 g of NaCl daily. The following year observations in an adrenalectomized dog (LOEB 1933) showed that Na^+ had been lost by excretion in the urine, and it was clear that the benefit derived from treatment with NaCl was due to replacement of Na^+. Further studies of patients with adrenal insufficiency showed that K^+ excretion did not accompany excess Na^+ excretion, although it did in other circumstances in which Na^+ was lost in the urine (LOEB et al. 1934). They demonstrated also that Na^+ excretion was not the consequence of acidosis, since there was no increase in the excretion of ammonium ion or in titratable acidity of the urine in a patient as acute signs of insufficiency developed. Because of these findings they suggested that the kidney might be the locus of action of a sodium-regulating factor of the adrenal. In 1935, LOEB et al. summarized their observations on the relationship of dehydration and acidosis to low plasma Na^+ and Cl^- in patients with adrenal insufficiency. Emphasis was placed on the fact that Na^+ loss was greater than that of either Cl^- or water. They found that daily treatment with 10 to 15 ml of the adrenocortical extract then available was less efficacious than the addition of NaCl to the diet.

ZWEMER (1934) also reviewed the evidence that the dehydration and acidosis of adrenal insufficiency was a consequence of loss of sodium; he was able to effect repair of the electrolyte disturbance in adrenalectomized cats with either sodium salts or adrenocortical extract. SWINGLE et al. (1934b) were unable to obtain satisfactory results in dogs unless extract was given in addition to salt. Probably excessive quantities of salt (1 to 2 g/kg) provoked the vomiting which occurred in these adrenalectomized dogs and were responsible for further losses of electrolytes and water.

V. Water distribution

SILVETTE and BRITTON (1932) found that in adrenalectomized cats the water content of muscle and liver was increased and that of the blood decreased. Similar results were obtained in rats, the analyses being done on samples of the ground carcass. They assumed that a shift of water into the tissues would lead to increased osmolarity in the blood and suggested that increased Cl^- excretion might serve as a partial compensation. Later studies demonstrated decreased glycogen and Cl^- in tissue, and the proposal was made that increased tissue water was an osmotic consequence of decreased tissue glycogen (SILVETTE 1934, SILVETTE and BRITTON 1936).

GILMAN (1934) *discussed the necessity of a shift of water from extra- to intracellular space when Na^+ is lost from the extracellular fluid.* He demonstrated that dogs that had lost Na^+ into the peritoneal space after intraperitoneal injection of glucose withstood hemorrhage as poorly as did adrenalectomized animals. GILMAN felt that renal retention of water would be necessary in order to reconstitute extracellular fluid volume and composition in adrenal insufficiency. SWINGLE et al. (1934a) made careful studies on adrenalectomized dogs allowed to develop severe insufficiency after the withdrawal of adrenal extract and then given large doses of extract (103 ml to a 12.1 kg dog) while food and fluid were withheld. It was evident that substantial recovery, accompanied by diuresis, occurred under these circumstances, so that at least some of the water and salt missing from the circulation must have been present somewhere in the animals themselves. HARROP et al. (1935) confirmed repair of adrenal insufficiency in dogs treated with adrenal extract despite a negative balance for Na^+ and water during the first day of recovery. SWINGLE et al. (1936a) found that adrenal extract could alleviate the symptoms and restore the blood pressure of adrenalectomized dogs maintained on a diet low in NaCl. Increase of extracellular volume was reflected in a fall of urea nitrogen without a change in Na^+ concentration.

Despite the evidence cited above, GROLLMAN (1936) stated that the water content of the tissues was not increased in adrenal insufficiency, but rather that osmotically active solutes were lost by damaged renal tubules. He apparently thought that no change in osmotic pressure occurred and commented: "The view that the adrenal cortical hormone in some way controls 'osmotic pressure' is too fantastic to require comment."

VI. Potassium metabolism

BAUMANN and KURLAND (1927) *were the first to describe elevated* K^+ (as well as decreased Na^+) *concentration in the plasma of adrenalectomized cats.* They emphasized the marked alteration in the plasma Na^+/K^+ ratio (20 in the normal and only 12 in adrenalectomized animals). HASTINGS and COMPERE (1931) found that the serum K^+ of dogs rose markedly (from 3 to 20 meq) 48 hours after adrenalectomy and suggested that hyperkalemia might be related to the cause of death. The occurrence of hyperkalemia in patients with Addison's disease was recognized by LOEB (1932, 1933), at the same time that depression of Na^+ concentration and repair of the electrolyte abnormalities by administration of Na^+ salts were noted. LOEB pointed out that hyperkalemia in the presence of hyponatremia was unique to this disease.

ZWEMER (1934) and ZWEMER and TRUSKOWSKI (1936, 1937) placed great emphasis on the elevation of K^+, believing that it induced Na^+ diuresis and hypoglycemia. They were able to reproduce some of the signs of adrenal insufficiency by intraperitoneal administration of KCl to normal cats, and found that adrenalectomized cats were killed by one-fifth the dose tolerated by intact animals and that cortical extract could reduce induced hyperkalemia as well as that of adrenal insufficiency. The observation of increased K^+ excretion during the repair of adrenal insufficiency in dogs treated with adrenal extract was interpreted by HARROP et al. (1935) as possibly reflecting prior cellular damage and alteration in cellular permeability. Analysis of electrolyte distribution between serum and cells (NILSON 1937) showed an increase of K^+ in both compartments in adrenalectomized dogs when a crisis was induced either by a low Na^+ diet or by a gradual increase of K^+ supplement to a low Na^+ diet.

While the idea grew that potassium might be the toxic factor so long thought to be present in adrenal insufficiency, INGLE et al. (1937) demonstrated that in adrenalectomized-nephrectomized rats plasma K^+ concentration could be reduced by administration of an adrenocortical extract. In animals in which muscle work was forced by faradic stimulation, death occurred at K^+ concentrations lower than those found in animals that were not stimulated (8.5 meq/l compared to 10.5 meq/l). Evidently the absolute concentration of potassium was not the only factor leading to death.

ALLERS et al. (1936) found that K^+ was a toxic substance for adrenalectomized dogs, while WILDER et al. (1937) showed that patients with Addison's disease excreted more Na^+ when K^+ intake was high. The detailed renal function studies carried out by HARRISON and DARROW (1939) demonstrated that in adrenal insufficiency K^+ excretion by the kidney was low in relation to plasma concentration. The maximal concentration of K^+ in the urine in relation to plasma was 10, while urine to plasma ratios as high as 75 were observed in normal fasting dogs given excess K^+.

Reviews by KENDALL (1937, 1948) and by SWINGLE and REMINGTON (1944) summarize the development of understanding of changes in electrolyte distribution in adrenal insufficiency.

VII. Carbohydrate metabolism

1. Hypoglycemia

Porges (1909/10) first observed hypoglycemia in adrenalectomized dogs, 13 hours after surgery, and in 3 patients thought to have adrenal insufficiency. He concluded that the low blood sugar was not a consequence of diminished blood pressure but that it probably resulted from damage to nerves rather than being a simple result of absence of the adrenals. Barlow and Ellis (1924) found significantly low blood sugars in adrenalectomized cats. G. Stewart (1924) reported that the blood sugar of adrenalectomized cats, rabbits and dogs was not significantly different from that of intact animals. Although Rogoff and Stewart (1926b) found diminution of blood sugar in adrenalectomized animals, they regarded it as a terminal finding without significance. Banting and Gairns (1926) also thought that the adrenal cortex played no direct role in carbohydrate metabolism, since glucose prolonged the lives of their adrenalectomized dogs for only a few hours and since pancreatectomy following adrenalectomy led to an increase in blood sugar.

Bøggild (1925) *concluded that the hypoglycemia of adrenalectomized dogs must be attributed to absence of the cortex.* Hypoglycemia was observed in patients with Addison's disease (Rowntree 1925) and in adrenalectomized dogs (Yonkman 1926); Yonkman believed that the anorexia and lassitude in adrenal insufficiency were related to low blood sugar as well as to acidosis. Swingle (1926/27) observed hypoglycemia in adrenalectomized cats, and regarded this as a pathological change distinct from "acid intoxication," which he considered to be the cause of death. His animals could be revived only temporarily by the administration of glucose; since he gave 10 units of insulin with the glucose he may have inadvertently caused death in hypoglycemia. That hypoglycemia also occurred in adrenalectomized rats was established by Wyman and Walker (1929), who chose this animal deliberately with the thought that the frequent occurrence of accessory adrenals would allow them to study chronic adrenal insufficiency. Their studies demonstrated that the cortex was of much greater importance than the medulla in maintaining blood sugar, since hypoglycemia was not observed in those animals with substantial cortical accessory tissue. The extent of knowledge in 1930 in regard to carbohydrate metabolism in adrenal insufficiency was summarized by Britton.

In 1934 Zwemer and Sullivan reported extensive experience with determinations of blood sugar in adrenalectomized cats. They found essentially no difference in the average blood sugar of intact cats and those from which one adrenal medulla, an entire adrenal, or an entire adrenal plus the other medulla had been removed (average 100 mg%), while in those totally adrenalectomized the blood sugar was diminished (average 78 mg%). These authors found that adrenocortical extract was able to elevate the blood sugar of adrenalectomized animals; its effect was of lesser magnitude but greater duration than that following injection of epinephrine. These observations established that the adrenal cortex plays an important role in carbohydrate metabolism.

2. Liver glycogen

In 1910 Schwartz reported that the livers of adrenalectomized rats were almost entirely depleted of glycogen, and that a single feeding of glucose effected a partial restoration. Porges (1910) found low glycogen contents in the livers of adrenalectomized dogs whether they were fed meat or glucose prior to surgery. Kuriyama (1918) thought the low liver glycogen of adrenalectomized rats, fed 12 hours before sacrifice, reflected their general emaciation; he found the blood sugars to be within the normal limits of variation. BORNSTEIN and HORNEMANN (1923) attributed the decreased liver glycogen they found in adrenalectomized dogs to a disturbance in respiration.

A relationship between hypoglycemia and the decreased liver glycogen was established by CORI and CORI (1926). They demonstrated that adrenalectomized rats could replenish liver glycogen, depleted by a 24 hour fast, from glucose administered by stomach tube. They suggested that new formation of carbohydrate from protein might be disturbed by adrenalectomy. COREY and BRITTON (1937) observed that hypophysectomized rats, fed *ad libitum*, were able to maintain normal blood sugar and liver glycogen, while subsequent adrenalectomy led to a decline in both. Their results suggested to them that the changes in carbohydrate metabolism subsequent to hypophysectomy resulted principally from changes in the adrenals. RUSSELL and BENNETT (1937) demonstrated clearly that although hypophysectomized rats maintained normal blood sugar and liver and muscle glycogen when fed *ad libitum*, even an 8 hour fast led to a decrease in these values. Since hypophysectomized rats deposited less liver glycogen than intact animals when fasted after high carbohydrate feeding, it seemed clear that their rate of glucose utilization must have been greater.

3. Effect of adrenals on glycosuria

As early as 1908 FROUIN observed that in dogs which had been partially adrenalectomized, subsequent removal of three-quarters of the pancreas led to much less severe glycosuria than was found in similarly semi-pancreatectomized dogs with intact adrenals. Since FROUIN's experimental animals survived pancreatectomy for 16 and 24 days his results are probably meaningful.

The failure of phlorhizin to cause glycosuria in adrenalectomized dogs, and of epinephrine to cause glycosuria in patients with Addison's disease was interpreted as a result of absence of the adrenal medulla (EPPINGER et al. 1909). ROSENOW and JAGGUTIS (1922) showed that epinephrine did not cause hyperglycemia in patients with adrenal insufficiency.

In 1932 HOUSSAY et al. discovered that hypophysectomy abolished the diabetes of pancreatectomized dogs, and that glycosuria could be restored by the administration of anterior-lobe extract. Similar results were obtained for frogs and toads (HOUSSAY et al. 1936). HARTMAN and BROWNELL (1934) showed that the adrenal glands were essential for the occurrence of diabetes in pancreatectomized cats, but were puzzled by the failure of large amounts of "cortin" to restore glycosuria. LELOIR (1934) confirmed that total adrenalectomy led to modification of the intensity of pancreatic diabetes in dogs. BRITTON and SILVETTE (1937a) found that cats died within 24 hours of simultaneous adrenalectomy and pancreatectomy; when adrenalectomy was performed a week after pancreatectomy a 50% decline in blood sugar was noted within 15 hours.

4. Sensitivity to insulin

SUNDBERG (1923) found that bilaterally adrenalectomized rabbits were much more sensitive to insulin than were those subjected to laparotomy alone; he interpreted these results as indicative of the importance of the adrenal medulla because animals in which part of the adrenal was transplanted to the abdominal wall responded normally to insulin. Other reports of increased sensitivity of adrenalectomized dogs to insulin followed (J. LEWIS and MAGENTA 1925, HALLION and GAYET 1925). UNVERRICHT (1926) published a report about a young man with diabetes whose glycosuria disappeared and who began to have severe insulin reactions. Even when his dose of insulin had been decreased from 40 to 5 U daily his blood sugar was found to be 38 mg %. The diagnosis of Addison's disease, made on the basis of muscular weakness and fall of blood pressure from 125/50 to 90/50, was confirmed at autopsy.

5. Role of adrenal cortex in carbohydrate metabolism

In the period 1924–1940, there was considerable controversy about the relationship of the adrenal glands to carbohydrate metabolism. G. STEWART (1924) rejected as insignificant the observations of others that hypoglycemia and decreased liver glycogen occurred in adrenalectomized animals, suggesting that poor surgery, short survival periods, or general debility might be the explanation. Similarly he dismissed the finding of diminution in the severity of diabetes when both pancreatectomy and adrenalectomy had been performed. The vehemence of his views was clearly related to the general feeling that epinephrine played the major role in carbohydrate metabolism, and he was determined to establish that the deterioration following adrenalectomy was due to loss of the cortex rather than to the loss of the medulla. KENDALL (1938) suggested that changes in carbohydrate metabolism were secondary to changes in the distribution of K^+.

BRITTON and his co-workers were the strongest defenders of the viewpoint that the major abnormality in adrenal insufficiency was a disturbance in carbohydrate function. This concept was due in part to their studies on the opossum and marmot, in which plasma Na^+ and Cl^- both rise after adrenalectomy, while hypoglycemia occurs as in other species (BRITTON and SILVETTE 1934, SILVETTE and BRITTON 1936, 1938).

LONG (1937) could not accept the possibility that disturbed carbohydrate metabolism was secondary to changes in electrolytes, since administration of adrenocortical extract did not cause glycosuria to reappear in adrenalectomized-pancreatectomized rats. However, he found that diabetes could be restored in such animals by the implantation of adrenal tissue from young rats. He suggested that perhaps the dose of extract used was inadequate, or that the extract itself was incomplete. The increased availability of adrenocortical extract permitted further investigation of this point. KATZIN and LONG (1939) found a decrease in respiratory quotient from 0.86 to 0.78, accompanied by an increase in liver glycogen, in intact, fed rats that were given adrenal extract; the observations were interpreted as indicating decreased utilization of glucose. When extract was given to fasted animals (whether intact, hypophysectomized or adrenalectomized), increase in liver glycogen was accompanied by an increased rate of excretion of nitrogen in the urine, thus suggesting the formation of new carbohydrate from protein.

The following year the same group of workers (LONG et al. 1940) *offered a comprehensive discussion of the relationship of the adrenal cortex to carbohydrate metabolism and attempted to find an inclusive theory that could also explain disturbed electrolyte and water metabolism of adrenal insufficiency, since at that time it appeared that loss of one hormone was responsible for all the effects of adrenalectomy.* Using fasted, adrenalectomized mice and rats maintained on sodium salts they demonstrated that the depression of blood sugar and loss of liver glycogen and finally of muscle glycogen could be repaired by the administration of glucose or by feeding. Moreover, nitrogen excretion was 25% lower in fasting adrenalectomized rats than in fasting intact ones. An enormous increase in liver glycogen was found in intact animals treated with adrenocortical extract, dehydrocorticosterone or corticosterone. The source of the extra carbohydrate was identified as protein, since nitrogen excretion increased proportionally to the augmentation of blood glucose and liver glycogen.

LONG et al. also reported that in pancreatectomized, adrenalectomized rats, kept in good condition by increased NaCl intake, glycosuria was markedly diminished and could be increased by substituting for salt larger doses of adrenal extract or crystalline steroids than those necessary to maintain life. In hypophysectomized rats given adrenocortical hormones, similar new formation of glucose from protein was observed, but the increase in nitrogen excretion was greater and the deposition of glycogen less than in intact or adrenalectomized animals, indicating that a greater amount of glucose had been oxidized.

In the absence of the adrenal glands, a diabetogenic anterior pituitary extract did not augment the glycosuria of pancreatectomized rats. In those experimental animals in which deoxycorticosterone or progesterone was used, no diabetogenic effect was seen, although the doses used were large enough for life-maintenance of adrenalectomized animals. The authors suggested the difference might be merely a quantitative one, since they believed that the hormone active in carbohydrate metabolism was the same as that for maintenance of electrolyte balance and life. The evidence that cortical hormones caused an increase in the carbohydrate content of the liver, blood or urine in all the experimental conditions studied, and an increase in urinary nitrogen, indicated their protein catabolic effect and suggested that in any situation necessitating gluconeogenesis (such as fasting, pancreatectomy, exposure to cold, etc.) the adrenalectomized animal would demonstrate a fall in blood sugar.

This synthesis did not really allow for inclusion of electrolyte and water metabolism in the search for a common denominator. The explanation for this failure was shortly forthcoming, as it became clear that adrenocortical extract contained several compounds having different effects on electrolyte and carbohydrate metabolism.

THORN et al. (1940b) reviewed the situation with regard to carbohydrate metabolism in patients with Addison's disease. The respiratory quotients of such patients were elevated by feeding a high carbohydrate diet or by the intravenous administration of glucose. Treatment with adrenocortical extract led to a marked increase in the proportion of calories derived from protein. The abnormal respiratory quotient and the tendency to fasting hypoglycemia were corrected by the administration of adrenocortical extract, corticosterone or cortisol. The only effect observed from deoxycorticosterone therapy was an improvement in the oral glucose tolerance test, although rebound hypoglycemia and collapse (probably related to decreased plasma concentration of Na^+) were still observed 2 to 3 hours after the administration of glucose intravenously. The difference was interpreted as an indication that, when deoxycorticosterone was given, glucose was more readily absorbed from the gastrointestinal tract. E. ANDERSON and HERRING (1940) observed improved glucose absorption and glycogen deposition in adrenalectomized rats treated with either NaCl or deoxycorticosterone.

F. Replacement therapy

When a potent adrenal extract became available, studies of its mode of action were undertaken by various investigators. BRITTON and SILVETTE (1931a) expressed the opinion that the effect was entirely on carbohydrate metabolism, although in cats blood volume and non-protein nitrogen were restored to normal (BRITTON and SILVETTE 1931b). They found that saline-treated animals died with low blood sugar and normal plasma electrolytes, while when extract was given without saline supplement, improvement was evident before an alteration in Na^+ and Cl^- concentrations occurred (BRITTON and SILVETTE 1937b). THORN et al. (1937a, b) conducted balance studies in normal individuals and patients with Addison's disease which demonstrated an effect of "cortin" on the urinary excretion of electrolytes. SWINGLE (1937) emphasized that administration of extract to fasting and thirsting adrenalectomized dogs repaired their electrolyte disturbance. HARRISON and DARROW (1939) demonstrated clearly that the glomerular filtration rate and reabsorption of Na^+ and K^+ were restored to normal within an hour after the administration of adrenocortical extract to adrenalectomized dogs.

The argument concerning which of the various manifestations of adrenal insufficiency was of greatest importance began to be resolved as various crystalline components were separated from whole adrenocortical extract (KENDALL et al. 1934, KENDALL et al. 1935, H. MASON et al. 1936a, b, REICHSTEIN and von EUW 1938, REICHSTEIN and SHOPPEE 1943). It became evident that physiological actions could be separated into groups related to the chemical configuration of the compounds, and that no one compound corrected all of the disturbances found in adrenalectomized animals.

INGLE (1940) *showed that muscle efficiency was affected only by steroids which were oxygenated at C-11* (see p. 98). KENDALL (1937) reviewed the separation of compounds A (dehydrocorticosterone), B (corticosterone) and E (cortisone) from extracts of whole adrenal glands. Although he found that adrenalectomized dogs required larger amounts of cortisone (more than 7.5 mg daily) than of corticosterone (1.5 mg daily), no qualitative difference between the hormones was suggested.

When *deoxycorticosterone* was isolated and synthesized (REICHSTEIN and VON EUW 1938), its usefulness in the treatment of adrenal insufficiency was established almost immediately (LEVY-SIMPSON 1938, CLEGHORN et al. 1939, ZAHLER and LITZKA 1939, THORN 1939, THORN et al. 1939b), as well as the fact that *its action was essentially to cause the retention of Na^+ and excretion of K^+*, with a secondary restoration of plasma volume and blood pressure. The compound was essentially *without effect on carbohydrate metabolism* (THORN et al. 1940b, E. GORDAN 1940, CLEGHORN et al. 1941).

In 1939 HARTMAN et al. *described the separation from "cortin" of a factor that restored Na^+ concentration to normal.* The sodium-retaining factor was thought to be different from either deoxycorticosterone or corticosterone because animals became refractory to it while continuing to respond to crystalline material (HARTMAN and SPOOR 1940, HARTMAN et al. 1940). Although the refractory state was apparently due to antigens present in relatively impure preparations, the potency of the sodium factor was clearly different from that of the extract from which it had been separated and was greater than that of corticosterone, being approximately equivalent to that of deoxycorticosterone.

LONG et al. (1940), in their masterful synthesis of the role of the adrenal cortex in carbohydrate metabolism, found that deoxycorticosterone, dehydrocorticosterone and progesterone were essentially without effect on carbohydrate metabolism. They thought this inactivity was probably related to the doses employed, although they offered the possibility that there might be some qualitative differences among the various compounds. B. WELLS (1940) showed that the glycosuria of adrenalectomized rats treated with phlorhizin was accentuated by corticosterone or cortisone, while deoxycorticosterone and the amorphous fraction remaining after separation of the known compounds were ineffective.

B. WELLS and KENDALL (1940) demonstrated a similar separation of effects with relation to induction of atrophy of the adrenal cortex and thymus in intact rats. KENDALL (1940, 1942) *summarized the evidence for distinct separation of the activities of the 11-oxygenated corticosteroids (in gluconeogenesis and muscle efficiency), the 11-deoxycorticosteroids (in electrolyte and water distribution)* and the amorphous fraction (in renal function). The suggestion of a difference between the actions of deoxycorticosterone and the amorphous fraction, based largely on the much greater potency of the amorphous fraction, finally was resolved by the discovery of aldosterone (see Chapter 4, Part 1 of this Volume). However, even in 1944 it was remarked that the concept of a separate factor influencing renal function no longer seems justified (SWINGLE and REMINGTON 1944).

G. Etiology, diagnosis and treatment of Addison's disease

The cause of adrenal insufficiency usually was reported to be tuberculosis, with a few instances of metastases to the glands, atrophy, or infections such as syphilis (LEWIN 1890/91). In addition, reports of hemorrhages into the adrenals began to appear (VOELCKER 1895, p. 279, STILL 1898, A. GARROD and DRYSDALE 1898, ANDREWES 1898, E. TALBOT 1900). These disasters occurred in young children following an acute illness, which often was associated with purpura (LITTLE 1901). R. WATERHOUSE (1911) and FRIDERICHSEN (1917, 1918), whose names are attached to this syndrome, wrote about it some years after the first descriptions. Since no therapy was available at this time even for chronic adrenal insufficiency (or for infection), the relationship of damage to the adrenals to death from acute conditions could not be evaluated.

Progress in the diagnosis of Addison's disease was made only as advances were made in the understanding of adrenal function and the pathological physiology of adrenal insufficiency. Well into the twentieth century, physicians relied chiefly on the presence of characteristic pigment changes, combined with asthenia, weakness, and disturbances of digestion, and were able to verify the diagnosis only by postmortem examination. Later, roentgenographic evidence of adrenal calcification provided some help (ROWNTREE and SNELL 1931).

Gradually, some laboratory support for the diagnosis became available; acidosis, elevated blood urea nitrogen, and hypoglycemia were found, especially when patients were in crisis. As tests for renal function became available, it was noted that the ability to excrete a water load was limited (ROWNTREE 1925). ROWNTREE and SNELL (1931) found that administration of large amounts of water in tests of ability to excrete dilute urine might cause nausea and vomiting in patients with Addison's disease. The ability to excrete a water load became a common test of adrenal function (F. ROBINSON et al. 1941).

In 1931 ROWNTREE and SNELL suggested additional laboratory findings that might indicate Addison's disease, although none is specific. These included the finding of low basal metabolic rate, relative lymphocytosis, increased viscosity of the blood, and decreased blood volume in crisis. They also recommended skin biopsies as an aid in the differential diagnosis of abnormalities of pigmentation. They found a tintometer helpful for following changes in skin color.

As soon as adrenocortical extract became available, its use for diagnostic purposes became popular (as recommended by HARTMAN et al. 1932b) and led to an admonition by LOEB (1935) that it not be used indiscriminately in small doses for all who were "weary and heavy laden."

The understanding of changes in distribution of fluid and electrolytes provided a basis for understanding the vulnerability of patients to further loss of water or of base, and allowed the development of one group of procedures for evaluating adrenal function. The use of a salt-free diet was advocated as a diagnostic aid. HARROP et al. (1933c) found that on such a diet a patient with Addison's disease had a plasma Cl^- of 81 meq/l and a urine Cl^- of 40 meq/l, while a normal person maintained a plasma chloride of 101 and excreted very little Cl^-. It soon became clear that sodium deprivation could result in severe or even fatal circulatory collapse (KEPLER 1935, KLINE 1937, LILIENFELD 1938). CUTLER et al. (1938a, b) proposed a modification of the Na^+ restriction procedure which shortened the time required for the test but added to its danger. Potassium was added to the low Na^+ diet and the concentrations of Na^+, K^+, and Cl^- in plasma and urine were measured. The ability to excrete a water load of 20 ml/kg was tested when the electrolyte concentrations were measured. It was found that the least overlap between the results in normal individuals and those with adrenal insufficiency occurred in the excretion of Cl^- (high in the patients) and of water (low in the patients). GROLLMAN (1936) mentioned the occurrence of plasma Na^+/K^+ ratios of less than half normal when a crisis was impending.

CUTLER et al. (1938a) and HARROP et al. (1933c) felt that tests of carbohydrate metabolism were too nonspecific to be helpful. However, the insulin-sensitivity of patients with Addison's disease and the small rise in blood glucose after a test dose of glucose, sometimes terminating in hypoglycemia, were used as diagnostic aids by some (FRASER et al. 1941b).

Studies of steroid excretion in the urine were foreshadowed by the report of PERLA and MARMORSTON-GOTTESMAN (1931) of a substance in the urine of normal individuals which increased the resistance of adrenalectomized rats to histamine. GROLLMAN and FIROR (1933), using a different method of extraction, also found a substance in urine which prolonged the lives of adrenalectomized rats. CALLOW et al. (1940) described a chemical method for the measurement of 17-ketosteroids, androgens and estrogens in urine; the quantity excreted was decreased in patients with adrenal insufficiency. N. TALBOT et al. (1940) and FRASER et al. (1941 a) defined the limits to be expected in normal individuals. Assay of corticosteroid excretion by chemical means (N. TALBOT et al. 1945, HEARD et al. 1946) or by bioassay (VENNING and BROWNE 1947) provided an additional diagnostic tool.

When ACTH became available, the ability of the adrenal to respond to stimulation could be evaluated; this maneuver is especially reliable when increase in excretion of 17-ketosteroids or corticosteroids is used as the end point. (See p. 137ff.)

H. Treatment of adrenal insufficiency

Treatment, at first limited to general measures of diet and rest, was completely unsuccessful and the disease had to be regarded as uniformly fatal. The prognosis for patients in whom the diagnosis of Addison's disease had been made was regarded as grave indeed, yet suggestions for therapy were offered. Emphasis was placed on bed rest during paroxysms and avoidance of mental and bodily excitement during remissions. The danger to the patient of even minor stresses was recognized in the strict injunction against giving drastic purgatives.

Attempts at providing functional adrenal grafts were as unsuccessful in patients as in animals; the surgical procedure itself may have accelerated death in some instances.

Adrenal transplants. JABOULAY (1897) had extremely poor results when he attempted to treat Addison's disease with adrenal transplants; both his patients died within 24 hours. Transplantation experiments in animals had less disastrous results, but long-lasting function even of autotransplants was rarely achieved (BUSCH et al. 1908, BIEDL 1913, H. JAFFE and PLAVSKA 1926, OLDBERG 1929/30).

Adrenal tissue preparations. OSLER (1896/97) reasoned that extracts of the adrenal gland must be tried in therapy, since the disease was akin to myxedema in being due directly to loss of an internal secretion. Improvement of patients given hog adrenals, ground in glycerol, by mouth is difficult to interpret since many patients had tuberculosis and were benefited by bed-rest, good food and other nonspecific programs (W. JOHNSTON 1900). Other physicians (e. g., KINNICUT 1897) administered adrenal extracts by mouth on the premise that the adrenals either secreted or excreted active principles that served to destroy toxic substances in the circulation. KINNICUT attributed his meager results (improvement in only 6 of 48 patients) to the fact that many of them had active tuberculosis. The large number of patients diagnosed as having Addison's disease suggests that other ailments, which might not have responded even to the potent preparations available today, affected at least some of his patients. E.W. ADAMS (1903), in a review of 93 cases of Addison's disease, expressed the opinion that 16 patients had been permanently improved by oral administration of adrenal material; no criteria for establishment of the diagnoses were included.

Oral and rectal administration of dried adrenal glands and rectal or subcutaneous use of epinephrine remained the main "specific" forms of therapy in Addison's disease for some years. This program, called the *Muirhead regime,* had serious limitations in that rectal and gastric irritation often made it difficult for the patient to continue the medication (MUIRHEAD 1921, ROWNTREE 1922, ROWNTREE 1925). In 1931 ROWNTREE and SNELL reviewed their extensive experience with this form of therapy, and emphasized that sudden worsening and death might follow such stresses as the administration of a cathartic, anti-syphilitic treatment or surgical intervention for the removal of a tuberculous organ. Although blood transfusions and infusions of glucose were given to seriously dehydrated patients who were in shock, most of them died once a real crisis developed. ROWNTREE and SNELL felt that there was definite benefit from this program, but stated that "the limitations of the Muirhead treatment should be faced squarely." A few patients survived for several years.

Following the isolation of *ascorbic acid* from the adrenal cortex by SZENT-GYÖRGYI (1929) this substance was given to some patients with Addison's disease and there were some optimistic reports of its helpfulness (ROWNTREE and SNELL 1931).

The concept that accumulation of toxic substances was responsible for the symptoms of adrenal insufficiency provided the basis of the earliest attempts at *therapy with fluids.* It was no doubt fortuitous that SODDU (1898) used physiological saline to replace the blood he withdrew from adrenalectomized dogs, thinking the blood contained a toxic substance. Prolongation of the lives of adrenalectomized dogs by intravenous administration of Ringer's solution was interpreted by G. STEWART and ROGOFF (1925) as compatible with the idea that the treatment facilitated elimination of toxic substances via the intestinal mucosa. BANTING and GAIRNS (1926) found that the elevated concentrations of blood urea nitrogen and non-protein nitrogen of adrenalectomized dogs could be restored to normal by administration of serum and ample saline solution.

COREY (1926/27) administered a variety of solutions by stomach tube to adrenalectomized cats, because of their physical evidence of dehydration, and achieved prolongation of life especially with those mixtures containing sodium salts. The effectiveness of the various solutions cannot be compared easily since the volume and frequency of administration were not kept constant. COREY was undecided whether the effect was a temporary relief of dehydration or the induction of a diuresis, which flushed toxic substances from the blood stream. SWINGLE (1926/27) thought the temporary revival induced by the administration of acacia, glucose or Ringer's solution was the result of hemodilution. MARINE and BAUMANN (1927) *were impressed by the effectiveness in adrenalectomized cats of 0.9% NaCl given intraperitoneally.* ROGOFF and STEWART documented prolongation of life of adrenalectomized dogs by intravenous injection of Ringer's solution and continued to believe that success was due to washing out of poisons from the gastrointestinal mucosa (1926b) or adhering to the nervous system (1928a).

Dietary treatment of adrenal insufficiency acquired a rational basis as soon as the basic disturbance in Na^+ distribution had been recognized and definite prolongation of life of adrenalectomized animals had been achieved by administration of sodium salts. LOEB (1933) first treated a patient with Addison's disease *with NaCl,* given rectally and by mouth; the plasma electrolytes returned to normal. Further experience (LOEB 1935) showed that even administration of NaCl could be overdone, for one patient developed edema, only to become ill again when a salt-poor regimen was instituted. RUBIN and KRICK (1934) were able to keep adrenalectomized rats in good condition for months by providing drinking fluid containing NaCl (as well as small amounts of Ca^{2+}, Mg^{2+} and K^+), while SWINGLE et al. (1934b) had only limited success with dogs given rather large doses of NaCl by mouth. ZWEMER (1934) found that cats did well when part of the Na^+ was provided as $NaHCO_3$. Treatment with added NaCl (SNELL 1934, KEPLER 1935) or preferably NaCl and Na citrate (ALLERS 1935) was found very helpful in the management of patients with Addison's disease.

Curiously, the fact that an increased appetite for salt occurs in some patients with Addison's disease was not recognized until after the beneficial effect of treatment with Na^+ became known. ROGOFF (1932b) stated that "most patients with Addison's disease crave spicy or sour foods, such as sauerkraut and pickles, and that these articles of diet are usually well tolerated" but did not comment on their Na^+ content. RICHTER (1936), allowing rats access to both tap water and saline, found that after adrenalectomy the spontaneous intake of saline solutions was much increased. Studies on this subject were extended and confirmed (RICHTER 1941, 1942/43, W. CLARK and CLAUSEN 1943). WILKINS and RICHTER (1940) found salt craving in a young child with adrenal insufficiency, and in 1942 THORN first mentioned that about 20% of patients with adrenal insufficiency develop a marked interest in eating salty foods.

That adrenal insufficiency leads to increased sensitivity to K^+ was described for cats (ZWEMER and TRUSZKOWSKI 1936), dogs (ALLERS et al. 1936, MARENZI 1938), and human beings (WILDER et al. 1937). These observations formed the basis for restricting the K^+ content of the diet while providing extra Na^+ (WILDER et al. 1936). Under such a regimen patients did quite well and did not require adrenal extract except when an infection or other stress occurred. ALLERS and CRANDALL (1936) found that adrenalectomized puppies could grow well on a low K^+, high Na^+ diet, and ALLERS et al. (1936) were able to maintain adult dogs on a similar program if they were not fasted, so that hypoglycemia did not develop. SWANN (1937) found poor weight gain in adult rats maintained on 74 mg of Na^+ and 44 mg of K^+ daily, although they were able to survive adrenalectomy for some time. This ration provided more K^+ in relation to Na^+ than the diet ALLERS and CRANDALL used in dogs (200 mg K^+, and 8 to 20 g NaCl). Animals treated by dietary means alone maintained normal electrolytes and blood urea nitrogen, and normal blood sugar unless fasted, but had very poor resistance to stress of any kind.

After the value of sodium salts in the treatment of adrenal insufficiency had been recognized, it was found that the dose of extract could be smaller if sodium salts were also used. When NaCl was added to a diet restricted in K^+, and adrenal extract was given, the results were much better than with previous methods of therapy. Nevertheless it was difficult to provide adequate amounts of extract, and the maintenance of severely deficient patients remained a problem (LOEB 1935, C. GREEN 1937).

Shortly after the effectiveness of deoxycorticosterone was demonstrated for treating acute and chronic adrenal insufficiency, the need to use it with caution became clear. Excess weight gain, edema, hypertension and cardiac failure were noted (PENTSCHEW 1939, FERREBEE et al. 1939, HENI 1939), and some unexpected deaths occurred (WILDER 1940). The occurrence of hypoglycemia in the course of deoxycorticosterone therapy was attributed by some to overdosage *per se* (E. GORDAN 1940). However, KUHLMANN et al. (1939) showed that overdosage with deoxycorticosterone would lead to severe depletion of K^+ in normal dogs, and that addition of KCl to the regimen protected them from this toxic action. Their signs of muscle weakness and hypertension were similar to the findings in some patients receiving deoxycorticosterone. These observations were confirmed in adrenalectomized dogs (E. GORDAN 1940, REMINGTON et al. 1941). FERREBEE et al. (1941/42) demonstrated that, in intact dogs given deoxycorticosterone and a low K^+ diet, marked changes in muscle composition (decreased K^+ and elevated Na^+) were associated with the development of paralysis and death. DARROW and MILLER (1942) produced necrosis of myocardial fibers in rats, cats and dogs treated with deoxycorticosterone: they suggested that a serum K^+ concentration of less than 3.5 meq/l should be regarded as dangerous. When deoxycorticosterone was given to patients with Addison's disease, the low K^+ diet was shown to represent a hazard for them (WILDER 1940, TOOKE et al. 1940), since K^+ depletion developed and Na^+ retention was greater when K^+ intake was restricted.

Provision of deoxycorticosterone as pellets implanted subcutaneously was recommended by THORN et al. (1939a) and was found efficacious in dogs (BILLMANN and GRATHWOHL 1940, HENI 1941). Since the oral route of administration was unsatisfactory even in doses 10 times greater than those required parenterally, E. ANDERSON et al. (1940b) proposed sublingual therapy, the drug being dissolved in propylene glycol and absorbed into the blood stream. The effectiveness of this mode of therapy was confirmed for human beings (THORN et al. 1942) although it was not adequate for dogs (CLEGHORN et al. 1943), possibly because the animals swallowed the solution before it could be absorbed by the oral mucosa. EVERSOLE and GAUNT (1943) also found the oral route of administration to be of very little value and GROLLMAN (1941) observed only slight benefit to young adrenalectomized rats given deoxycorticosterone mixed with dry food.

Even with an appropriate dose of deoxycorticosterone, given by daily injection or provided by implantation of pellets, and with the addition of NaCl to the diet, patients with adrenal insufficiency continued to be subject to hypoglycemia (THORN et al. 1942) *and to be vulnerable to crises in the course of infections* (KEMPERER 1945).

It was pointed out by WILDER (1940) and by LOEB (1942) that *cortisone was effective under these circumstances, as were corticosterone or whole adrenocortical extract.*

It was not until the usefulness of cortisone in rheumatoid arthritis was described (HENCH et al. 1949) that cortisone was produced commercially in large enough quantities to allow it to be added to the routine treatment of patients with Addison's disease; this treatment has improved the prognosis substantially (see p. 41ff.). Before deoxycorticosterone was available, 80% of the patients died within 3 years of the onset of adrenal insufficiency (THORN 1944). Between 1938 and 1948, this figure was reduced to 30%.

Etiology of adrenocortical insufficiency

A. Naturally occurring insufficiency in man

I. Primary adrenal insufficiency

Adrenal insufficiency may conveniently be described as either acute or chronic. Such a subdivision is to some extent artificial, since a chronic process leading to destruction of adrenal tissue may first become evident as a crisis in the course of an infection or other severe stress. Acute damage to the adrenal glands, moreover, may be the cause of subsequent chronic insufficiency.

1. Acute adrenal insufficiency

a) Hemorrhage and severe infection

Acute destruction of the adrenal glands is most commonly due to hemorrhage and is found especially in patients who have died of an overwhelming infection. The term "Waterhouse-Friderichsen syndrome" is reserved by some for the association of adrenal hemorrhage with meningococcal septicemia; the same situation may be found in individuals who have died with septicemia due to other organisms, however. Descriptions of this condition are rather frequent (AEGERTER 1936, MARTLAND 1944, FRIDERICHSEN 1955). (See p. 45 for early reports.) It is difficult to prove that death was caused by involvement of the adrenals by hemorrhage or by pseudotubular degeneration of the cortical tissue (RICH 1944). The few measurements of plasma electrolytes or blood sugar which have been reported do not provide clear evidence of adrenal insufficiency (NITSCHKE and KRATSCHELL 1938, STENGER 1939, COSGRIFF 1944, D'AGATI and MARANGONI 1945). COSGRIFF (1944) has expressed the opinion that sepsis rather than adrenal insufficiency is the major problem. H. WILLIAMS (1942) summarized the findings in 38 patients who had meningococcemia. In 17 the illness followed a fulminating course, and in 9 of these bilateral adrenal hemorrhage was found. There were no differences in the clinical findings between these 9 and the other 8 patients; in all there was severe circulatory collapse and hyperpyrexia, with death occurring 5 to 36 hours after the onset of symptoms. Similarly, it cannot be assumed that a patient who survived did so because of substitution therapy. RUCKS and HOBSON (1943) claimed success in a patient who received adrenal extract in addition to intravenous fluids, blood and sulfathiazole. The dose they used (5 ml of adrenocortical extract every 6 hours) seems grossly inadequate to have accomplished this result. Since their patient had already survived for 36 hours before the extract was given, it seems likely that the fortunate outcome should be attributed to the other treatment he received. In one patient, M. SMITH (1945) observed rising blood concentrations of nonprotein nitrogen (from 36 to 72 mg%) and K^+ (4.7 to 7.9 meq/l) as CO_2-combining power dropped from 40 to 35 volumes per cent; the patient recovered from shock and infection following the use of adrenocortical extract as well as antibiotics and intravenous fluids. D. WRIGHT and REPPERT (1946) emphasized the importance of treatment for shock and infection in the recovery of 4 adult patients who were given deoxycorticosterone acetate and adrenal extract as additional therapy.

That adrenal insufficiency does occur in severe bacterial infection cannot be regarded as established (SPINK 1960). But the use of adrenal extract or cortisol in addition to antibiotics, fluid and blood is probably not unreasonable so long as the specific measures to combat infection and shock are given priority in the institution of therapy. See also section *e* (page 51) on adrenal insufficiency in shock.

b) Hemorrhage without obvious infection

Adrenal hemorrhage in the newborn may follow a difficult delivery or strenuous efforts at resuscitation (CORCORAN and STRAUSS 1924, SNELLING and ERB 1935, E. HALL and HEMKEN 1936, STEVENS and TOMSYKOSKI 1954). In many instances the hemorrhage is unilateral and presents either as a surgical emergency or concealed bleeding.

Hemorrhage and necrosis of the adrenals is an occasional complication of pregnancy or delivery (CRAWFORD 1951). PLOTZ (1953) ascribed about 4% of obstetrical deaths to this cause.

Trauma (SCHRADER 1929, GUTTMAN 1930) or severe burns (WEISKOTTEN 1917, SNELLING and ERB 1935) have been given as the cause of bilateral adrenal hemorrhage. MANDELSTAM et al. (1958) reported the occurrence of an episode of acute adrenal insufficiency 45 days after a severe burn, and postulated that adrenal dysfunction persisted, since the patient showed no response to ACTH although he had no symptoms of adrenal insufficiency. No convincing evidence of adrenal insufficiency was presented in any of these instances.

c) "Idiopathic" adrenal hemorrhage; reactions to ACTH

Bilateral adrenal hemorrhage, without evident precipitating factors, was described by PEARL and BRUNN (1928) and by N. MACLEAN and IRONSIDE (1958); in each instance the first symptom was severe abdominal pain. CHOKAS (1958) reviewed instances of adrenal hemorrhage occurring in patients receiving anticoagulant therapy.

Hemorrhage has been described in patients who were receiving ACTH for ulcerative colitis (WILSON and ROTH 1953), hypertension (GREENE 1953) or gastrointestinal bleeding which followed administration of heparin (MOOLTEN 1957). Although the relationship between the administration of ACTH and the hemorrhagic lesion of the adrenal is not established, it is worth noting that INGLE (1951) induced adrenal hemorrhage in rats by the prolonged administration of ACTH. WILBUR and RICH (1953) also found that ACTH could lead to anatomical damage to the adrenal cortex in rats.

It seems possible that hemorrhage of the adrenals under circumstances of severe physiological stress may be a consequence of excessivestimulation from the pituitary. (See also Part 1, Chapter 1, p. 129ff. and p. 63 of this Chapter.)

d) Bilateral thrombosis

HIRSCH and CAPPS (1927) described a patient who died in convulsions and who was found to have thrombosis of the inferior vena cava and of both adrenal veins. E. HALL and HEMKEN (1936) mentioned the development of thrombi in the adrenal veins in patients with pneumonia and with tonsillitis. GUTTMAN (1930) gave venous thrombosis as the cause of the adrenal insufficiency in 3 of 402 patients, and arterial embolus in one. Necrosis of the adrenals and hemorrhagic infarction of the kidneys were found in a man who developed anuria after translumbar aortography (CICCANTELLI et al. 1958); the authors felt that adrenal insufficiency was a factor in the patient's death.

e) Shock

Adrenal insufficiency has been considered as a factor in shock, and substitution therapy recommended without adequate evidence that such therapy is appropriate. RABINOVICI (1951) subjected rats to irreversible hemorrhagic shock and found that there was extensive hemorrhage of the adrenal cortex although its lipide content was normal; he suggested that ischemia might prevent release of hormones in the gland. FRANK (1953) indicated that hormonal therapy was ineffective in shock which did not respond to blood transfusions. FRANKSSON et al. (1954) found that the plasma corticoid concentration in patients in surgical shock was even higher than that observed in individuals after uncomplicated surgical procedures. MELBY et al. (1957) and MELBY and SPINK (1958) studied patients in shock due to infection, and found that plasma corticoid concentrations, which were already elevated, rose even further when ACTH was given. Thus *there is little reason to invoke adrenal insufficiency or inability to respond to ACTH in shock.*

2. Chronic adrenal insufficiency

The relative incidence of etiological factors has changed considerably since Addison's time, when the most common cause of adrenal insufficiency was tuberculosis of both adrenal glands, with an occasional instance of atrophy or of involvement of the adrenals by metastatic tumors.

a) Atrophy

In most cases no basis for atrophy of the adrenals can be found. BARKER (1929) reported 3 cases of atrophy among 28 patients with Addison's disease. BRENNER (1928) summarized 42 patients who died of Addison's disease; in 5 the adrenals were atrophic and in one no adrenal glands could be found. H. WELLS (1930) reported 6 instances of atrophic adrenals and suggested it be called "cytotoxic suprarenal contraction." He described infiltration with lymphocytes and commented that there was even more infiltration in the thyroid than in the adrenals. GUTTMAN (1930) found a 16% incidence of atrophy among 402 autopsy reports on Addisonians. DUFF and BERNSTEIN (1933) published 5 instances of atrophy, with lymphocytic infiltration of the adrenal and thyroid glands. J. CLARK and ROWNTREE (1934), in a review of 25,000 autopsies, found lesions of the adrenals described in 3%. Atrophy of the adrenal cortex occurred in 7 instances, and in one patient there had been symptoms of adrenal insufficiency. E. HALL and HEMKEN (1936) found atrophy as common as tuberculous involvement of the adrenals (5 cases of each among 15 individuals). SNELL (1935) commented that the incidence of atrophy seemed to be increasing, and suggested that improved therapy allowed more patients with tuberculosis to survive, since they usually had some remnants of functioning cortical tissue. DUFFIN (1943) reported necrosis or atrophy in 7 of 17 instances of adrenal insufficiency. O'DONNELL (1950) reviewed 9000 autopsies from 1928; among these there were 18 cases of Addison's disease, in only 6 of which were the adrenal glands atrophic. SLOPER (1955) summarized the pathology of Addison's disease as seen at the London Hospital between 1912 and 1951. He reported atrophy of the adrenals in 15 patients; the average weight of the two glands was 2.7 g, the range being 1.9–4.6 (normal 10–15 g, BOURNE 1961). LABHART (1957) found 4 instances of atrophy as compared to 8 of adrenal tuberculosis.

HADLOW (1953) reported 3 instances of spontaneous adrenal atrophy in dogs; in 2 there was lymphocytic infiltration of the cortical remnants and 2 also had small peptic ulcers.

Adrenal atrophy has also been described as part of a *pluriglandular syndrome* under circumstances ruling out the possibility of primary pituitary disease. ATRIA et al. (1948) reported adrenal atrophy in a patient with Turner's syndrome who died in shock following an appendectomy.

The combination of *hypothyroidism* and adrenal insufficiency was described by BERNSTEIN (1948) in a patient with diabetes mellitus. STATLAND and LERMAN (1950) described diminished adrenal function in 43 patients with myxedema; one appeared to have Addison's disease as well as hypothyroidism. BLOODWORTH et al. (1954) reviewed 35 patients in whom adrenal insufficiency was associated with hypothyroidism; autopsy examination in 12 showed atrophy of the adrenals in 9 and tuberculosis of the glands in 3. Definite lesions of the thyroid gland were found in 8 of the 12. The paper of PAULL and PHILLIPS (1954) suggests that the adrenal insufficiency of hypothyroidism may be functional rather than anatomical, since adrenal function became normal on thyroid replacement therapy.

The combination of *thyrotoxicosis* and adrenal insufficiency has been discussed by RUPP and PASCHKIS (1957) as well as by THORN (1951). This combination is apparently unusual.

Adrenal insufficiency occasionally develops in patients with *idiopathic hypoparathyroidism* (LEONARD 1946, PAPADATOS and KLEIN 1954, PERLMUTTER et al. 1956).

CRAIG et al. (1955) and WHITAKER et al. (1956) have reviewed the condition of *adrenal atrophy associated with superficial moniliasis*, which occurs especially in children. Adrenal insufficiency or hypoparathyroidism, or both, may develop many years after the first manifestations of steatorrhea and of moniliasis involving the oral mucosa and nails; there is a definite tendency for this syndrome to be familial (MORSE et al. 1961). These authors suggested that the syndrome might be a variant of mucoviscidosis. Although the role of the monilia infection (usually *Monilia albicans*) is not known, it appears possible that a cytotoxic factor leads to atrophy of the adrenal cortex or parathyroids. No monilia organisms have ever been cultured from the involved endocrine organs. CRAIG et al. (1955), who reviewed previous reports, pointed out the frequent occurrence of liver damage in children dying of this disease and suggested a positive correlation with hepatitis. There is frequently marked lymphocytic infiltration of many organs, including the adrenal cortices.

The association of *adrenal atrophy and diffuse cerebral sclerosis*, which occurs especially in young males, has recently been reviewed by HOEFNAGEL et al. (1962). Earlier reports include those of SIEMERLING and CREUTZFELDT (1923) and GAGNON and LEBLANC (1959). Addison's disease associated with *spastic paraplegia* may also be familial (HARRIS-JONES and NIXON 1955, PENMAN 1960); the onset is usually in adult life, rather than in childhood. Reports of *familial Addison's disease* without known neurological complications are reviewed by HARRIS-JONES and NIXON (1955) and by MEAKIN et al. (1959). (See also lipoid hyperplasia, p. 54.)

b) Tuberculosis

The incidence of tuberculosis as a cause for adrenal insufficiency has declined substantially and may be expected to decrease further as therapy for tuberculosis becomes increasingly effective. Whereas the incidence of tuberculosis in Addisonians was at one time as high as 70 to 80% (BARKER 1929, GUTTMAN 1930, JAUDON 1946b) it now appears to be between 40 and 60% (O'DONNELL 1950, LABHART 1957). The figures differ, of course, in various parts of the world in relation to the frequency of tuberculosis in the population.

c) *Other infections*

WEINER (1936) described a patient whose adrenals were small (2 and 4 g); the small amount of cortex contained many large cells, some with 2 or 3 nuclei and acidophilic *nuclear inclusions* suggestive of a viral infection. No such inclusion bodies were found in other tissues. GUTTMAN (1930) mentioned 2 instances of *pyogenic* infection of the adrenals and ZOLOTAREVA and GISSINA (1956) ascribed adrenal insufficiency to chronic suppuration consequent to asymptomatic suppurative endometritis following an abortion.

Other specific organisms have been implicated in the production of adrenocortical insufficiency. *Histoplasmosis* was given as the cause of one case of adrenal insufficiency by O'DONNELL (1950). Four cases with this disease were reported by CRISPELL et al. (1956); postmortem evidence for histoplasmosis of the adrenal was definite in one patient, while in the others the diagnosis was based on the occurrence of adrenal insufficiency associated with pulmonary histoplasmosis. MALONEY (1952) reported 2 patients with Addison's disease due to chronic disseminated *coccidiomycosis*. GUTTMAN (1930) listed one instance of *syphilis* involving the adrenal glands. RICKARDS and BARRETT (1954) cited a variety of inflammatory processes in the adrenals including *giant-cell granuloma* (without identification of the organism), *coccidiomycosis*, *torulosis*, *moniliasis* and *syphilis*[1].

d) *Infiltrations*

Amyloid apparently is the most common substance infiltrating the adrenal glands. In some instances there has been recognizable cause for amyloidosis, such as tuberculosis. MCCUTCHEON (1922) described one patient, as did HUNTER and RUSH (1926/27). PHILPOTT (1927/28) described another patient and mentioned that adrenal involvement in generalized amyloidosis is rare, having occurred to some extent in only 9 of 118 patients. GUTTMAN (1930) attributed adrenal insufficiency to this cause in 7 of his 402 cases. O'DONNELL (1950) accepted adrenal insufficiency as the cause of death in only 3 of 29 patients in whom amyloidosis involved the adrenals. The case reported by SNAPPER and CH'IN (1940) is difficult to accept as proved. LEVINSON et al. (1947) gave chemical evidence of adrenal insufficiency in a patient who had generalized amyloidosis following severe thermal burns. In this patient, as well as in many of the others reported, the amyloid infiltration was most pronounced in the inner layers of the cortex.

[1] J. K. FRENKEL, whose studies on intracellular infection of adrenocortical cells by microorganisms were cited in Chapter 1, Part 1, p. 132 and p. 141, has published further experimental studies, as well as two reviews (1960a, b, 1961). These make clear that intracellular infection by various bacteria, fungi and protozoans is enhanced by corticoids only when resistance has previously been developed. In other words, the glucocorticoids here cause a reversal of the immune state, as elsewhere in the body (see also Chapter 7, p. 16). In viral infections where immunity does not appear to depend on antibodies, there is no tendency for adrenal infection.

In his 1960(b) review, FRENKEL gives the following therapeutic advice: Patients with active adrenal infection should receive exogenous hormone treatment to suppress ACTH output and hence depress local cortisol synthesis. Since most available glucocorticoids have a direct stimulatory effect on the adrenal infection when given in doses that depress ACTH output, as much mineralocorticoid (deoxycorticosterone; 2-methyl-9α-fluorocortisol) as possible should be used to reduce the amount of exogenous glucocorticoid required. Some weak glucocorticoids, such as corticosterone, also serve to depress ACTH output more than to enhance the intracellular infection. SU 4885 may also prove useful. — H. W. DEANE.

Hemochromatosis involving the adrenal cortex has been described in one case (Clinicopathologic Conference 1950); the portion of the cortex which looked normal may not have been adequate for life.

e) Metastases

Although the possibility that metastases to the adrenal gland can be so extensive as to lead to adrenal insufficiency has been subject to considerable doubt, in some instances this etiology seems to be well established. GLOMSET (1938) reviewed the autopsy findings in 445 patients who had malignancies with metastases. In 110 cases (13%) there were metastases in one or both adrenal glands, but in no instance had symptoms of adrenal insufficiency been recognized. BULLOCK and HIRST (1953) described 224 instances of adrenal metastases among 2,833 cases of carcinoma but did not comment on their functional implications. BLACK (1926) reported bilateral endothelioma of the adrenals with metastases to the lungs and pleura and CAWADIAS (1946) described bilateral adrenal tumor with metastases to the kidney; in both instances it seems possible that the adrenal lesions were secondary rather than primary. BUTTERFLY et al. (1952) and WALLACH and SCHARFMAN (1952) each reported one clear instance of adrenal insufficiency due to metastases from bronchiogenic carcinoma; and metastases from a hepatoma (Case Records Massachusetts General Hospital 1953) and gastric carcinoma (LEARY and MASTERS 1957) have also been described.

f) Lipoid hyperplasia

PRADER and GURTNER (1955) described marked adrenal hypertrophy with conspicuous quantities of lipide in the cells, in an infant who died with symptoms of adrenal insufficiency. In this instance the child, who appeared to be a girl, excreted normal amounts of 17-ketosteroids and was found to have bilateral inguinal testes. This syndrome differs from virilizing congenital adrenal hyperplasia, which can be considered as overactivity of the cortex and will be discussed in Part 3 of this Volume. Since 1955 a total of 6 instances of adrenal insufficiency with lipoid hyperplasia of the adrenals have been described from Switzerland (summarized by PRADER and SIEBENMANN 1957); 3 of the children were girls, and 3 were male pseudohermaphrodites. There is a suggestion that the disorder may be genetically determined, as there was consanguinity in 2 of the families with affected children. The authors have suggested that there may be an enzymatic disturbance interfering with the synthesis of adrenal steroids leading to accumulation of cholesterol and intermediate hormone precursors in the cortical cells. The pathology has been described in detail by SIEBENMANN (1957) (see also Chapter 1). SANDISON (1955) reported "lipoidosis" of the adrenal cortex in a girl who died at 3 months of age; no information on adrenal function was available. An additional case report and a good discussion of the syndrome can be found in DHOM (1956/57).

The association of male pseudohermaphroditism, adrenal hyperplasia and scanty sexual hair in an adult was described by HEALEY and GUY (1931). This syndrome may represent a form of lipoid hyperplasia of the adrenals. (See also pseudo-adrenal insufficiency, p. 57.)

g) Calcification

MARINE (1926) described the finding of calcification of the adrenals in 64 of 257 cats. The abnormality was most marked in the zona fasciculata; he believed the lesion might be a consequence of distemper, and said that, when severe, the signs were similar to those seen after adrenalectomy. In some instances of adrenal insufficiency in man, the presence of calcification may justify the suspicion that previous hemorrhage was the underlying disorder (VICTOR 1921, ATRIA et al. 1948). This possibility finds support in the observation of SNELLING and ERB (1935) that the adrenals may calcify within 8 to 10 days after trauma or hemorrhage. GARDNER (1957) observed radiological evidence of calcification in 2 children, one of whom definitely suffered trauma during delivery. All these patients manifested adrenal insufficiency.

h) Cystic degeneration

This pathological finding may also be the result of trauma and hemorrhage at delivery. F. MOORE and CERMAK (1950) found cystic adrenals in an infant who died despite temporary improvement following treatment for adrenal insufficiency.

i) Aplasia

Aplasia of the adrenal glands has been occasionally reported, but seems doubtful. For example, this diagnosis seems highly unlikely in a patient who had survived into adult life in apparent good health (CROSBY 1930).

j) Vascular lesions

SILBERMAN et al. (1950) described a patient with thromboangitis obliterans who had adrenal insufficiency which improved; after death he was found to have small scarred adrenals in which many vessels had been thrombosed and recanalized. It also seems possible that the small adrenals found in a patient whose adrenals had been denervated because of recurrent duodenal ulcer should be ascribed to interference with their blood supply (SNELL et al. 1936).

k) Pemphigus

There have been some suggestions that patients with acute pemphigus may suffer from adrenal insufficiency. The detailed study of TALBOTT et al. (1940) does not provide evidence to this effect; rather it indicates that a disturbed electrolyte equilibrium results in those patients whose skin lesions are exudative.

II. Secondary adrenal insufficiency

The early observations on the relationship between the anterior pituitary and adrenal cortex have already been discussed (see p. 33). In man, spontaneous adrenal insufficiency as a consequence of inadequate secretion of ACTH is most frequently associated with a destructive lesion of the pituitary and occasionally with failure of development of the pituitary gland. Reviews by ESCAMILLA and LISSER (1942), PERKINS and RYNEARSON (1952) and VAN BUCHEM (1955) give some idea of the varieties of disturbed pituitary function. In the past decade, however, treatment with corticosteroids has become the most common cause of inadequate secretion of ACTH.

1. Abnormalities in pituitary development

BENDER and LERI (1903), MEYER (1912) and KOHN (1922) described the adrenal cortex as small in anencephalic fetuses; KOHN (1922) suggested that the difficulty might lie in the absence of neurohypophysial and cerebral connections to the anterior pituitary. COSMETTATOS (1920) described adrenal atrophy in a hydrocephalic infant who died at one day of age, and EDMONDS (1950) found small adrenals in fetuses with cyclopia.

Deficiency in weight of the adrenal glands is frequently associated with absence of the fetal cortex at an age when substantial remnants should still be found (DEAMER and SILVER 1950, GEPPERT et al. 1950, PROVENZANO 1950, WELSH and MEHLIN

1954, A. WILLIAMS and ROBINSON 1956). ANGEVINE (1938) analyzed the findings in 18 anencephalic infants, and pointed out that the adrenals were smallest in the heaviest fetuses, which suggests that atrophy occurred in the latter part of gestation. BENIRSCHKE et al. (1956) emphasized that the weight deficit of the adrenals is due to atrophy of the fetal cortex beginning at about the 20th week of gestation. They suggested that the underlying factor may be maldevelopment of the hypothalamus, as adrenal atrophy was noted in hydrocephalic as well as in anencephalic fetuses. Whether there is a functional deficiency in such adrenals has not been determined, since the cerebral anomaly is not compatible with life. J. NICHOLS et al. (1958) found that plasma corticoid concentrations in such fetuses at term were not different from those of normal newborn infants although the plasma content of 17-ketosteroids was extremely low, suggesting that the adrenals may have been underactive.

Adrenal hypoplasia associated with agenesis or hypoplasia of the pituitary may occur without other obvious malformations of the central nervous system. BLIZZARD and ALBERTS (1956) described an infant in whom no trace of a pituitary gland could be found; the adrenals were very small (combined weight less than 1 g) and entirely without fetal cortex. MOSIER (1956) reported the association of hypoplasia of the pituitary and the adrenals in an infant who died in the neonatal period; REID (1960) reported an instance of congenital absence of the pituitary gland and also discussed cases reported by others. On the other hand, HARLEM and MYHRE (1957) described a boy who died at 31 days of age; the hypoplastic adrenals consisted of fetal cortex only, although the pituitary gland appeared normal.

2. Pituitary atrophy

In 1918 SIMMONDS recognized that atrophy of the adrenal cortex and other target glands is found when there is atrophy or destruction of the anterior lobe of the pituitary. ESCAMILLA and LISSER (1942) reviewed many cases of Simmonds' disease, emphasizing disturbances in gonadal and thyroid function. However, in 6 of 7 patients so studied, there was reduced tolerance to sodium restriction. The probable explanation of this symptom after restriction of sodium intake is an accentuation of water intoxication (see p. 132). In 43% of the patients, the fasting blood sugar was less than 60 mg %. In 11 of 21 patients in whom glucose tolerance tests were done, carbohydrate tolerance was increased; in a few instances coma was induced by an insulin-tolerance test. (See p. 134.)

3. Pituitary necrosis

The term Sheehan's syndrome is commonly reserved for hypopituitarism following hemorrhage or shock as a complication of childbirth. Following the first clear-cut description (SHEEHAN 1932) additional reports appeared (SHEEHAN 1937, SHEEHAN and MURDOCH 1938, SHEEHAN and SUMMERS 1949). A history of postpartum hemorrhage may be obtained in a large proportion of women with hypopituitarism. VAN BUCHEM (1955) gave this as the etiology in 4 of 8 such patients.

Pituitary necrosis may also occur in men (MARZULLO and HANDELSMAN 1951) without obvious cause, or following infection (W. GORDAN and SHIMKIN 1943, ZOELCKLER and ORBISON 1955).

4. Destructive lesions

The etiology of hypopituitarism may be a craniopharyngioma (FARBER et al. 1941), acromegaly (RUMMELL et al. 1951), intrasellar tumor (SHEEHAN and SUMMERS 1949), skull fracture (VAN BUCHEM 1955), giant-cell granuloma (DONAICH and WRIGHT 1951, RICKARDS and HARVEY 1954), a pituitary cyst (MCKAY et al. 1950), degenerative lesions of the central nervous system (JACOB 1923) or other cerebral lesion (CHRISTY 1960). Destruction of the pituitary by metastases from carcinoma has also been reported (Clinical Pathological Conference 1957).

5. Selective pituitary failure

There are a few well-documented instances of apparent failure of ACTH production only (STEINBERG et al. 1954) and other degrees of partial rather than generalized hypopituitarism (MADDOCK et al. 1953).

6. Suppression of ACTH secretion

In the past, suppression of ACTH production was most common in patients with functioning tumors of the adrenal cortex. In these, the remaining adrenal had become atrophic and death occurred within a few hours after surgical removal of the tumor (PAUL 1931, JOSEPHSON 1938). In some instances, the atrophy was so profound that congenital absence of the contralateral adrenal was suggested (LUKENS et al. 1937). PRATHER (1933) and WALTERS et al. (1934) proposed that patients having adrenal tumors removed be treated as if they had acute adrenal insufficiency, already discussed (p. 48). INGLE and KENDALL (1937) and INGLE et al. (1938) put forth the hypothesis that the amount of circulating adrenal hormone influences the pituitary in its control over adrenocortical activity. Although adrenal tumors causing Cushing's syndrome are those most likely to suppress ACTH production, atrophy of the normal or uninvolved adrenal cortex has occasionally been noted in patients with virilizing adrenal tumors who had not received cortisone therapy (SOBEL et al. 1953).

In recent years the most common cause of reduced ACTH secretion is treatment with corticosteroids for such diseases as asthma and rheumatoid arthritis. Failure to recognize that patients receiving steroid therapy need an increased supply when undergoing surgery or other stress may lead to fatal collapse (L. LEWIS et al. 1953, SALASSA et al. 1953, H. WEST 1953). Subnormal response to ACTH may be evident for some time after steroid therapy is discontinued, but some individuals respond promptly within a few days after a prolonged course of therapy (CHRISTY et al. 1956, BIRKE et al. 1956, YOUNG et al. 1957, HOLUB et al. 1959, LIDDLE et al. 1959).

Relative adrenal insufficiency may be a factor in thyroid storm, where a markedly enhanced need for adrenal hormones can be expected. Since therapy for this condition necessarily includes a wide variety of measures, the specific effect of adrenal substitution therapy in recovery cannot be evaluated (MCARTHUR et al. 1947).

III. Pseudo-adrenal insufficiency

This diagnosis may be suggested when it appears that the peripheral tissues fail to respond to normal amounts of hormone (ALBRIGHT et al. 1942b).

1. Absence of sexual hair

Failure of sexual-hair development occurs in some individuals with extreme degrees of male pseudohermaphroditism; androgens and 17-ketosteroid precursors are produced (GOLDBERG and MAXWELL 1948), the hair follicles are present, but no hair growth occurs in response to large doses of androgen (WILKINS 1957, p. 276). This situation, associated with adrenal hyperplasia, was first reported by HEALEY and GUY (1931). The possibility of a relationship between testicular feminization with failure of sexual hair growth, and lipoid hyperplasia of the adrenal cortex, in which severe male pseudohermaphroditism also occurs, is discussed by PRADER and GURTNER (1955) and by PRADER (1957).

2. Pseudohypoaldosteronism

Pseudohypoaldosteronism has been documented by DONNELL et al. (1959) in an infant who was severely dehydrated, with elevated plasma K^+ and reduced Na^+ concentrations. The child did not respond to deoxycorticosterone or other mineralocorticoids. He was found to have markedly elevated aldosterone excretion (800 μg per 24 hours), which fell to normal as the child improved when treated with large amounts of salt.

IV. Transient adrenal insufficiency

JAUDON (1946b, 1948) has written on this subject, discussing adrenal insufficiency in patients under 16 years of age. In some the signs of adrenal insufficiency were associated with virilism (see Part 3 of this Volume), whereas in others there were disturbances of nutrition, which may have been the primary problem. *There is no substantial evidence that transient adrenal insufficiency actually exists.*

B. Experimentally induced insufficiency*

I. Adrenalectomy

Although the surgical procedures employed for adrenalectomy are much the same for all mammals, minor problems peculiar to a given species may be encountered. One must know, for example, the degree of proximity of the right gland to the inferior vena cava, the desirability of performing a two-stage operation, the need for the resection of ribs, and so on.

The techniques of adrenalectomy in a number of vertebrate species have been described by HARTMAN and BROWNELL (1949). In the following table, more recent reports of operative techniques are listed: some describe adrenalectomy in animals hitherto unstudied; others consist of new reports upon species long in use for the investigation of adrenocortical insufficiency.

Table 1. *Techniques for adrenalectomy in various species*

Species	Reference
Wallaby	BUTTLE et al. (1952)
Guinea pig . . .	MORRISON (1954)
Rabbit	ZAK et al. (1957)
Goat	COWIE and STEWART (1949)
Sheep	MÅNSSON and OBEL (1958)
Bull	MÅNSSON and OBEL (1958)
Rhesus monkey .	GREEP et al. (1952)
Man	LAPIDES (1958)

The problem of anesthesia during adrenalectomy in the human being has been reviewed by BELLVILLE et al. (1958).

The removal of adrenocortical or interrenal tissue from lower vertebrates presents difficulties not encountered in mammals. This tissue may not be present in a single pair of organs but may exist in "islands" oriented along the long axis of

* In this and most subsequent sections of this chapter, the form of adrenal insufficiency considered is primary; secondary insufficiency and relative insufficiency are not evaluated or characterized. — Sub-Eds.

the organism. (See Chapter 1.) Much of this tissue occurs in very close proximity to blood vessels or excretory organs. It has therefore been found expedient to cauterize such tissue rather than attempt its removal. Descriptions of some such procedures and citations of original reports of studies in lower vertebrates will be found in the books by HARTMAN and BROWNELL (1949) and CHESTER JONES (1957).

II. Chemical inhibitors of the adrenal cortex

There has been great interest in recent years in the possibilities of producing partial or total adrenocortical insufficiency by the use of chemical agents. This interest stems from the discovery of a number of compounds that exert a relatively specific inhibitory effect on adrenocortical tissue, such as amphenone B; 2,2-bis (2-chlorophenyl-4-chlorophenyl)-1,1-dichloroethane (*o, p'* DDD), and 2-methyl-1,2-bis-3-(pyridyl)-1-propanone (Su 4885). The pharmacological effects of these compounds are discussed in Chapters 1, 3 and 4, Part 1 of this Volume.

III. Vitamin deficiencies

Many attempts have been made to elucidate the role of vitamins in adrenocortical function. The usual experimental approach to this problem has been to study vitamin-deficient animals for manifestations of adrenocortical insufficiency. The vitamin deficiencies have been produced by feeding a diet devoid of the vitamin, by administering a metabolic antagonist of the vitamin or by the simultaneous use of both procedures. Only situations in which the water-soluble vitamins are lacking have been studied; the roles of lipide-soluble vitamins in adrenocortical function remain unexplored.

The initial response to all vitamin deficiencies studied thus far appears to be one of adrenocortical hyperactivity. In most vitamin deficiencies, this hyperactivity appears to persist until the onset of the preterminal period; whereas in some, the initial response appears to be followed by the gradual development of adrenocortical hypofunction.

Many of the findings in this area are difficult to interpret for a variety of reasons. In many of the studies, indirect criteria of adrenocortical insufficiency, such as sensitivity to insulin or increased blood eosinophil concentrations, have been employed; direct measurements of the hormonal output of the gland are few and relatively recent. When indications of adrenocortical insufficiency are obtained, it is difficult to know if the disturbance has occurred within the adrenal cortex itself or is to be attributed to a failure of the anterior pituitary gland to secrete adrenocorticotropic hormone (ACTH). The situation is further rendered difficult of analysis by the disturbances that occur in nonendocrine tissues when a vitamin is lacking, for such disturbances may modify the effects of adrenocortical hormones. The fact that investigators have made observations upon animals at varying periods of time after the institution of measures designed to produce a vitamin deficiency makes the comparison of findings from various laboratories difficult.

The earlier findings in this area of vitamin-adrenocortical interrelationships have been reviewed briefly by RALLI (1950) and more extensively by MORGAN (1951).

1. Ascorbic acid deficiency

The unusually large quantities of ascorbic acid that are present in the adrenal gland of most species (HARTMAN and BROWNELL 1949) have provoked many investigations of the role of this vitamin in adrenocortical function. The following discussion will be confined to reports of studies in scorbutic mammals. The earlier findings in this area have been reviewed by MEIKLEJOHN (1953).

Only a few investigators have reported signs of apparent adrenocortical insufficiency in scorbutic animals. BANERJEE and DEB (1952) observed that serum potassium concentrations are elevated in the scorbutic guinea pig while the urinary excretion of this cation is subnormal; serum sodium concentrations were found to be subnormal. These authors also concluded that the synthesis of adrenocortical hormones may be impaired, for the concentration of cholesterol in the adrenal gland of the scorbutic guinea pig is unusually low (BANERJEE and DEB 1951).

Such findings are in the minority, for most investigators have reported signs of adrenocortical hyperactivity in states of scurvy; this hyperactivity appears to persist until the structure of the gland begins to break down in the preterminal period. C. STEWART et al. (1952) found that monkeys are unusually resistant to the hypoglycemic effects of insulin early in the period of ascorbic acid deficiency. DONE et al. (1953) found that the concentration of plasma 17-hydroxycorticosteroids in the scorbutic guinea pig is approximately 10 times greater than normal. BURSTEIN et al. (1955) reported that scorbutic guinea pigs excrete more cortisol in the urine than do normal animals. Unusually high titers of formaldehydogenic substances (presumably corticosteroids) were found in the urines of scorbutic guinea pigs by NADEL and SCHNEIDER (1952). In both scorbutic guinea pigs and human beings, CLAYTON et al. (1954) measured abnormally large quantities of urinary 17-ketosteroids and 17-ketogenic steroids. BACCHUS and HEIFFER (1953) found that the elevated urinary titers of 17-ketosteroids and 17-hydroxycorticosteroids fell to subnormal levels only when necrotic changes could be observed in the adrenal glands of scorbutic guinea pigs.

The adrenocortical response to ACTH likewise seems unimpaired in states of ascorbic acid deficiency. CLAYTON and PRUNTY (1953) found that the excretion of urinary 17-ketosteroids showed an increase after the administration of ACTH to scorbutic guinea pigs. In scorbutic human beings, TREAGER et al. (1950) noted that the administration of ACTH produced an eosinopenia and an enhanced excretion of urinary 17-ketosteroids.

CLAYTON and PRUNTY (1951) were unable to influence the course of scurvy in guinea pigs by the administration of ACTH or cortisone.

It can thus be seen that the adrenal cortex is apparently capable of hyperactivity in the presence of ascorbic acid deficiency; this indicates that the vitamin is not indispensible to adrenocortical function. The concensus has been well put by MEIKLEJOHN (1953): "In summation, it seems fair to say that to date there is no undisputed evidence that ascorbic acid plays any part in assisting the synthesis or secretion of the adrenocortical hormones." (See also Chapters 1, 3.)

2. Pantothenic-acid deficiency

The strongest evidence that a vitamin may play an important role in adrenocortical function has been obtained in instances of pantothenic-acid deficiency. The earlier findings in this area were reviewed by RALLI and DUMM (1953) and NOVELLI (1953).

In addition to feeding a diet devoid of this vitamin, attempts have been made by some investigators (e. g., BEAN et al. 1955) to induce pantothenate deficiency by the administration of a supposed metabolic antagonist, ω-methylpantothenic acid. It should be noted that the many physiological disturbances that follow the administration of this agent are not readily reversed by pantothenic acid. One may therefore surmise that the effects of this agent cannot be attributed wholly to interference with the utilization of pantothenic acid (A. WHITE et al. 1959, p. 1012).

DEANE and McKIBBIN (1946) described in detail the morphological and histochemical changes that occur in the adrenal gland of the rat in the course of pantothenic acid deficiency. The ultimate picture is one of focal necrosis and hemorrhage within the gland.

Many indications of adrenocortical insufficiency have been observed in states of pantothenic-acid deficiency. Sensitivity to the hypoglycemic effects of insulin is increased in human beings (BEAN and HODGES 1954, BEAN et al. 1955) and rats (WINTERS et al. 1952b). There are other signs of disturbances in intermediary metabolism that may reflect adrenocortical insufficiency. In pantothenate-deficient dogs, forced excercise resulted in a collapse characterized by hypoglycemia; such a condition could be prevented by the prior administration of lipide extracts of adrenocortical tissue (ARNRICH et al. 1956a). Pantothenate-deficient rats did not exhibit the normal response of hyperglycemia and enhanced hepatic glycogenesis when exposed to atmospheric pressures equivalent to an altitude of 20,000 feet for 24 hours, even though adrenocortical hypertrophy had developed. Again a normal response occurred in such animals if they were pretreated with adrenocortical extracts (HURLEY and MORGAN 1952). Pantothenic-acid therapy is efficacious also if instituted before the final stages of "adrenal exhaustion."

ERSHOFF (1953) found that no pantothenate-deficient rats were able to survive for 72 hours at 2° C., whereas all of the control animals were able to withstand this stress. However, it is of interest to note that the survival of the experimental animals in the cold was not prolonged by the administration of a variety of hormones, including anterior pituitary extracts, cortisone acetate, thyroxine, growth hormone or ACTH. The syndrome of pantothenic-acid deficiency obviously consists of more than adrenocortical hypofunction.

Both GAUNT et al. (1946) and GUGGENHEIM (1954) have reported that pantothenate-deficient rats excrete a water load at a subnormal rate and have an enhanced susceptibility to water intoxication. This disturbance can be corrected by the administration of ACTH or cortisone (GUGGENHEIM 1954). McQUEENEY et al. (1947) found that pantothenate-deficient rats were able to conserve sodium while on a sodium-deficient diet. Furthermore, steroids with "mineralocorticoid" activity do little to reverse the changes wrought by pantothenic-acid deficiency (HURLEY and MORGAN 1952, GUGGENHEIM 1954).

Other findings in pantothenate deficiency that are compatible with a diagnosis of adrenocortical insufficiency are, in human beings, hypochlorhydria (BEAN and HODGES 1954) and postural hypotension (BEAN et al. 1955). Pantothenate-deficient dogs do not exhibit the eosinopenia after exercise that normal animals do (ARNRICH et al. 1956b).

The *in vitro* synthesis of adrenocortical steroids is definitely subnormal in glands obtained from pantothenate-deficient rats (EISENSTEIN 1957), as is the output of urinary 17-ketosteroids by pantothenate-deficient human beings (BEAN et al. 1955). The adrenal content of cholesterol is low in the pantothenate-deficient rat (WINTERS et al. 1952c, DUMM et al. 1953, PERRY et al. 1953) but normal in the pantothenate-deficient dog (ARNRICH et al. 1956b). Moreover, in such rats, repletion of adrenal cholesterol after a stress takes an unusually long period of time (DUMM et al. 1953). These findings have prompted the suggestion that the biogenesis of steroid hormones by the adrenal cortex is impaired in pantothenate-deficient rats because of a disturbance in the ability of the gland to synthesize cholesterol (WINTERS et al. 1952c, DUMM et al. 1953, FIDANZA and BONOMOLO 1956).

Reports on the effects of ACTH, endogenous and exogenous, in pantothenate deficiency are in conflict. Pantothenate-deficient rats exhibit a normal depletion of adrenal ascorbic acid when placed in the cold (PERRY et al. 1953). They likewise exhibit depletion of adrenal ascorbic acid and cholesterol and a lymphopenia when given ACTH (ERSHOFF and PARROTT 1953, ERSHOFF et al. 1953). The response to epinephrine, however, is delayed and may be abnormal (i. e., a lymphocytosis develops instead of a lymphopenia); it has therefore been suggested that the adrenocortical insufficiency of pantothenate deficiency may stem from a disturbance in the secretion of ACTH (ERSHOFF and PARROTT 1953, ERSHOFF et al. 1953).

However, many reports of subnormal responsiveness to ACTH in pantothenate deficiency have been made. In pantothenate-deficient rats, the administration of ACTH does not produce the customary eosinopenia and lymphopenia (WINTERS et al. 1952a), there is no increase in liver glycogen stores or in the urinary excretion of glucose and nitrogen (WINTERS et al. 1952b), and the fat content of the liver does not increase significantly (MORGAN and LEWIS 1953). In pantothenate-deficient guinea pigs, only a small increase in the urinary excretion of 17-hydroxycorticosteroids occurs (EISENSTEIN and WENNEKER 1958). Little or no eosinopenia follows the administration of ACTH to pantothenate-deficient human beings (BEAN and HODGES 1954).

These findings of a subnormal response to ACTH suggest that the state of insufficiency might be ascribed to the adrenal cortex itself and not to the pituitary. In point of fact, SCHULTZ et al. (1952) concluded that abnormally large quantities of ACTH are released in pantothenate deficiency. It was their view that, as the output of adrenocortical hormones wanes in the absence of pantothenic acid, the anterior pituitary gland is prompted to secrete ever-greater quantities of ACTH. They felt that the adrenocortical necrosis that occurs in pantothenate deficiency is a reflection of glandular exhaustion from overstimulation by ACTH. (See Chapter 1.) They were able to accelerate the process of adrenocortical necrosis in pantothenate-deficient rats by the administration of ACTH and to retard or prevent it by the administration of cortisone. This concept is not supported by the report of BUTLER and MORGAN (1955) that normal quantities of ACTH were found in the anterior pituitary gland and in the venous blood of pantothenate-deficient rats.

Taken in sum, the evidence indicates that pantothenic acid is required for normal adrenocortical function. There are only a few findings not consonant with this view. Atrophy of the thymus occurs in pantothenate-deficient rats (MCQUEENEY et al. 1947, MORGAN and LEWIS 1953) and mice (MELAMPY et al. 1951). Perhaps this is a reflection of the need for pantothenic acid for normal thymic growth. The other inconsistent finding was made by ERSHOFF et al. (1953); they observed that the ability of rats to survive intoxication by egg white was unimpaired by pantothenate deficiency. Such a response would not be anticipated if any significant degree of adrenocortical insufficiency were present (HARTMAN and BROWNELL 1949).

3. Riboflavin deficiency

Adrenocortical function appears to be impaired only in the later stages of riboflavin deficiency. ARNRICH et al. (1956a, b) could detect no signs of adrenocortical insufficiency in the response of riboflavin-deficient dogs to the stress of exercise, and the adrenal content of cholesterol was found to be normal in these animals. FORKER and MORGAN (1954) observed that rats on a riboflavin-deficient diet for 3 weeks gave evidence of abnormally enhanced gluconeogenesis when subjected to anoxia; however, when they had been on the diet for 8 weeks, this response was essentially abolished. GAUNT et al. (1946) likewise noted that delayed diuresis after a water load was exhibited by rats in states only of severe riboflavin deficiency. ERSHOFF (1954) has noted that the riboflavin-deficient rat may exhibit the same degree of sensitivity to the hypoglycemic effects of insulin as does the adrenalectomized rat.

It would thus seem that, in riboflavin deficiency, the initial response is one of adrenocortical hyperfunction. As the severity of the deficiency state increases, evidence of impaired adrenocortical function may be seen. Such a sequence of events suggests that riboflavin may play no vital role in the synthesis and secretion of adrenocortical hormones.

4. Pyridoxine deficiency

There is little or no evidence to suggest that pyridoxine plays an indispensible role in normal adrenocortical function. In pyridoxine-deficient rats, the adrenal glands are hypertrophied, and, by histological and histochemical criteria, appear to be hyperactive (STEBBINS 1951 b, COMBRIDGE 1956). The processes of steroidogenesis and the response to ACTH appear to be normal (EISENSTEIN 1959).

The pyridoxine-deficient rat excretes a water load at a subnormal rate (STEBBINS 1951 a, GUGGENHEIM 1954), but such a response may be a reflection of disturbances other than adrenocortical insufficiency.

IV. Ionizing radiations

Adrenocortical insufficiency can be produced by the cytotoxic effects of ionizing radiations. The initial response to lethal, whole-body irradiation appears to be one of adrenocortical hyperactivity. Thus, in mice, tolerance for potassium was increased in the immediate post-irradiation period (W. SMITH 1951). The response of the perfused calf adrenal to ACTH was enhanced at this time (ROSENFELD 1958). This may be followed by a period of relative normalcy with respect to adrenocortical function, and then the preterminal decline commences (ROSENFELD 1958).

A decrease in the gland's output of hormones may be detected before degenerative changes in the structure of the adrenal cortex can be observed by light microscopy (UNGAR et al. 1955). Despite the gradual waning of steroidogenesis within the gland in the preterminal period, it is of interest to note that the concentrations of cortical steroids in the peripheral blood of the calf at this time are at or above normal levels.

V. Adrenal exhaustion

It has occasionally been demonstrated that, if the adrenal gland is subjected to prolonged, intense stimulation by ACTH, it becomes "exhausted" and adrenocortical insufficiency ensues. In this situation, the gland not only becomes "exhausted" in the sense of its capacity for hormonal secretion, but portions of it actually undergo hemorrhagic disintegration.

INGLE (1951) was able to produce necrotic changes in the inner cortical zones of the adrenal gland of the rat by the administration of large doses of ACTH for 21 days. Three instances of adrenal hemorrhage during the course of ACTH therapy in human beings have been reported: in two of the cases, hemorrhages occurred in both glands (GREENE 1953, WILSON and ROTH 1953) and in the third case, only one gland was involved (GALIN 1958). It remains to be determined, however, to what extent ACTH therapy played a causative role in the adrenal hemorrhages encountered in these clinical situations. (See also p. 50.)

Large doses of mammalian ACTH will also produce pathological lesions in the adrenal glands of lower vertebrates. M. MILLER (1953) was able to produce such changes in the glands of an amphibian, *Triturus torosus*, and also in the glands of a lizard, *Xantusia vigilis* (M. MILLER 1952).

Manifestations of adrenocortical insufficiency

A. Ability to survive

I. Introduction

States of adrenocortical insufficiency are usually found to have a fatal termination (HARTMAN and BROWNELL 1949, THORN 1951). However, any statement as to the inevitability of death in the absence of functional adrenocortical tissue must be subject to some qualification. Certain investigators have reported instances of "indefinite" survival after adrenalectomy (HARTMAN and BROWNELL 1949). Some of these findings can be readily explained in terms of the presence of accessory adrenocortical tissue, the composition of the diet supplied, or the intervention of other factors known to prolong survival after adrenalectomy. A few of these findings (e. g., AGATE and ZWEMER 1935, HARTMAN et al. 1943), however, cannot easily be incorporated into our general concepts of adrenocortical insufficiency. Indeed, they seem to indicate that animals can survive for a normal span of life without adrenocortical tissue (see p. 66 ff.). To be sure, it has never been demonstrated unequivocally that the animals employed in these studies were wholly free of any tissue capable of secreting adrenocortical hormones. However, unless these findings can be explained in ways consonant with our present general concepts of adrenocortical insufficiency, we cannot reject the view that, under certain circumstances, an organism may survive in the absence of adrenocortical hormones and in the absence of any supportive measures.

Additionally, the actual length of the period of survival after the onset of adrenocortical insufficiency can be influenced by a number of factors. Earlier investigators of adrenocortical insufficiency recognized the existence of these extraneous factors, and their findings have been well summarized in previous publications (HARTMAN and BROWNELL 1949, GAUNT and EVERSOLE 1949, NOBLE 1950, CHESTER JONES 1957). In the paragraphs to follow, these factors will be enumerated; the relatively few recent findings in this area will be cited under the appropriate headings.

II. The influence of various factors upon survival

1. The degree of adrenocortical insufficiency

Varying degrees of adrenocortical insufficiency can exist that presumably reflect varying decreases in the secretory capacity of the adrenal cortex. Such states of partial adrenocortical insufficiency can best be observed in human beings in whom the condition is of spontaneous origin (THORN 1951). The degree of adrenocortical insufficiency can be estimated, in part, by the signs, symptoms and clinical course of the disease as well as in the requirements for maintenance therapy (see section on Diagnosis, p. 131). On occasion, the pathological process appears to encroach upon only a portion of the adrenocortical tissue and then to become arrested (ROWNTREE 1940). *The milder forms of adrenocortical insufficiency are not incompatible with survival and pose a threat to life only in times of great stress,* as during a major surgical procedure (THORN 1951).

In the absence of accessory adrenocortical tissue, bilateral adrenalectomy results in a state of total adrenocortical insufficiency; such a state may also arise spontaneously (see section on Etiology, p. 49 ff.). The organism subsequently manifests many, if not all, of the physiological and biochemical derangements described in subsequent portions of this chapter. In most individuals, death is preceded by a preterminal crisis that may last for hours or days. Ultimately, the moribund organism is no longer capable of responding even to very large doses of adrenocortical hormones, and death follows shortly (HARTMAN and BROWNELL 1949).

2. The rate of development of adrenocortical insufficiency

It has been demonstrated that certain species are better able to tolerate adrenocortical insufficiency if it develops slowly rather than suddenly. For example, SILVETTE and BRITTON (1936) removed both adrenal glands from opossums in one operation. Their animals survived, on the average, for 6 days thereafter. HARTMAN et al. (1943) adrenalectomized opossums in the following fashion: they first removed one gland and then, after an interval of some days, removed the other. Moreover, they injected their animals with adrenocortical extracts for the first 3 days after removal of the second gland. Many of the animals survived for weeks or months and a few seemed destined to live out a normal life span. There seemed to be no explanation for these findings other than the surgical schedule that they observed; with single-stage adrenalectomies, their animals had the same short period of survival as did those of SILVETTE and BRITTON.

Findings such as these have prompted many investigators to perform adrenalectomy in two stages. For example, it is commonly done in the case of the dog (NOBLE 1950), but does not significantly prolong the postoperative period of survival of many species (HARTMAN and BROWNELL 1949).

3. Species

The period of survival after adrenalectomy is very much a function of species (HARTMAN and BROWNELL 1949, NOBLE 1950, 1955, CHESTER JONES 1957). The survival of some birds after adrenalectomy is reckoned in terms of hours (NOBLE 1950). The sloths *(Choloepus hoffmanni* and *Bradypus griseus)* survive for only 24 hours after adrenalectomy (BRITTON et al. 1938); this constitutes the shortest period of survival reported for any mammal. As described above, the opossum *(Didelphis virginiana)* may survive for months or even a year after adrenalectomy (HARTMAN et al. 1943). These findings have been interpreted as a reflection of varying degrees of dependence of the different species upon adrenocortical hormones (HARTMAN and BROWNELL 1949).

4. Age

It is generally held that younger animals succumb more readily to the effects of adrenalectomy than do older ones (HARTMAN and BROWNELL 1949), but this variable has been systematically investigated only in the case of the laboratory rat.

GILLMAN and GOLDBERG (1942) found that rats weighing from 10 to 20 grams survived, on the average, for 5.6 days after adrenalectomy, whereas rats weighing from 100 to 200 grams had an average survival period of 8.4 days. They also noted that a higher percentage of younger rats died within 48 hours after adrenalectomy; they seem less able to withstand the stress of surgery. These were excluded in calculating the average.

COWIE (1949a) performed similar experiments on rats of varying ages. He found that the period of survival after adrenalectomy increased with age up to the age of 4 months; in animals 4 to 12 months of age, the periods of survival were essentially identical.

5. The presence of accessory adrenocortical tissue

The presence of functional masses of ectopic adrenocortical tissue that may occur within the organism has sometimes proved to be a hindrance to the elucidation of adrenocortical function (see p. 30). On the other hand, the specter of such tissue can be overemphasized and perhaps has been invoked too often as an explanation for otherwise inexplicable periods of prolonged survival after adrenalectomy (GAUNT and EVERSOLE 1949).

The magnitude of the problem posed by accessory adrenocortical tissue is a function of species. (See also Chapters 1 and 4.) It has been reported that 70% of all rabbits examined possess ectopic cortical tissue (KOJIMA 1929). In a series of 100 consecutive autopsies performed upon unselected human beings, GRAHAM (1953) found that 32 had accessory adrenal tissue, usually in the area of the celiac plexus. Both cortical and medullary tissue could be detected in half of these 32 instances. Accessory adrenal tissue is often postulated to occur in the strains of rats used in biological research but it is difficult to find even after adrenalectomy (GAUNT and EVERSOLE 1949). In a study of several hundred rhesus monkeys, SCHMIDT (1956) discovered accessory adrenal tissue in only one animal. It was imbedded in the peripheral cortex of the kidney; histologically, it possessed glomerulosal, fascicular and reticular zones but no medullary tissue.

6. Environmental factors

The environmental conditions found to have the greatest influence on the length of the period of survival after adrenalectomy are ambient temperature and the composition of the diet.

Adrenalectomized mammals react adversely to extremes of heat and cold. While investigators might naturally avoid placing adrenalectomized animals in a cold environment, a similar avoidance of elevated temperatures is not often made nor is it always possible to escape them, especially in the summer. The effect of heat upon the survival of adrenalectomized rats was assessed by LANGLEY et al. (1956); they found the period of survival to be significantly decreased when the animals were placed in rooms having ambient temperatures of 34–38°. The fact that cold temperatures likewise hasten the demise of the adrenalectomized rat is well known (HARTMAN and BROWNELL 1949, NOBLE 1950).

The response of poikilothermic animals after adrenalectomy to alterations in temperature is somewhat different from that of the homeothermic mammals. MAES (1937) found that adrenalectomized frogs kept at 5° survived approximately 7 times as long as similar animals kept at 28°. [See also CHESTER JONES (1957, pp. 149–167).]

The time of the year at which adrenalectomy is performed appears to exert a significant influence upon the duration of survival only in poikilothermic animals and in those animals that undergo hibernation. The marmot (groundhog) adrenalectomized in the summer survives only about as long as the cat; adrenalectomized during the winter, the marmot will survive throughout the normal period of hibernation (HARTMAN and BROWNELL 1949).

The inorganic constituents of the diet consumed in states of adrenocortical insufficiency may profoundly influence the length of the period of survival. A diet rich in sodium is well-known for its beneficial effects in many species (KENDALL 1948); a diet rich in potassium will, however, shorten the period of survival. However, sodium therapy does not materially prolong the survival of rhesus monkeys (HOFMANN et al. 1954) or wild Norway rats (RICHTER et al. 1950) after adrenalectomy.

The effects of various organic constituents of the diet upon the survival period after adrenalectomy have been investigated only slightly. INGLE et al. (1943) found that the period of survival of adrenalectomized rats was not significantly influenced by variations in the relative amounts of carbohydrate, protein or fat in the diet.

Booker et al. (1955) found that rather large doses of glutathione prolonged the survival of adrenalectomized mice in the cold.

An outstanding example of the influence of environmental conditions upon the period of survival after adrenalectomy was reported by Agate and Zwemer (1935). One group of rats in this study was made up of animals raised in the authors' colony where careful attention was paid to diet, temperature and cleanliness. Only 2% of the animals in this group were dead by 60 days after adrenalectomy. A second group of rats was obtained from outside sources; they subsequently received the same care as the animals in the first group. By 60 days after adrenalectomy, 52% of these animals were dead. In summarizing the literature of the time, the authors noted that other investigators had reported an average mortality of 89% of rats by 60 days after adrenalectomy. The factor that seemed to be of significance in the findings of Agate and Zwemer was the particular care given to the environment of their animals. The authors could not find accessory adrenocortical tissue in most of their animals. It seems possible that, in a carefully regulated environment, some rats may survive for long periods of time without adrenocortical hormones (Gaunt and Eversole 1949).

Some 19th and early 20th century investigators reported remarkably short periods of survival for animals after adrenalectomy; the survival periods for cats, dogs and rats were often measured not in days, as they now are, but in hours. With our present knowledge, we can say that many of these reports probably reflected poor surgical techniques and inadequate care of the animals in the postoperative period (G. Stewart 1924).

In any determination of the average period of survival of a group of animals after adrenalectomy, one would obviously arrive at a falsely low figure if data from animals succumbing principally to the stress of surgery were included in the computations. It has therefore become conventional to exclude from consideration any animals dying within 48 hours after adrenalectomy (Gillman and Goldberg 1942); such a statement naturally applies only to those species known to survive adrenalectomy for approximately 5 days or more.

As organisms in adrenocortical insufficiency are unusually susceptible to a wide variety of noxious influences, such as bacterial infections, low temperatures, etc. (see p. 129), the period of survival after adrenalectomy will be reduced to the extent that these factors are operative in the experimental situation. If one wishes to observe the picture of uncomplicated adrenocortical insufficiency, care should be taken to protect adrenalectomized animals from all influences to which they are known to be particularly sensitive (Hartman and Brownell 1949).

7. The influence of other endocrine glands

a) Reproductive organs

The influence of ovarian secretions upon the length of the period of survival after adrenalectomy was widely studied in the period before 1945 (Parkes 1945); the more recent studies consist of repetitions and extensions of these earlier investigations.

In confirmation of earlier findings (Parkes 1945), Cowie (1949a) found that female rats live for a significantly longer period of time after adrenalectomy than do male rats; such a distinction between the sexes cannot be made, however, in prepubescent animals. The same difference in the survival period after adrenalectomy has been observed in male and female golden hamsters (Snyder and Wyman 1951). The gonadectomized female rat lives for a shorter period of time after adrenalectomy than does the rat with ovaries intact (LANGLEY et al. 1956). Implants of ovarian tissue into the ears, spleens or testes of male or female rats (LICHTON et al. 1953) or mice (R. HILL 1948) enable the animals to live for a somewhat longer period of time after adrenalectomy than would otherwise be expected.

The survival of adrenalectomized rats can be prolonged by the administration of progesterone, is usually unaffected by the administration of androgenic hormones and may be unaffected or reduced by the administration of estrogenic hormones (PARKES 1945).

Pregnant or pseudopregnant dogs and ferrets survive for a longer period of time after adrenalectomy than do non-pregnant animals. Whether or not pregnancy or pseudopregnancy prolongs the period of survival after adrenalectomy in rats and cats remains an unresolved dispute (PARKES 1945). This matter is discussed further in the section on the reproductive system (pp. 118–122).

b) *Thyroid gland*

The influence of thyroidal secretions upon the period of survival is discussed in a subsequent portion of this chapter (pp. 122–126).

c) *Anterior pituitary*

The survival period after adrenalectomy in hypophysectomized rats is shorter than it is after simple adrenalectomy (GREEP 1950). In one study, the hypophysectomized rat lived, on the average, 4.8 days after adrenalectomy; the non-hypophysectomized rat had an average postoperative survival of 10.5 days (HOFMANN 1952).

B. Growth

I. Weight

Most studies of growth in adrenalectomized animals are based on rate of *gain in weight, which may reflect changes in water or fat content of the animal rather than changes in amount of protoplasm.* Most such studies have been directed towards evaluating the effectiveness of various forms of therapy. Discussion of the poor weight gain of operated animals has usually failed to take into account differences in the amount of food consumed, even when control animals were fed the same diet as the adrenalectomized ones (GAUNT et al. 1935, E. ANDERSON et al. 1940a, KUIZENGA et al. 1943, EVERSOLE 1945, SEGALOFF 1946, GAUNT et al. 1951).

II. Body composition

COHN et al. (1955) analyzed the carcasses of force-fed rats and found that the increase in nitrogen and fat content was no different in adrenalectomized and intact animals. Both groups of force-fed animals gained more fat than did intact animals allowed to eat *ad libitum.*

III. Skeleton

1. Man

Little information is available on the growth of human beings with Addison's disease, since onset in childhood is unusual. JAUDON (1946a) gave growth data for a girl whose signs appeared at 6½ months of age. Her height was always within normal limits; treatment consisted of deoxycorticosterone and the addition of salt to her diet. N. TALBOT and SOBEL (1947a, b) found that children with adrenal insufficiency grew normally as long as electrolyte and water balance was satisfactory. EHRENGUT (1956) analyzed the data on the growth of 16 children with Addison's disease; stature was below average in only one of this group.

2. Animals

a) Linear growth

The information available from carefully conducted studies of skeletal growth indicates that adrenalectomized animals grow more rapidly than intact ones. ALLERS and CRANDALL (1936) reported *radiological* evidence of growth of the skeleton and teeth in adrenalectomized puppies but offered no quantitative data. WADE and HAZELWOOD (1941) mentioned that 50-day-old adrenalectomized rats averaged 15.1 cm in *length*, compared to 16.1 cm for control animals. A low K^+ and high Na^+ diet was provided, but the animals were not pair-fed and the operated animals weighed 90 g while the intact ones weighed 100 g. INGALLS and HAYES (1941) found demineralization and decrease in the columnar cartilage of the epiphyses of adrenalectomized rats; only 3 of 11 animals survived for more than 10 days. The results were probably influenced by "small doses of cortin" given for the first 10 days after adrenalectomy.

WYMAN and TUM-SUDEN (1945) concluded that the decreased width of the tibial epiphyseal cartilage of adrenalectomized rats could be related to their decreased food intake, because the width of the cartilage of intact animals on restricted diets was no greater than that of the adrenalectomized animals. By 2 to 4 weeks after operation the epiphyseal cartilage in adrenalectomized animals was definitely wider than in the control animals. MAASEN (1952) studied pair-fed rats and found that the tibias of adrenalectomized animals were longer than those of controls, although the weights and volumes of these bones were smaller. The results were attributed to absence of growth-inhibiting corticosteroids in the operated animals. BUFFETT and WYMAN (1955) adrenalectomized male and female rats ranging in age from 80 to 100 days, and supplied them with physiological saline and food *ad libitum* in the postoperative period. At the end of the third week there was a significant increase in the width of the epiphyseal cartilage of the tibia; the widths were the same in males as in females. As the width of the cartilage is somewhat greater in the intact male rat than in the female, a relatively greater growth of cartilage occurred in the female after adrenalectomy than in the male. Histologically, enhanced chondrogenesis was evident after adrenalectomy (BUFFETT and WYMAN 1956); these animals were maintained on saline drinking fluid.

M. SIMPSON et al. (1950/51) found that hypophysectomized, adrenalectomized rats showed the same epiphyseal response to growth hormone as did those merely hypophysectomized.

b) Hypercalcemia

Some disturbance of calcium metabolism after adrenalectomy is indicated by the high Ca^{2+} concentrations in the plasma of dogs, although there is considerable variation from dog to dog (ROGOFF and STEWART 1928c). It is not clear to what extent the increased level (about 15 mg per 100 ml, compared to a normal of 10 mg) reflects an increased loss from the skeleton and to what extent it should be attributed to adrenal insufficiency. Hypercalcemia has occassionally been reported in human beings with Addison's disease (LABHART 1957, PRADER et al. 1959).

c) Mineralization

The findings of PUGSLEY and COLLIP (1936) indicated that a greater amount of Ca^{2+} was released from bone after the administration of parathyroid extract in the adrenalectomized rat than in the intact animal. The various reports of demineralization of the skeleton after adrenalectomy suggest that the ability of bone to maintain its Ca^{2+} content is subnormal.

INGALLS and HAYES (1941) observed widespread and marked demineralization of the entire skeleton in adrenalectomized animals receiving no supportive therapy. That loss of bone salts cannot be attributed to malnutrition is indicated by the observations of MAASEN (1952), who found that the weight and volume of the tibias of adrenalectomized rats were less than in pair-fed controls, although the length of these bones was greater in the adrenalectomized animals. APPLEBAUM and SEELIG (1955) noted a pronounced loss of mandibular bone supporting the molars in the rat after adrenalectomy.

It seems clear that an increase in chondrogenesis after adrenalectomy reflects the absence of the growth-inhibiting effect of the glucocorticoids (see Chapter 7). Abnormalities of mineralization possibly reflect the reduction in pH of the extracellular fluid (see p. 79).

C. The functions of various organs in adrenocortical insufficiency

I. The kidney

Renal dysfunction in adrenocortical insufficiency has been widely recognized but only imperfectly studied. In studies made before 1940, a number of the upsets occasioned by adrenalectomy, such as changes in the composition of the blood, were often attributed to the kidney without any measurement of renal function being made. From approximately 1940 onward, modern techniques for the evaluation of renal function began to be utilized by investigators of adrenocortical insufficiency. Although these studies yielded much information, the scope of our knowledge of renal function in adrenocortical insufficiency is still smaller than it need be; in particular, there has never been full use made of the techniques available for the study of renal tubular function.

Disturbances in renal hemodynamics and in renal tubular function occur in states of adrenocortical insufficiency. Briefly, it has been found that *glomerular filtration rate and renal blood flow are reduced.* In addition, *alterations in the reabsorptive and secretory activities of the renal tubular cells have been described.* The investigations upon which these statements are based will be reported in detail in the paragraphs to follow. The nature of the defect (or defects) in the function of the renal tubular cells remains largely unknown. The transport processes within these cells ordinarily show fluctuations in activity in accordance with the needs of the moment for the maintenance of homeostasis. In adrenocortical insufficiency, the capacity of these processes to accommodate to the needs of the organism (e. g., to conserve sodium or to excrete potassium) is limited. How the kidney loses some of its ability to maintain the constancy of the internal environment in adrenocortical insufficiency remains an unanswered question of fundamental importance.

It is unknown to what extent the distortions in the composition and volume of the extracellular and intracellular fluids that occur in adrenocortical insufficiency can be attributed to altered renal function, and to what extent extra-renal factors are operative in this situation.

1. Anatomical changes

Striking changes in the anatomy and chemical composition of the kidney are not consistently encountered in states of adrenocortical insufficiency.

Few investigations have been made of the structure of the kidney in adrenalectomized animals; in these studies, some investigators have reported the occurrence of degenerative changes in the kidney after adrenalectomy, while others have reported that no morphological alterations occur. These findings were reviewed by HARTMAN and BROWNELL (1949).

Similarly, there have been no consistent findings of alterations in renal structure in Addison's disease. This was the conclusion of TALBOTT et al. (1942) based upon their own findings and a review of the literature. SORKIN (1949) found some pathological changes in the kidneys of 8 of 21 patients studied. Leukocytic infiltration, usually in the cortex, occurred in these 8 instances. Furthermore, in 4 of the 8, focal cortical atrophy was present and, in an additional case, subcapsular atrophy.

From a consideration of all of these findings, it seems reasonable to conclude, as did TALBOTT et al. (1942), that the renal dysfunction that occurs in adrenocortical insufficiency cannot be readily accounted for on the basis of structural changes in the kidney, as determined by light microscopy. In any event, functional changes can be detected within a day or two after adrenalectomy; it does not seem likely that gross anatomical lesions will have developed in this short period of time. It is possible, however, that the situation in some species is subsequently aggravated by anatomical damage (LUFT and SJÖGREN 1949).

2. Glomerular filtration rate and renal blood flow

Test:

The concept that the renal clearance of an appropriate substance could be used as a measure of glomerular filtration was first expressed by REHBERG (1926), but underwent its greatest development at the hands of HOMER SMITH and his associates (H. SMITH 1951). The substances under study must be capable of being filtered completely at the glomerulus and must not be secreted or reabsorbed by the renal tubule. Of the substances meeting these specifications, the one most commonly employed for the measurement of *glomerular filtration rate (GFR)* is the polysaccharide inulin. If the urinary concentration of inulin is divided by its concentration in plasma and the result multiplied by the rate of urine formation, the GFR for the organism (usually expressed in ml/min) can be obtained.

In adrenocortical insufficiency, the GFR is commonly found to be reduced. Representative examples of these findings are presented in Table 2.

Table 2. *The glomerular filtration rate in adrenocortical insufficiency*

Species	Clearance of:	GFR (ml/min)			Reference
		Control	Adrenocortical insufficiency	% Decrease	
Rat	Inulin	0.33[1]	0.04[1]	88	STONER and DEXTER (1954)
Dog	Creatinine	65	14—27	70	HARRISON and DARROW (1939)
Dog	Inulin	71	27	62	GAUDINO and LEVITT (1949)
Dog	Creatinine	65	40	38	O. GARROD et al. (1955)
Man	Inulin	120	77	36	TALBOTT et al. (1942)
Man	Inulin	117 (women)	64	45	SANDERSON (1948)
		131 (men)	73	41	
Man	Mannitol	90[2] (women)	50	45	C. WATERHOUSE and KEUTMANN (1948)
		150[2] (men)	80	47	

[1] Per 100 g body weight (for rat only).
[2] Per 1.73 square meters of body surface.

It is not an easy matter to determine why the GFR is subnormal in adrenocortical insufficiency, for alterations in the rate at which plasma is filtered at the glomerulus can be produced by a variety of disturbances. It seems unlikely that adrenocortical hormones play any primary role in the maintenance of the normal GFR, since ROEMMELT et al. (1949) demonstrated in the dog that for periods up to

3 months following adrenalectomy the GFR could be maintained at the preoperative level if the animals were permitted free access to salt and water. The absence of adrenal medullary secretions likewise does not seem to be an important causative factor. STONER and DEXTER (1954) found the GFR to be unchanged by adrenal demedullation in the rat, whereas adrenalectomy, as recorded in Table 2, was followed by a profound drop in the GFR. It would seem that the reduction in the GFR is a secondary reflection of some primary disturbance produced by adrenocortical insufficiency.

In a recent review of the influence of the adrenal cortex on renal function, NABARRO (1956) concluded that the drop in the GFR occurring in adrenocortical insufficiency is caused by a number of factors. (1) Sodium depletion and the reduction in extracellular fluid volume are clearly contributory; the findings of ROEMMELT et al. (1949) cited above make such a conclusion inescapable. However, in adrenalectomized human beings and dogs, the GFR may be subnormal despite maintenance of a normal volume of extracellular fluid (MENDELSOHN and PEARSON 1955, O. GARROD et al. 1955). (2) As the pressure that effects filtration in the glomerulus derives from the arterial blood pressure, the cardiovascular status of the organism must be considered. Later it will be shown that hypotension is a common finding in adrenocortical insufficiency (p. 88 ff.). It would seem unwise to deny that this hypotension can contribute to the reduction in the GFR, particularly in instances such as the preterminal state when a shock-like picture may be seen. However, when the blood pressure and the extracellular fluid volume are normal but the GFR is subnormal, the possible intervention of additional factors must be considered. (3) Of primary importance for filtration pressure in the glomerulus is the effective diameter or "tone" of the afferent and the efferent glomerular arterioles (H. SMITH 1951). In general, the filtration pressure can be reduced by an increased tone in the afferent arteriole or by relaxation of the efferent vessel. In Addisonian patients having a normal blood pressure and extracellular fluid volume, TALBOTT et al. (1942) found that the GFR was reduced more than was renal plasma flow. They concluded that such a finding could be a reflection of efferent arteriolar dilation; SANDERSON (1948) came to a similar conclusion.

In summary, the reduction in the GFR, so commonly encountered in adrenocortical insufficiency, seemingly cannot be wholly accounted for on the basis of a single causative factor. Oligemia, hypotension and vasomotor changes within the kidney probably play contributory roles. As the GFR falls, the excretory capacity of the kidney becomes increasingly impaired. The abnormally high concentrations of some substances in the blood, such as non-protein nitrogen, do reflect, in part, this inadequacy.

The drop in the GFR should not be regarded as having only deleterious effects for the organism. As PITTS (1952) points out, the Na^+ losses that occur in adrenocortical insufficiency would proceed at a faster rate if the GFR were not reduced. THOMPSON and PITTS (1952) have shown that experimental reductions in renal arterial pressure are followed by reductions in the GFR and Na^+ excretion. As the capacity of the renal tubular cells to reabsorb Na^+ is impaired in adrenocortical insufficiency (see p. 74), the maintenance of a normal GFR would lead to a more rapid Na^+ depletion and the survival period might be correspondingly shortened.

Test:

The blood flow through the kidney can be determined by measuring the renal clearance of a substance subject to filtration at the glomerulus and secretion by the renal tubular cells. To be suitable for this purpose, the substance in question must be excreted so efficiently as to be essentially absent from the renal venous blood. For this measurement, *p*-aminohippurate, diodrast or diodone are commonly employed. The value obtained from such a measurement is an expression of *effective renal plasma flow (ERPF)*. It is approximately 90—92% of true renal plasma flow, for some of the blood entering the kidney flows through perirenal fat and inert tissue in the renal capsule, pelves and calyces (H. SMITH 1951); thus the clearance of *p*-aminohippurate or diodrast is less than 100% only to the extent that blood entering the kidney does not perfuse tubular excretory tissue. Additionally, such measurements are valid only when low plasma levels of *p*-aminohippurate or diodrast obtain; if the plasma level is so high as to exceed the capacity of the renal tubular cells to secrete the substance, then obviously the plasma will not be wholly cleared of the substance in a single passage through the kidney. The *effective renal blood flow* may be calculated by dividing the ERPF by (1 — hematocrit).

In adrenocortical insufficiency, renal plasma flow has been found to be subnormal. However, it is usually depressed less than is the GFR, and, as a consequence, the filtration fraction (GFR/ERPF) in adrenocortical insufficiency is reduced. Examples of the reduction in ERPF are presented in the following table.

Table 3. *Effective renal plasma flow in adrenocortical insufficiency*

Species	Clearance of:	ERPF (ml/min)			Reference
		Control	Adrenocortical insufficiency	% Decrease	
Dog	PAH[1]	227	110	52	GAUDINO and LEVITT (1949)
Dog	PAH	159	128	19	O. GARROD et al. (1955)
Man	Diodrast	690	465	34	TALBOTT et al. (1942)
Man	Diodone	697 (men)	430	38	SANDERSON (1948)
		594 (women)	347	42	
Man	PAH	550—800 (men)	250—580	34	C. WATERHOUSE and KEUTMANN (1948)
		500—700 (women)	250—480	39	

[1] PAH = *p*-aminohippurate.

A comparison of the data in Tables 2 and 3 reveals that the depression in the GFR may be as much as twice the depression in the ERPF. The oligemia and impaired cardiovascular function of adrenocortical insufficiency no doubt are related causally to the reduction in the ERPF as well as to the drop in the GFR. The fact that the GFR is always found to be reduced somewhat more than is the ERPF suggests the operation of an additional factor in the glomerulus. Not only is the glomerulus being perfused with less blood per unit time and at a lower pressure but, as noted above, both TALBOTT et al. (1942) and SANDERSON (1948) have suggested that the efferent arterioles are dilated to an abnormal degree. Such a dilation by itself could cause some decrease in GFR.

3. Renal tubular reabsorptive and secretory activities

a) Inorganic electrolytes

α) Sodium. No other aspect of adrenocortical insufficiency has received greater attention from investigators than has the loss of sodium. PITTS (1952) has stated, "Inadequacy of renal tubular absorption of salt is probably the most clear-cut renal dysfunction in adrenal insufficiency." The defect seems to be resident in those renal tubular processes responsible for the reabsorption of sodium from the glomerular filtrate.

The failure of the kidney to conserve Na^+ becomes evident most dramatically in those instances in which the onset of adrenocortical insufficiency is abrupt, as it is following adrenalectomy. Within a matter of a few days, perhaps sooner, the organism is usually in negative Na^+ balance. With the passage of time, the amount of Na^+ in the body becomes progressively smaller. The magnitude of the renal defect, in relative terms, is not a great one. It may mean a failure to reabsorb only

2% of the Na^+ filtered at the glomerulus (PITTS 1952). In time, however, the cumulative effects of such losses result in a marked depletion of Na^+ in the body. In the rat, for example, within 5 to 7 days after adrenalectomy the quantity of Na^+ in extracellular fluid and tissues is significantly reduced, as is its concentration (H. WHITE and ROLF 1955). The *loss of Na^+ from blood* means that insufficient osmotically active solute is available for the maintenance of a normal blood volume, and, as *water is excreted* with the Na^+, *oligemia ensues*. Manifestly, more Na^+ is lost than water, for the concentration of Na^+ in extracellular fluid decreases. This may reflect the enhanced renal reabsorption of water that occurs in adrenocortical insufficiency (see below). At the same time, the K^+ concentrations in some tissues rise (H. WHITE and ROLF 1955) and water moves accordingly from extracellular to intracellular spaces. This means a further reduction in extracellular fluid volume and a further reduction in the intracellular concentration of Na^+ remaining in the tissues. The reasons for the increases in intracellular K^+ concentrations will be discussed shortly.

The problem of Na^+ losses from the body is aggravated by the loss of appetite usually encountered in adrenocortical insufficiency (see p. 113) as this means a reduced Na^+ intake. Many physiological upsets secondary to Na^+ depletion can be partially or wholly reversed if Na^+ intake is commensurate with the urinary losses of this ion (KENDALL 1948). Such secondary upsets are described at various places in this chapter and delineate clearly the widespread and serious consequences of the renal failure to conserve Na^+ in a normal fashion. The defect in the renal tubules with regard to Na^+ reabsorption cannot be remedied by salt supplements, however, and apparently reflects directly the absence of adrenocortical hormones, the most important of which is aldosterone.

Sodium is probably reabsorbed from the glomerular filtrate by several renal tubular processes (H. SMITH 1951). Some progress has been made in identifying those reabsorptive processes for Na^+ that are disturbed in adrenocortical insufficiency. The work of PITTS and his collaborators tends to suggest that the processes in the distal convoluted tubule are the ones that are deranged. This view is confirmed by the findings of VANDER et al. (1960). In the distal tubule, one mechanism of Na^+ reabsorption involves an ion-exchange reaction. The tubular cells secrete H^+ or K^+ into the tubular fluid; in exchange Na^+ is simultaneously absorbed by the tubular cells and restored to the extracellular fluid. The secretion of these two cations is subnormal in the adrenalectomized dog (PITTS 1952). The decreased secretion of H^+ means not only that Na^+ reabsorption is impaired but also that the maintenance of acid-base balance is more difficult to achieve. This problem will be discussed in subsequent paragraphs.

PITTS (1952) held that other mechanisms involved in Na^+ reabsorption may also be affected adversely by adrenocortical insufficiency. The natriuresis that follows the administration of relatively large doses of Pitressin is not accompanied by changes in the acidity of the urine or by changes in the rate of excretion of ammonium ions. Pitressin, therefore, is presumably acting on other mechanisms than the ion-exchange reactions described above. About 80% of the Na^+ loss that follows Pitressin administration can be prevented by the prior injection of deoxycorticosterone (SARTORIUS and ROBERTS 1949). The natriuresis produced by mercurial diuretics likewise seems to reflect the inhibition of Na^+-reabsorptive mechanisms other than the ion-exchange reactions. PITTS (1952) found that these Na^+ losses could be prevented by the prior administration of adrenocorticotropic hormone. PITTS believed that these experiments suggest the existence of renal tubular processes for Na^+ reabsorption that can be inhibited by Pitressin and mercurial diuretics and that can be stimulated by adrenocortical hormones.

In terms of specifically defined renal tubular processes, one can, to date, implicate only the ion-exchange reactions in the Na^+ losses that attend adrenocortical insufficiency. It is not known whether the magnitude of the depression is sufficient to account for all of the Na^+ lost. In any event, it should be appreciated that adrenocortical insufficiency results in only a relatively small impairment in the kidney's capacity to conserve Na^+. It is only with passage of time that the cumulative effects of this small impairment result in a state of Na^+ depletion that has serious secondary consequences.

In states of adrenocortical insufficiency, it seems probable that the renal tubular mechanisms for the reabsorption of Na^+ are operating to the full extent of their somewhat impaired capacity. If a Na^+ load is now suddenly imposed upon the organism, the response of the kidney is abnormally slow. For some time after the imposition of the load, the kidney reabsorbs greater quantities of Na^+ than is the case when the normal kidney is faced with a similar excess of Na^+. This is one manifestation of the kidney's reduced capacity to preserve homeostasis in adrenocortical insufficiency, and has been likened to its similar inability to excrete excess water rapidly (ROEMMELT et al. 1949). (See Chapter 4, Part 1, p. 542.) Observations of enhanced Na^+ reabsorption in the presence of Na^+ loads have been made in adrenalectomized dogs (KOTTKE et al. 1942, ROEMMELT et al. 1949, MADSEN 1955, SHARE and HALL 1955), adrenalectomized rats (KELLOGG and BURACK 1954) and in patients with Addison's disease (WILLSON and SUNDERMAN 1939, REFORZO-MEMBRIVES et al. 1945, BURNETT et al. 1953).

The time factor in these studies appears to be an important consideration. The renal inability to adjust rapidly to a Na^+ load is most evident when hypertonic NaCl is administered intravenously. The normal dog may respond in this situation not only by decreasing its renal tubular reabsorption of Na^+ but also by increasing the GFR (SHARE and HALL 1955). The adrenalectomized dog shows a similar response but of smaller magnitude. Only upon the passage of a greater period of time will the adrenalectomized dog excrete the excess Na^+ (MADSEN 1955). If the Na^+ load is imposed more gradually (i. e., by mouth), the difference between the normal response and that in adrenocortical insufficiency is less evident. The patient with Addison's disease, upon gradual Na^+ loading, is quite capable of achieving a daily excretion of Na^+ comparable to that of the normal individual (BURNETT et al. 1953). In this instance, the adaptation seems to occur wholly in the renal tubular cells, since the GFR is, if anything, slightly lowered during the period of increased salt intake.

The normal individual usually displays a diurnal rhythm in certain aspects of renal function. A greater volume of urine and therefore a greater amount of inorganic electrolytes are excreted during the day in comparison with the night. In Addison's disease, this rhythm may be abolished or reversed (BURNETT et al. 1953, NABARRO 1956).

β) Potassium. The phenomenon of "potassium retention" has likewise engaged the attention of many investigators of adrenocortical insufficiency. However, as a phrase, it has often been misused. In many instances, its use seems to derive from the finding of an increased concentration of K^+ in the plasma, and the phrase is employed as a synonym for hyperkalemia. "Potassium retention," interpreted literally, means an increase in the total quantity of K^+ in the body. Careful balance studies indicate that, in certain species, the reverse can be true in adrenocortical insufficiency (H. WHITE and ROLF 1955). Until these carefully designed studies were carried out, it was often assumed that, in adrenocortical insufficiency, a reciprocal relationship *always* obtained with regard to Na^+ and K^+, and that the content of Na^+ in the body decreased when that of K^+ increased.

As careful studies have not been carried out in many species, the applicability of the following statements is limited largely to the rat and dog. Shortly after adrenalectomy, two disturbances develop with regard to the distribution and excretion of K^+. The organism loses, to some extent, its ability to regulate the distribution of K^+ between the extracellular and intracellular fluid compartments. It also loses, to some extent, its ability to react to hyperkalemia with an increased renal excretion of K^+. These two impairments can account for the alterations in K^+ distribution and excretion encountered in adrenocortical insufficiency, but the precise nature of the cellular disturbances that give rise to these impairments is not known.

Very soon after adrenalectomy, the concentration of K^+ in the plasma begins to rise. In the dog, by the second postoperative day, it may have increased by as much as 50% (HASTINGS and COMPERE 1931). In the rat, by the fifth postoperative day, according to one study (H. WHITE and ROLF 1955), it rises from a normal value of 3.72 meq/l of extracellular fluid to 6.25. While the rats suffered, on the average, a net loss of 2.63 meq of Na^+ from their bodies in the first 5 days after the operation, they also suffered a net loss of 1.73 meq of K^+. It can be seen, therefore, that *the plasma concentration of K^+ can rise at a time when the rat is in negative K^+ balance*. This indicates an impairment in the ability of the organism to maintain K^+ normally and largely sequestered in intracellular fluid compartments. The increased K^+ in the urine and that in the extracellular fluid is probably derived partially from dietary sources and partially from tissue catabolism. In the study of H. WHITE and ROLF (1955), body weight, in this time period, fell from an average of 250 g to 215. Skin, muscle and viscera were all reduced in mass. The fat concentration of these tissues (expressed as g/kg of fresh tissue) was reduced 50 to 100%; this finding probably accounts in part for the fact that the water concentration in these tissues (in g/kg of fresh tissue) was elevated slightly at this time along with tissue concentration of fat-free solids. The concentration of K^+ in muscle, skin and viscera, if expressed in terms of meq/kg of fresh tissue or meq/100 g of fat-free solids, rises, but, if the concentration of K^+ is expressed in terms of meq/kg of intracellular water, it rises in muscle and viscera and falls in skin. Such changes are the mirror images of the alterations that occur in the water content of these tissues when expressed as g of water/100 g of fat-free solids. Thus, increases in the K^+ concentration of certain tissues do not necessarily mean that the organism is in a state of "potassium retention." It is obviously a complex situation for, in adrenocortical insufficiency, the cellular content of water, fat and fat-free solids is subject to change and the changes are not all in the same direction for all tissues. While concentrations of K^+ are doubtless of physiological significance, it must be borne in mind that units of reference for concentrations (i. e., volumes, masses, etc.) may be changing along with the absolute K^+ content of a cell.

In the study of H. WHITE and ROLF (1955) the total quantity of K^+ in the rat fell from 18.4 meq to 17.2 by the fifth day after adrenalectomy; this finding agrees rather closely with the external balance data reported above and shows clearly that no "potassium retention" *per se* occurred in these animals. Renal excretion of K^+ coupled with a reduced dietary intake eventually resulted in a state of negative K^+ balance that presumably would have been even greater had the kidney possessed a normal capacity for secreting K^+. The significance of the reduced intake is indicated by the fact that, in pair-fed controls, intracellular changes in Na^+, K^+ and water occurred that, in many instances, were in the same direction as those observed following adrenalectomy. Such animals, however, maintained normal concentrations of Na^+ and K^+ in extracellular fluid. The reviews of HARTMAN and BROWNELL (1949) and NOBLE (1950) clearly reveal that earlier investigators were aware that there was no necessary relationship between plasma K^+ concentration and the K^+ balance of the organism, and that the quantity of K^+ in a tissue could decrease after adrenalectomy. "Potassium retention" was not an inevitable consequence of adrenocortical insufficiency. Despite such findings, the specters of increased plasma K^+ concentration, increased sensitivity to K^+ (NOBLE 1950) and "potassium retention" continued to dominate the thoughts of many working in this field. Only with a complete understanding of adrenocortical insufficiency can the phenomenon of "potassium retention" be placed in a proper perspective.

A true state of "potassium retention" can occur in adrenocortical insufficiency (HILLS et al. 1953), in which the total quantity of K^+ in the body is increased. To establish a positive K^+ balance, however, the dietary intake of K^+ must exceed the ability of the kidney to excrete this cation.

The importance of the dietary intake of K^+ is indicated by the finding of ALLERS and KENDALL (1937) that appreciable increases in the plasma K^+ concentration after adrenalectomy can be prevented by restricting the intake of this cation. There is some evidence that cats (NOBLE 1950) and dogs (NILSON 1937) will reject a diet rich in K^+ after adrenalectomy, although such a diet will be readily consumed by intact animals. However, if the K^+ is contained in some food highly-prized by dogs and cats, such as meat, they will eat it eagerly after adrenalectomy. Eventually, they may refuse all food but meat (KENDALL 1948). In such situations the intake of K^+ may be adequate for the development of a positive K^+ balance.

It is not possible at this time to say precisely why the renal excretion of K^+ is impaired in adrenocortical insufficiency. It is currently believed that urinary K^+ is largely, if not wholly, a secretory product of renal tubular cells (MUDGE 1958). According to this view, K^+ filtered at the glomerulus undergoes virtually complete reabsorption; excretion of K^+ is regulated by variations in the rates at which renal tubular cells secrete this ion. At least one such process seems to be impaired in adrenocortical insufficiency. As noted above, an ion-exchange mechanism whereby Na^+ is reabsorbed from the tubular fluid and K^+ secreted in exchange operates subnormally in adrenocortical insufficiency (PITTS 1952). It is not known if this mechanism is depressed to an extent that could account wholly for the inability of the kidney to excrete K^+ at a normal rate. Other secretory processes for K^+, as yet unidentified, may also be implicated. It should be noted that there is no total loss of the ability to excrete K^+. When there is a reduced intake of K^+, the kidney is still able to excrete K^+ to an extent that will permit the development of a negative K^+ balance. In the words of PITTS (1952), "No single discrete renal function present in the normal animal is absent or lost in the adrenalectomized animal or in the patient with Addison's disease. The adrenal cortical hormones merely modify rates of certain processes which proceed whether the hormones are present or not."

Normally, the organism preserves the K^+ concentration in extracellular fluid within certain physiological limits. The excretory defect of adrenocortical insufficiency accounts, in part, for the hyperkalemia encountered in this state. It is to be doubted, however, that hyperkalemia *per se* is cause of death in many species (HARTMAN and BROWNELL 1949, NOBLE 1950).

γ) Inorganic anions. The Na^+ losses that occur in adrenocortical insufficiency require the loss of equivalent amounts of anions. The two most prevalent inorganic anions in extracellular fluid are chloride and bicarbonate ions, and it is these anions that are found in the urine in the greatest quantities in adrenocortical insufficiency (HARTMAN and BROWNELL 1949, NOBLE 1950).

Conversely, there is usually a small increase in the plasma concentrations of sulfate and phosphate (HARTMAN and BROWNELL 1949, NOBLE 1950). Both of these anions are subject to glomerular filtration and, to a limited extent, are reabsorbed from the filtrate by renal tubular cells (H. SMITH 1951). A number of factors probably contribute to the increased plasma concentrations of these ions in adrenocortical insufficiency. A decreased volume of extracellular fluid coupled with a reduction in the GFR would be contributory. In addition, the continuing process of cellular catabolism probably more than makes up for the urinary losses of phosphate, by the release of anion originally contained in intracellular fluid. One may infer from the scanty evidence available that the tubular reabsorptive processes for these anions are not seriously disrupted by adrenocortical insufficiency.

b) Water

The failure to excrete excess water at a normal rate has been widely studied in adrenalectomized animals and in patients with Addison's disease. When a water load is administered by mouth to such subjects, it is excreted at a markedly subnormal rate (GAUNT 1951). Most of the water is eventually excreted but the organism has been exposed to the risk of water intoxication for a dangerously long period of time. This danger exists because the lag in the absorption of water from the intestinal tract that occurs in adrenocortical insufficiency is not large enough to explain the delayed diuretic response (SHIPLEY 1945, SLESSOR 1951, GAUNT 1951, NABARRO 1956). After the bulk of the water has been absorbed into the body, there follows a period in which the minute urine volume is low, hemodilution occurs and the risk of excessive intracellular hydration is great. It is in this period of time that the normal kidney would markedly increase its output of urine. In adrenocortical insufficiency, the renal response is slow and the period of danger is protracted.

It is currently believed that the most important factor in the delayed excretion of water in adrenocortical insufficiency is sluggishness of response on the part of the kidney. The reductions in the GFR and ERPF mentioned above doubtlessly interfere with an increase in the flow of urine but, within limits, they do not seem to be critical factors. There is no good correlation between reduction in the GFR and the extent to which the diuretic response to water is decreased (LOTSPEICH 1949). Furthermore, adrenocortical hormones can exert a favorable response upon water diuresis without markedly increasing the GFR (NABARRO 1956, KLEEMAN et al. 1959). *The principal disturbance seems to be in the renal tubular cells. When a water load is imposed, these cells do not seem to be able to decrease the rate at which they reabsorb water as rapidly as do normal tubular cells. Hence diuresis is delayed.* Two explanations of this phenomenon have been offered.

(1) The antidiuretic hormone (ADH) of the neurohypophysis is responsible for the delay. In adrenocortical insufficiency, the kidney becomes unusually sensitive to the effects of ADH, or the rate of inactivation of this hormone is reduced. In any event, a relative or absolute excess of ADH is present. As adrenocortical steroids normally promote water diuresis while ADH opposes it, hormonal imbalance occurs in adrenocortical insufficiency. One manifestation of this imbalance is a delayed excretion of water (GAUNT et al. 1949, GAUNT 1951, SLESSOR 1951, GINSBURG 1954, DIAMOND 1956). In opposition to this concept are the findings that blood levels of ADH are not increased after adrenalectomy in the rat (AMES and VAN DYKE 1952), nor are the blood levels of antidiuretic substances in patients with Addison's disease correlated with the ability to excrete water (CHALMERS and LEWIS 1951). When normal and adrenalectomized rats are exposed to an ambient temperature of 35°, increases in the blood concentrations of antidiuretic substances occur in both types of animals and the responses are in-

distinguishable (K. Robinson and Macfarlane 1956). In the patient with Addison's disease, the renal sensitivity to ADH and the duration of its effect is unaffected by doses of cortisone that make possible a normal diuretic response to water loads (Chalmers and Lewis 1951).

(2) Adrenocortical hormones like cortisol are thought to decrease the permeability of the renal tubular epithelium to water (Raisz et al. 1957, Kleeman et al. 1958, 1959). When a water load is imposed, a normal organism presumably increases its output of adrenocortical hormones. They, in turn, act upon renal tubular cells in such a way that the rate of reabsorption of water is reduced and the minute urine volume increases accordingly (Chalmers and Lewis 1951, Nabarro 1956). In adrenocortical insufficiency, rapid changes in the permeability of the renal tubular epithelium to water are not possible.

It is not easy at this time to decide which concept has the greater merit. The concept involving ADH has been somewhat vitiated by the findings reported above, and by the criticism that some of the important supporting data were obtained under inappropriate experimental conditions or by the use of unsatisfactory bioassay methods for ADH (Nabarro 1956). The "permeability" concept, on the other hand, suffers from a lack of evidence. It seems a likely explanation for the delayed diuresis but has not, as yet, been firmly buttressed by experimental findings.

c) *Acid-base balance*

As the pH of the urine rarely falls below 4.5, the organism can excrete excess hydrogen ions only to a limited extent in the form of strong acid; they must be incorporated largely into salts of strong or weak acids. This is made possible by renal tubular processes that secrete hydrogen ions or ammonia. A concomitant event in these reactions is the reabsorption of Na^+. These ion-exchange reactions are adversely affected by adrenocortical insufficiency. Accordingly, there is a contemporaneous impairment in Na^+ reabsorption and the regulation of acid-base balance.

In adrenocortical insufficiency, the pH of the blood is commonly lower than normal and the pH of the urine higher. A number of factors contribute to this derangement of acid-base balance. Altered tissue metabolism and cellular catabolism may result in the release of abnormally large quantities of "fixed" acid[1] into the extracellular fluid. Bicarbonate ions are displaced from their association with Na^+ so that the "fixed" acids may be neutralized, and the pH of extracellular fluid falls as a consequence. A normal kidney would be able largely to compensate for these changes. In adrenocortical insufficiency, however, disturbances in renal tubular function limit the ability of the kidney to combat acidosis.

The carbonic acid-bicarbonate system constitutes an important buffer in extracellular fluid. The integrity of this system depends upon the efficient renal reabsorption of bicarbonate ions and the normal operation of other renal acidification mechanisms. Preliminary to the reabsorption of HCO_3^-, renal tubular cells must synthesize H_2CO_3 from CO_2 and H_2O. This reaction is catalyzed by the enzyme carbonic anhydrase; from the response to acetazoleamide, it has been concluded that the activity of this enzyme is not impaired in adrenocortical insufficiency (Hamwi and Brown 1956). The H_2CO_3 dissociates to H^+ and HCO_3^-. The H^+ is secreted by the renal tubular cells into the tubular lumen in exchange for the Na^+ in $NaHCO_3$ filtered at the glomerulus. This exchange reaction occurs at a subnormal rate in adrenocortical insufficiency; the inability of the kidney to secrete hydrogen

[1] The term "fixed" acid refers to those other than carbonic acid, i.e., nonvolatile.

ions for this purpose is probably one of the fundamental reasons for the reduction in the body's content of bicarbonate in adrenocortical insufficiency (SARTORIUS et al. 1952).

True acidification of the urine involves similar renal tubular processes for H^+ secretion. Again the requisite preliminary step is the formation of H_2CO_3 within the renal tubular cell with the aid of carbonic anhydrase so that a supply of H^+ is made available. The secretion of such ions permits the conversion of Na_2HPO_4 within the tubular fluid to NaH_2PO_4. Thus H^+ is excreted and simultaneously Na^+ is reabsorbed. The Na^+ is restored to the extracellular fluid in association with the HCO_3^- formed in the renal tubular cell. This process may be severely depressed in adrenocortical insufficiency (DUMM et al. 1955). SARTORIUS et al. (1953) found that the ability of the adrenalectomized rat to excrete titratable acid in the urine amounted to but 15 to 25% of that of the normal animal.

A third renal defense against systemic acidosis involves the formation of ammonium ions. Again, the preliminary formation of H_2CO_3 with the aid of carbonic anhydrase appears to be mandatory. Additionally, the renal tubular cells must produce ammonia. This may be obtained by the deamidation of glutamine. This acidification mechanism comes into play to compensate for the presence of strong acids in the body. Such acids form Na^+ salts in the body that are filtered at the glomerulus. The secretion of H^+ by the renal tubular cells converts such salts into strong acids. Simultaneously, Na^+ is reabsorbed and returned to extracellular fluid as $NaHCO_3$. The strong acids, in turn, react with ammonia secreted by the renal tubular cells and are excreted as ammonium salts in the urine.

Amino acids (such as alanine and glutamic acid) are deaminated oxidatively by the kidney less rapidly after adrenalectomy in the rat (RUSSELL and WILHELMI 1941). In this finding we may have the explanation as to why the adrenalectomized rat excretes ammonia at only 50 to 60% of the normal rate (SARTORIUS et al. 1953).

Thus it appears that the acidosis that may occur in adrenocortical insufficiency cannot be antagonized effectively by the kidney because the renal tubular cells cannot secrete hydrogen ions and ammonia (HARRIS et al. 1952) at normal rates. Concomitantly, Na^+ conservation is impaired, since both Na^+ and H^+ participate in common ion-exchange reactions.

d) Organic substances

The concentration of certain organic substances in the blood may be altered by adrenocortical insufficiency. In particular, the azotemia so often observed in this state has been the subject of many studies. In adrenalectomized animals otherwise untreated, the blood levels of non-protein nitrogen (comprised largely of urea) serve as a fairly accurate index of the severity of the insufficiency (NOBLE 1950). Following adrenalectomy, there is usually a steady rise in the blood concentration of urea throughout the survival period (NOBLE 1950). Several factors contribute to the development of this condition. The non-protein nitrogen in the extracellular fluid is probably being augmented constantly as a result of cellular catabolism, and reductions in the ERPF and the GFR lead to a reduction in the rate at which urea and other nitrogenous substances are filtered at the glomerulus. However, the concentration of these substances in the blood of an adrenalectomized dog rises even though the GFR is normal (H. WHITE et al. 1949). These studies were made in dogs maintained on deoxycorticosterone acetate, and in which the urea clearance: inulin clearance ratio remained normal. One must conclude, however, that the origin of the azotemia in adrenocortical insufficiency remains a matter for speculation.

The conservation of certain physiologically important organic substances, such as glucose and amino acids, is accomplished by processes of renal tubular reabsorption. It is believed that such processes are active transport systems that consume energy; they are limited with regard to the amount of solute that can be reabsorbed from the tubular fluid in any period of time. This limitation is represented by the concept of a tubular "transfer maximum," or Tm (H. SMITH 1951). The values for the Tm of any substance are usually expressed in mg/min. The mechanisms underlying these rate-limited transport processes are not known but are currently under intensive investigation (TAGGART 1958). Little is known of the impact of adrenocortical insufficiency upon these active reabsorptive processes. It is known only that, in Addison's disease, the Tm for glucose is reduced (TALBOTT et al. 1942, SANDERSON 1948).

The renal tubular cells are likewise capable of secreting a number of organic substances into the tubular fluid. These processes are also believed to be active transport systems for which a Tm can be determined. The capacity of the renal tubular cells to secrete certain organic substances is not known to have any physiological significance. Nevertheless, the Tm for such substances as diodrast or *p*-aminohippurate can serve as a measure of the quantity of functional tubular tissue (H. SMITH 1951).

The maximal rate at which diodrast (Tm_D) or *p*-aminohippurate (Tm_{PAH}) is secreted in adrenocortical insufficiency is usually, but not always, found to be reduced. The Tm_{PAH} may be reduced in the rat as soon as one day after adrenalectomy (FRIEDMAN et al. 1948). The findings of GAUDINO and LEVITT (1949) indicate that, following adrenalectomy in the dog, the Tm_{PAH} may be 50% of normal. TALBOTT et al. (1942) found the Tm_D in patients with Addison's disease to be only slightly subnormal, while SANDERSON (1948) found, in his patients, the Tm_D was 61 to 92% of normal. C. WATERHOUSE and KEUTMANN (1948) found the Tm_{PAH} to be subnormal in female Addisonian patients, whereas it was normal in males unless hypertensive disease had previously developed. However, hypertensive disease may itself be characterized by a reduction in Tm_{PAH} so the contribution of Addison's disease to the subnormal Tm_{PAH} in these men cannot be assessed easily (H. SMITH 1951).

From the data above, it would seem (1) that a reduction in the secretory Tm occurs more readily in acute adrenocortical insufficiency after adrenalectomy than when the onset is more gradual, as it is in Addison's disease; or (2) that changes of secretory Tm in adrenocortical insufficiency are a function of species. A reduction in the secretory Tm may be brought about in several different ways. There may be a true reduction in the mass of secretory tubular tissue in the kidney (as in the instance of some localized destructive process) or there may be a generalized reduction in the secretory activity of all of the renal tubular cells. A reduction in secretory Tm will also follow a reduction in renal blood flow of sufficient magnitude. In this instance, the tubular secretory processes may be capable of achieving normal rates of transport, but renal blood flow is inadequate and the supply of diodrast or *p*-aminohippurate presented to the tubular cells per unit time is insufficient for the attainment of a normal Tm. In adrenocortical insufficiency, the ERPF is usually subnormal, as indicated earlier, and may account largely, if not wholly, for the reductions in Tm reported above.

ADDIS et al. (1950) studied spontaneous and induced proteinurias in the rat before and after adrenalectomy. The male rat spontaneously excretes protein at a rate that is about twice that of the female. Adrenalectomy (with subsequent maintenance on saline) did not affect the spontaneous proteinuria in females but effected a 50% reduction in proteinuria in males. In both sexes, after adrenalectomy, the excretion of human serum albumin, injected intraperitoneally, was reduced to 5 to 10% of normal. In all instances, these animals were maintained on saline and so, presumably, the GFR and the ERPF were maintained within normal limits (FRIEDMAN et al. 1948, LOTSPEICH 1949). It does not seem likely that slowed absorption from the injection site would account for such a gross secretory defect. The reduction in protein excretion after adrenalectomy could represent a decrease in the filtered load of protein or enhanced tubular reabsorption.

4. Alterations in extracellular fluid: renal and extra-renal factors

In the preceding paragraphs some of the distortions in the composition and volume of extracellular fluid that occur in adrenocortical insufficiency have been described. Some of the earlier investigators held that such distortions were wholly a reflection of renal dysfunction (LOEB et al. 1933, HARROP et al. 1933a). While not denying the presence of renal dysfunction, other investigators (e. g., SWINGLE et al. 1936b) felt that adrenocortical hormones also regulate the distribution of water and electrolytes between the extracellular and intracellular fluid compartments. They suggested that some of the distortions in extracellular fluid that occur in adrenocortical insufficiency must be attributed to the absence of this hormonal regulation. Observations in adrenalectomized animals indicated that shifts of Na^+, K^+ and water between the fluid compartments take place that cannot be explained wholly on the basis of renal intervention. For example, adrenalectomized dogs, adjudged to be in states of severe insufficiency on the basis of the composition of extracellular fluid, showed within 48 to 60 hours after the administration of 2 α-methyl-9 α-fluorocortisol, restoration of normality in the composition of the extracellular fluid even though deprived of food and water (SWINGLE et al. 1957). Not only were the plasma concentrations of Na^+ and K^+ restored to normal but also plasma volume (SWINGLE et al. 1959). Not all of the adrenocortical steroids are alike in their ability to effect such shifts of water and electrolytes. The so-called glucocorticoids, such as cortisol, are most potent in this respect. Aldosterone, a so-called mineralocorticoid, is less effective, particularly with regard to the restoration of a normal plasma volume; it is presumed to act primarily upon the kidney (SWINGLE et al. 1959). It is the view of SWINGLE et al. (1959) that, in adrenocortical insufficiency, the content of Na^+ and water in extracellular fluid is reduced not only because of urinary losses but also because Na^+ and water move from extracellular to intracellular fluid; such Na^+ and water re-enter the extracellular fluid when appropriate adrenocortical hormones are administered.

These views have been supported by other experimental evidence. FLANAGAN et al. (1950) showed that, following adrenalectomy in the dog, more Na^+ was lost from extracellular fluid than can be accounted for by the excretory losses of this ion. When deoxycorticosterone acetate was administered to these animals, the increase in the quantity of Na^+ in extracellular fluid was greater than the dietary intake would provide. GAUDINO and LEVITT (1949) reported that a marked increase in the volume of intracellular fluid occurs following adrenalectomy in the dog. In studies upon patients subjected to subtotal adrenalectomy, HILLS et al. (1953) likewise presented evidence for the internal transfer of Na^+ and water. They concluded, "... external loss of sodium and chloride appears to be overshadowed by the internal transfers of these ions which accompany the condition." In extensive studies on adrenalectomized and nephrectomized rats, WOODBURY (1953) likewise concluded that, "... adrenocortical hormones influence electrolyte metabolism in all tissues of the body, and that this influence is exerted independently of the known effects of these hormones on the kidney." Additional evidence is to be found in the reports of STERN et al. (1951), LEDINGHAM (1954) and LUFT et al. (1954).

The available evidence indicates strongly that adrenocortical hormones regulate the distribution of certain ions within the body. In particular, they appear to play an important role in the maintenance of Na^+ as the predominant inorganic cation of extracellular fluid and K^+ as the predominant cation of intracellular fluid. *The situation with regard to water shifts is less clear.* Conventionally, it is assumed that shifts in water between extracellular and intracellular fluid passively follow shifts in ions. Evidence suggesting that adrenocortical hormones can effect shifts in water independent of shifts in ions is present in the experiments cited above, but it is less than decisive.

Some of the experiments devised to study the regulatory effects of adrenocortical hormones upon the distribution of inorganic electrolytes within the body have resulted in the creation of some knotty problems. In the main, these problems relate to the quantity of Na^+ within the body and its distribution between extracellular and intracellular fluid during adrenocortical insufficiency. For example, there is the matter of the "missing sodium." FLANAGAN et al. (1950) reported that more Na^+ is lost from extracellular fluid after adrenalectomy in the dog than can be accounted for by excretory losses of this ion. [This finding posed a problem, since often (H. WHITE and ROLF 1955) but not always (GROLLMAN 1954) the Na^+ content of soft tissues decreases after adrenalectomy.] The explanation FLANAGAN et al. (1950) offered was that perhaps the "missing sodium" could be found in bone. While it can be demonstrated that some of the Na^+ in bone is exchangeable and not wholly inert (STERN et al. 1951, G. NICHOLS and NICHOLS 1956), it has also been found that, in adrenocortical insufficiency, the Na^+ in bone is reduced (STERN et al. 1951, H. WHITE and ROLF 1955). As the "missing sodium" apparently does not reside in intracellular spaces, the possibility must be entertained that it represents a miscalculation. To determine the quantity of Na^+ in extracellular fluids, some measure of the volume of this space must be made. A number of compounds, such as mannitol, inulin, thiocyanate and sucrose, have been employed to measure the extracellular space. They must be utilized properly and with a full awareness of possible error; such an awareness was possibly not present in all of the studies reported above. For example, not all of the normal channels of perfusion may have been open. The possibility of capillary stasis in adrenocortical insufficiency was pointed out many years ago (REMINGTON 1940). In such a situation, a substance like inulin might not be distributed throughout the entire volume of extracellular fluid. Consequently, the calculated volume of extracellular fluid, and the calculated total quantity of Na^+ contained therein would be erroneously small. It is possible that errors in technique or the existence of closed perfusion channels can account for the "missing sodium" reported in studies upon dogs and human beings. It is of great interest that ROLF et al. (1959), using sucrose to measure the volume of extracellular fluid, reported that, in adrenalectomized rats, the loss of Na^+ from extracellular fluid very closely approximates the excretory losses of this ion. Furthermore, NETRAVISESH and WHITE (1953) reported that the amount of Na^+ lost by adrenalectomized dogs during a period of adrenocortical insufficiency closely approximated the amount retained upon subsequent replenishment and recovery. In adrenocortical insufficiency, Na^+ does not move into cells, but rather loss of intracellular Na^+ usually occurs. This, coupled with the studies reported above in which the Na^+ lost from extracellular fluid is wholly accounted for in excretory losses, makes it seem likely that no significant quantity of "missing sodium" exists. It is possible that the findings of SWINGLE and his co-workers can be explained without the postulate of shifts of Na^+ into the cells. A not-inconsiderable quantity of Na^+ still remains in intracellular fluid after adrenalectomy (GROLLMAN 1954, H. WHITE and ROLF 1955). Even with no dietary intake of Na^+, it may be possible to increase the Na^+ in extracellular fluid by the administration of appropriate adrenocortical hormones that will effect the movement of Na^+ from intracellular to extracellular fluid. This would entail an even greater depletion of the intracellular content of Na^+. An omission in the studies of SWINGLE is any assessment of intracellular Na^+ before and after hormonal administration.

In conclusion, from the foregoing data it can be seen that the distortions in the volume and composition of extracellular fluid that occur in adrenocortical insufficiency are the resultant of two disturbances: (1) renal dysfunction and (2) some loss of the ability to regulate the distribution of electrolytes and water between extracellular and intracellular fluid. It is not possible to say which disturbance is of the greater consequence.

II. Cardiovascular system

The disturbances in cardiovascular function that occur in adrenocortical insufficiency are numerous. In the view of some (SWINGLE et al. 1959), death in adrenocortical insufficiency results primarily from circulatory collapse.

It has been known for some time that, *in adrenocortical insufficiency, the heart is small in size and its beat feeble* (ADDISON 1849); *it seems unable to react normally to increases in venous return or to increases in systemic arterial pressure. The blood volume is reduced, as is peripheral vascular resistance; hypotension is commonly present.* The distinctly abnormal status of the circulation in adrenocortical insufficiency is a reflection of the collective impact of these disturbances. There is little or no dispute today about the findings listed above, and they have been thoroughly treated in previous reviews (SWINGLE and REMINGTON 1944, HARTMAN and BROWNELL 1949, NOBLE 1950, 1955, RAMEY and GOLDSTEIN 1957). Controversy presently centers about attempts to explain the origins of these disturbances, especially the alterations in vasomotor function. In the paragraphs to follow, an attempt will be made to present the various pertinent findings in these matters.

1. The heart

a) Anatomical changes

α) Heart size. The size of the heart is commonly found to be subnormal in untreated cases of adrenocortical insufficiency. This decrease can be visualized during life by the use of roentgenographic techniques (THORN et al. 1942, REMINGTON 1951). It is likewise a common finding at autopsy. In Addison's disease the heart is almost always found to be smaller than normal (THORN 1951). However, in patients receiving hormonal therapy (particularly deoxycorticosterone) prior to death, the heart may be abnormally large (SOMERVILLE et al. 1951).

In adrenalectomized dogs, G. HALL and CLEGHORN (1938) found dilatation of the right half of the heart, and described the organ as being soft and flaccid. In the same species, REMINGTON (1951) and BROWN and REMINGTON (1955) described the heart at death as being "small and fixed." Within one week after adrenalectomy of rats, N. SOLOMON et al. (1959) found that the dry weight of the heart had decreased by 10%.

RAMEY and GOLDSTEIN (1957) concluded that the decrease in the size of the heart after adrenalectomy may reflect "a chronic diminution of tonic autonomic stimuli to the heart."

β) Pathological anatomy. Except for descriptions of size and weight, the anatomy of the heart in adrenocortical insufficiency has not been thoroughly investigated. With respect to Addison's disease, the few studies available were reviewed by SOMERVILLE et al. (1951); these reports indicate that, upon autopsy, many hearts were found to be normal. Except for "brown atrophy," no one pathological finding was consistently encountered in the abnormal hearts; many of the lesions described were deemed to be incidental in occurrence and unrelated to Addison's disease. In patients receiving deoxycorticosterone prior to death, focal myocardial necrosis was a common finding (SORKIN 1949).

G. HALL and CLEGHORN (1938) did an intensive study of the heart in dogs that had died from adrenocortical insufficiency. Upon the surface of the heart they observed both congested areas and pale infarcted areas. They could detect subendocardial hemorrhages and, occasionally, subepicardial hemorrhages. In some medium and small arteries, the endothelial lining was broken or absent, and there was evidence of edema in the walls. Some coronary arteries were completely occluded by thrombi, and in others spasm could be seen. Capillary congestion was described as being wide-spread. These findings were not compared with those occurring in dogs dying from other causes.

b) *Electrocardiogram*

The electrical activity in the heart is commonly found to be abnormal in adrenocortical insufficiency. The alterations in the electrocardiogram are usually observed to become progressively greater as the severity of the insufficiency increases.

In the adrenalectomized dog, bradycardia is observed (NICHOLSON and SOFFER 1935, G. HALL and CLEGHORN 1938, REMINGTON 1951, ROBERTS 1952); in many dogs, the heart rate was found to be less than 65 beats per minute, and rates below 40 were not uncommon (NICHOLSON and SOFFER 1935, G. HALL and CLEGHORN 1938). There is some dispute as to how much this bradycardia reflects vagal dominance of the heart. REMINGTON (1951) believed that it is a manifestation of a "prominent and unopposed vagal reflex," since he found that vagotomy reverses the bradycardia. G. HALL and CLEGHORN (1938), however, found that the heart rate increases for only a few minutes following the administration of atropine to adrenalectomized dogs and then returns to its initial low rate.

According to ROBERTS (1952), the first alteration in the electrocardiogram following adrenalectomy of the *dog* is an increase in the height of the T wave; this change occurs while serum concentrations of Na^+ and K^+ are still normal. The duration of the T wave may then become shorter (SWINGLE et al. 1952) or longer (ROBERTS 1952). The S—T segment becomes depressed (SWINGLE et al. 1952). The P wave becomes progressively flatter and may disappear entirely (NICHOLSON and SOFFER 1935, ROBERTS 1952, SWINGLE et al. 1952). The disappearance of the P waves and the appearance of "f" waves (G. HALL and CLEGHORN 1938) have been interpreted as manifestations of auricular fibrillation (NICHOLSON and SOFFER 1935). The remaining ventricular complexes may become irregular in shape and rhythm (NICHOLSON and SOFFER 1935, ROBERTS 1952). G. HALL and CLEGHORN (1938) felt that the ventricular complexes gave evidence of marked myocardial ischemia.

Following adrenalectomy of the *golden hamster*, bradycardia ensues, the T wave becomes flatter and may be inverted, and the Q—T interval is prolonged (WYMAN et al. 1953a).

The electrocardiographic findings in Addison's disease have been reviewed by SOMERVILLE et al. (1951). In a study of 90 *patients*, the electrocardiogram was found to be normal in half of the cases and abnormal in the other half. The most common findings were flattening or inversion of the T wave, prolongation of the Q—T, P—R and QRS intervals, and depression of the RS—T segment. However, the authors did not feel that there is any specific electrocardiographic pattern for Addison's disease.

The alterations in the electrocardiogram that occur in adrenocortical insufficiency have most often been attributed to changes in the concentrations of electrolytes in extracellular fluid, particularly the increases in K^+ (NICHOLSON and SOFFER 1935, ROBERTS 1952, SWINGLE et al. 1952, WYMAN et al. 1953a). While it is true that the alterations reported above are similar to those encountered in instances of hyperkalemia without adrenocortical insufficiency (ROBERTS 1952), it is of interest that DAVANZO et al. (1958) observed that electrocardiographic alterations persist in the adrenalectomized dog after plasma K^+ levels have been returned to normal by replacement therapy. At this time, however, plasma Mg^{2+} concentrations are still elevated; these authors felt that Mg^{2+} is implicated in the electrocardiographic disturbances of adrenocortical insufficiency. Since cortisone usually effects some improvement in the electrocardiogram of the patient with Addison's disease whereas deoxycorticosterone does not, SOMERVILLE et al. (1951) attributed the alterations in electrical activity to disturbances in carbohydrate metabolism. Since many biological effects are exerted by cortisone, such reasoning has its limitations. It is of interest to note that no correlation between abnormality of heart size and normality or abnormality of electrocardiogram could be demonstrated in Addison's disease (SOMERVILLE et al. 1951).

c) Cardiac output

The findings of bradycardia and a decrease in heart size in adrenocortical insufficiency would lead one to expect that there would be a contemporaneous reduction in cardiac output. In the adrenalectomized rat, cardiac output has been found to be only 60 to 70% of that of the intact animal (HOELSCHER 1954, N. SOLOMON et al. 1959). In the adrenalectomized dog the cardiac index (cardiac output per square meter of body surface) is reduced to 75% of the value obtained in intact animals (REMINGTON 1951).

Cardiac output is the product of heart rate and stroke volume, and the effects of adrenocortical insufficiency upon these two factors must be discussed separately. In the adrenalectomized rat, the *heart rate* is actually slightly elevated at a time when cardiac output is significantly reduced (HOELSCHER 1954). In the adrenalectomized dog, however, a profound bradycardia may occur (see above). Moreover, the heart rate is not increased by substances like epinephrine and norepinephrine that produce a tachycardia in intact dogs (BROWN and REMINGTON 1955). It seems likely that the reduction in cardiac output in the dog may be attributed, in part, to bradycardia. The situation in other species is unanalyzed.

Stroke volume may be influenced by a number of factors: heart rate, the rate of return of blood to the heart, ventricular distensibility and the force of the ventricular contraction. From the evidence available, it seems unlikely that stroke volume is generally limited in adrenocortical insufficiency by a rapid heart rate, although rapid rates are observed occasionally (REMINGTON 1951). Conversely, in the adrenalectomized dog, the heart has been observed actually to become smaller in size as the length of the cycle increased (REMINGTON 1951).

The irregularities in cardiac rhythm detected on the electrocardiogram after adrenalectomy in certain species (see above) might well result in a reduction in stroke volume. An abnormal spread of the wave of excitation over the ventricle or the influence of ectopic foci within the ventricle could result in myocardial contractions wholly or partially ineffective with respect to the ejection of blood from the ventricle. As simultaneous measurements of electrical and mechanical events within the heart have rarely been made, it is difficult to know if the pumping capacity of the heart is markedly impaired before serious abnormalities can be seen in the electrocardiogram. In the terminal stages of adrenocortical insufficiency in the dog, both electrical and mechanical events within the heart are markedly abnormal (G. HALL and CLEGHORN 1938, REMINGTON 1951).

No direct measurements have been made of venous return in states of adrenocortical insufficiency. It might well be reduced if oligemia and pooling of blood in peripheral beds (see p. 92) are present. Despite maintenance of a normal plasma volume, HECHTER (1945) found that, upon exsanguination, a subnormal volume of blood is obtained from adrenalectomized rats that have been maintained on a 1% NaCl solution. He attributed this reduction in "bleeding volume" to reductions in venous return and cardiac output. Venous pressure may be abnormally high in the adrenalectomized dog at a time when cardiac output is reduced (REMINGTON 1951). This observation suggests that cardiac output is unable to keep up with venous return.

The reduction in heart size in adrenocortical insufficiency suggests that the chambers of the heart are less distensible. Such a change in the relaxation curve of the ventricle (BROWN and REMINGTON 1955) would definitely limit stroke volume. A reduction in cardiac distensibility seems to occur in adrenocortical insufficiency but the cause is unknown.

Several investigators have attributed certain findings in adrenocortical insufficiency to a decreased force of myocardial contraction. FOWLER and CLEGHORN (1942), for example, attributed the weaker pressor responses to Pitressin and $BaCl_2$ in adrenalectomized cats to an inability to maintain cardiac output in the presence of an increased systemic blood pressure. The fact that a normal pressor response to epinephrine was obtained was attributed to the cardiotonic effects of this agent. Additional indirect evidence can be derived from the finding of a decrease in heart size. Should this decrease impose a sufficiently great reduction upon diastolic volume in the heart, then, by Starling's Law of the Heart, the force of myocardial contractions would be reduced correspondingly.

In a heart-lung preparation, one may regulate the venous pressure and the resistance offered to the ejection of blood by the left ventricle. N. SOLOMON et al. (1959) made heart-lung preparations of both intact and adrenalectomized rats that had been maintained on saline for 2 to 3 weeks postoperatively. They calculated a left-ventricular work index (expressed in g-meters/100 mg dry wt of heart) that measured systemic, but not coronary, perfusion. At a mean arterial pressure (or resistance) of 80 mm of Hg, the left-ventricular work index in the adrenalectomized rat was 28% of that of the intact rat; at a mean arterial pressure of 110 mm of Hg, it was only 11%. Such findings indicate that the pumping capacity of the left ventricle is severely impaired by adrenalectomy in the rat. Such an impairment could reflect a limited myocardial distensibility that would limit diastolic volume and hence the force of myocardial contraction. It might also reflect myocardial asthenia; in this instance, for any given diastolic volume the force of myocardial contraction would be abnormally weak. This situation would be represented by a downward shift in the conventional plot of the force-volume curve for the heart. It is possible that both of these aberrations occur in the heart after adrenalectomy; the scanty evidence available favors the former explanation. N. SOLOMON et al. (1959) reported that a period of at least 2 weeks must elapse between adrenalectomy and testing before a significant reduction in work capacity can be demonstrated in the heart-lung preparation. At this time, myocardial atrophy is also definitely established. REIDENBERG et al. (1958) made direct measurements of the force of myocardial contractions in the dog. On the second day after completion of two-stage adrenalectomy, they found that the myocardial response to the positive inotropic effects of norepinephrine was essentially normal.

As the diastolic blood pressure is often subnormal in adrenocortical insufficiency (see p. 88), the perfusion of the coronary vascular bed may be inadequate. Such a reduction in coronary blood flow, if it were of sufficient magnitude, would certainly impair the work of the heart. As measurements of coronary blood flow have not been made in adrenocortical insufficiency, it must remain a matter for conjecture.

The presently available evidence does not permit any firm conclusions as to why cardiac output is reduced in adrenocortical insufficiency. However, two factors seem to be most contributory: (1) the marked bradycardia that occurs in some species, and (2) the subnormal myocardial distensibility that limits stroke volume and, thereby, the force of myocardial contractions. It is not known to what extent deficiencies in the activity of the sympathetic nervous system are implicated in this situation.

d) The chemical composition of the heart

Table 4 lists some of the changes in the concentrations of water and electrolytes that have been reported to occur in the myocardium after adrenalectomy.

It is apparent that no very great changes occur in the intracellular concentrations of electrolytes in the myocardium after adrenalectomy. And yet, in the rat, in which no changes in intracellular K^+ can be detected, plasma K^+ rises from 5.1 meq/l to 7.4. This means that the ionic gradient from intracellular to extracellular fluid with respect to K^+ decreases by one third following adrenalectomy; it remains to be determined if this decrease in any way impairs myocardial function.

Table 4. *Influence of adrenalectomy on myocardial water and electroytes*

Species	Reference unit	Status	Na^+(meq)	K^+(meq)	H_2O(g)	Reference
Dog	kg of fresh tissue	Intact	—	75.6	773	CLARKE and CLEGHORN (1942)
		Adrx.	—	83.7	782	
Cat	100 g fat-free solids	Intact	25.2	40.2	420.5	DARROW et al. (1939)
		Adrx.	24.2	44.0	442.7	
Cat	100 g fat-free solids	Intact	19.3	39.7	371	ROBERTSON and DUNIHUE (1954)
		Adrx.	18.7	42.9	394	
Rat	kg wet tissue	Intact	—	85	—	
		Adrx.	—	83	—	ST. GEORGE et al. (1955)
	kg dry tissue	Intact	—	35	—	
		Adrx.	—	33.3	—	

Adrenalectomy does not produce any significant change in the concentration of glycogen in the heart of the rat (RUSSELL and BLOOM 1956); this is true of both fasted and unfasted rats. The cytochrome *c* concentration in the heart is likewise unaffected by adrenalectomy in the rat, although the cytochrome oxidase activity does decrease slightly (TIPTON 1944).

The few studies that have been made of the chemical composition of the heart do not provide any ready explanation for the cardiac dysfunction that occurs in adrenocortical insufficiency.

2. Blood vessels

a) Pressure, flow and peripheral resistance

Systemic arterial hypotension occurs in states of adrenocortical insufficiency. Adrenalectomy is followed by a progressive fall in systemic blood pressure; the degree of hypotension parallels the severity of insufficiency (HARTMAN and BROWNELL 1949, NOBLE 1950). In the terminal stages of adrenocortical insufficiency, the blood pressure is at a shock-like level. It is not known, however, if circulatory collapse *per se* is responsible for death in adrenocortical insufficiency, or whether it is merely another manifestation of the body-wide derangements of tissue function that make death the usual result of this state.

In untreated patients with Addison's disease, THORN et al. (1942) found the average blood pressure to be 100/66. One might interpret such readings to indicate that the hypotension derives somewhat more from a reduction in cardiac output than from a decrease in peripheral resistance. The blood pressure may be normal or higher than normal in Addison's disease in states of excitement or in patients with a history of antecedent hypertension (THORN 1951).

Similar findings of hypotension have been made in adrenalectomized animals. Blood pressure measurements in animals not uncommonly are made while the subject is under the influence of some anesthetic agent (often a barbiturate). An anesthetic agent may interfere with the regulation of blood pressure by direct effects upon the central nervous system, or the heart and peripheral vessels, and indirectly by depressing respiration (PRICE 1960). Following adrenalectomy, the susceptibility of animals to drugs that depress the central nervous system (such as the barbiturates) is enhanced (REMINGTON 1951, WOODBURY 1958). Consequently, in any consideration of the findings about to be reported, it must be borne in mind that one cannot know to what extent the deviations from normality represent the effects of adrenocortical insufficiency upon cardiovascular function and to what extent they represent an enhanced susceptibility to anesthetic agents.

Hypotension may occur in the dog very soon after adrenalectomy. On the second day after completion of two-stage adrenalectomy, REIDENBERG et al. (1958) found that the pressure in the femoral artery had fallen from a preoperative level of 121/77 to 71/35; the pressure in the pulmonary artery had fallen from 15/5 to 9/< 1. However, ROGOFF and DOMINGUEZ (1927) reported that no significant change in blood pressure occurred in the dog following adrenalectomy until the onset of terminal symptoms.

In his studies on cardiovascular function in adrenalectomized dogs, REMINGTON (1951) noted, at the outset, that the animals were very sensitive to the effects of pentobarbital sodium: "Invariably, an adrenalectomized dog could be relaxed with anesthetic doses which would not even give drowsiness to intact dogs. Enough drug was given merely to place the animals in a stuporous condition. They required no more anesthetic, the stupor gradually increasing until death." Blood pressure in these animals initially was nearly normal and relatively stable. (These animals were taken off hormonal replacement therapy only from 1 to 4 days prior to these observations and were believed still to be in good condition.) The onset of the circulatory decline was signaled by a sharp decrease in blood pressure and peripheral resistance. No rise in the cardiac index ensued as it would in an intact dog after the same hemodynamic changes. Blood pressure fell to 92/62, then to 56/36 and, just before respiratory failure, to 40/20.

LEATHEM and DRILL (1944) found that the average systolic blood pressure in rats 6 days after adrenalectomy was 60 mm of Hg; the comparable value in intact animals was 110 to 120 mm of Hg.

Very few studies have been made of the effect of adrenocortical insufficiency upon the blood flow through specific vascular beds. As a first approximation, it can be said that flow varies directly with the perfusing pressure and inversely with resistance to flow. The pressure generally drops in adrenocortical insufficiency; the hypotension reported might lead one to speculate that flow suffers accordingly. Such an assumption has limited value in the absence of knowledge concerning the degree of peripheral vascular resistance. In adrenalectomized dogs initially maintained on replacement therapy in the postoperative period but taken off it 1 to 4 days prior to the following observations, REMINGTON (1951) found that blood pressure and peripheral resistance were essentially normal at first. Subsequently, both fell and flow either remained constant or declined. In a later report (BROWN and REMINGTON 1955), it was noted that, in adrenalectomized dogs with a similar postoperative history, arterial blood pressure apparently declined more rapidly than peripheral resistance; blood flow was decreased accordingly. If intact dogs were bled to the point where systemic arterial pressure fell to 90 mm of Hg, the blood flow in that portion of the hind leg served by the femoral artery was twice that observed in an adrenalectomized dog having the same systemic arterial pressure.

In conclusion, the hemodynamic status of the organism in adrenocortical insufficiency can be characterized as follows: some degree of hypotension is present. The magnitude of the hypotension depends upon the extent of the reductions in cardiac output, peripheral vascular resistance and plasma volume. (In the terminal stages, increases in blood viscosity may also play a role.) The hypotension may result in a decreased flow of blood, but the magnitude of the decrease depends largely upon the changes in peripheral resistance in the various vascular beds. The regulation of peripheral resistance (i. e., peripheral arteriolar "tone") will be considered in subsequent paragraphs dealing with vasomotor function.

b) Vasomotor function

Peripheral vascular resistance is determined, in a large measure, by the diameter or "tone" of the various arterioles in the body. Some degree of tone is intrinsically resident in all arterioles during life. The intrinsic tone is subject to modification by external chemical influences. Constrictor or dilator substances may be liberated at the endings of postganglionic autonomic nerves, or may be in the blood, as are epinephrine and histamine. Arteriolar tone is also influenced by the tensions of oxygen and carbon dioxide in the blood.

Additionally, peripheral vascular resistance may be modulated by changes in the tone of metarterioles and precapillaries. These smaller vessels possess contractile elements capable of responding to external stimuli, such as epinephrine. They also manifest slow spontaneous relaxation and contraction; this phenomenon is termed vasomotion (Zweifach et al. 1953).

Net systemic peripheral vascular resistance determines, in part, the work of the heart and the systemic arterial pressure. The peripheral vascular resistance of any vascular bed determines, in large measure, the rate of blood flow through that bed. The distribution of blood throughout the organism is influenced, in part, by the peripheral resistances of the various vascular beds. Assuming that all other relevant factors remain constant, the organism may then increase the perfusion of a given vascular bed by lowering the peripheral resistance in that bed while simultaneously increasing it in another bed. Thus, in exercise, the blood flow through muscle may be increased at the expense of the volume of blood flowing through the splanchnic vessels. Such circulatory adjustments are not readily effected in adrenocortical insufficiency, in part because the organism loses, to a significant degree, its ability to regulate peripheral resistance in the various vascular beds (Ramey and Goldstein 1957). A manifestation of this disturbance is the postural hypotension that often occurs in Addison's disease (Thorn 1951). The occurrence of this inability to regulate peripheral resistance, along with hypovolemia and cardiac hypodynamia, helps explain why circulatory collapse is so often observed in adrenocortical insufficiency.

Peripheral vascular resistance usually decreases in adrenocortical insufficiency (see above), particularly as death approaches. At present, controversy centers about the reason for this dilation of the small vessels in the arterial tree. Do the smooth muscles in the walls of these vessels lose their ability to respond to stimuli that would ordinarily produce contraction in them? Or is this a neurogenic phenomenon, characterized either by a failure of nerve endings to release their neurohumors or a derangement of those centers in the central nervous system that regulate circulatory homeostasis? Also to be considered is the possibility that some potent vasodilator substance in the blood is the cause of the decrease in peripheral resistance. The experimental findings in this complex area will be presented in the paragraphs to follow; it will be seen that much of the evidence is contradictory.

The concept that in adrenocortical insufficiency the contractile elements in the smaller blood vessels respond subnormally to vasoconstrictor agents has been championed by RAMEY and GOLDSTEIN (1957), and there is much evidence to support their view of a myogenic disturbance. However, in interpreting these findings, certain considerations should be borne in mind. A normal response to a pressor agent (as measured by an increase in the systemic arterial pressure) can be obtained only if certain factors in the total situation remain normal. (1) There must be a sufficient volume of blood in the circulatory tree so that, when its capacity is reduced as a result of vasoconstriction, the blood pressure can rise to an appreciable extent. (2) The heart must also be able to cope with an increase in systemic arterial pressure; if it cannot, because of muscular asthenia or an inadequacy of venous return, cardiac output will fall and the pressor reponse will be diminished accordingly. (3) It is not known if one can justifiably expect the same magnitude of pressor response in the hypotensive adrenalectomized animal as one sees in the normotensive intact animal. Neither the arterial distensibility curve (REMINGTON and HAMILTON 1945) nor the pressure-flow relation governing arterial drainage (H. GREEN et al. 1944) is linear with respect to pressure. A failure to obtain a pressor response of normal magnitude following adrenalectomy may then reflect, in part, the intervention of one or more of the above factors rather than resulting from inadequate arteriolar constriction alone. It is not known to what extents these factors were operative in some of the experimental findings reported below.

The pressor response to the stimulation of certain nerves is often reduced or absent following adrenalectomy (T. ELLIOTT 1914, 1915, SECKER 1939, 1949, BROWN and REMINGTON 1955). This has been interpreted as a failure of the nerves to liberate their neurohumors; the nictitating membrane of the adrenalectomized cat, for example, will respond to epinephrine at a time when motor nerve stimulation produces no response (SECKER 1939). This concept has not enjoyed popular favor, for many investigators believe that the failure to respond is not a neurogenic, but rather a myogenic, phenomenon (RAMEY and GOLDSTEIN 1957).

Numerous workers (FOWLER and CLEGHORN 1942, CLEGHORN et al. 1950, RAMEY et al. 1951, REMINGTON 1951, BROWN and REMINGTON 1955) have demonstrated that systemic arterial blood pressure in adrenalectomized dogs does not rise to a normal extent when a variety of vasoconstrictor agents (norepinephrine, epinephrine, $BaCl_2$, Pitressin, etc.) are injected. One group of investigators (RAMEY et al. 1951, RAMEY and GOLDSTEIN 1957) attributed the decreased pressor responses in adrenocortical insufficiency primarily to an increased threshold of the arterioles for response to vasoconstrictor substances. Other workers (FOWLER and CLEGHORN 1942, BROWN and REMINGTON 1955) believed that the inability of the heart to maintain cardiac output in the presence of increased peripheral vascular resistance is the more important factor.

The use of a technique whereby blood vessels in the mesoappendix of the rat are directly visualized has provided some interesting information with respect to vasomotor function in the adrenalectomized rat. Arteriolar tone is subnormal and vasomotion in the finer vessels tends to decrease and finally disappear. The reactivity of the arterioles, metarterioles and precapillaries to epinephrine and norepinephrine is reduced. The duration of the contractile response becomes shorter and finally disappears altogether. Just before death, the finer blood vessels are widely dilated, and either a very sluggish flow or complete stasis occurs in the capillary bed (FRITZ and LEVINE 1951, ZWEIFACH et al. 1953).

The blood vessels in the cheek pouch of the golden hamster may also be directly observed. Hamsters usually die within 5 to 7 days following adrenalectomy. Until 24 hours before death, blood flow through the cheek-pouch vessels is normal although the arterioles become abnormally constricted, beginning on the third to fourth postoperative day. These vessels show enhanced sensitivity to electrical stimuli that cause vasoconstriction but decreased sensitivity to norepinephrine. Blood flow becomes very sluggish on the day of death, and circulatory stasis can be seen in some of the finer vessels. These diminutions in flow are believed to reflect, in part, a decreased cardiac output (WYMAN et al. 1953b).

Not all investigators have found that the response to vasoconstrictor agents is subnormal following adrenalectomy. On the second postoperative day, REIDENBERG et al. (1958) found normal pressure increases in the femoral and pulmonary arteries of adrenalectomized dogs following the injection of norepinephrine. In adrenalectomized dogs, on the fourth postoperative day, SALMOIRAGHI and McCUBBIN (1954) observed normal pressor responses following the injection of epinephrine, norepinephrine and angiotonin, but a reduced or wholly absent pressor response to renin. They noted that in the terminal, shock-like state responses to all vasoconstrictor agents were markedly reduced. D. SMITH (1953) measured by means of an angioplethysmokymographic technique the changes in tone in isolated segments of the carotid arteries obtained from adrenalectomized swine. He found that the duration of the response of these vascular segments to epinephrine was normal, but the response to histamine was prolonged.

Taken in sum, the above observations indicate that the present controversy regarding the origin of the decrease in peripheral vascular resistance will not be resolved until more evidence is available. The situation is rendered more complex because of the varying experimental conditions that different investigators have employed. Some used dogs, others rats; the numerous factors that can contribute to circulatory collapse may be of different quantitative significance in various species. Some used animals immediately after adrenalectomy, others maintained their animals on saline for some time prior to observing them. Some supplied their animals with adrenocortical hormones until just shortly before making observations, others made no observations until their animals were overtly moribund. Some employed large doses of vasoconstrictor agents, others used small ones. Limited recognition has been given to the possibility that the state of anesthesia may modify the results. Different anesthetic agents were employed by different investigators; some conflict in findings may arise from this fact alone. It is not uncommon to see only the bare results of various studies being compared, with limited recognition being given to the experimental conditions under which they were obtained. Such presentations perpetuate rather than resolve controversies.

c) The capillaries

Some investigators have presented evidence that capillary permeability is increased in states of adrenocortical insufficiency; these findings include descriptions of increased protein concentration in lymph and increased movements of dyes out of the capillaries (NOBLE 1950, 1955, SPECTOR 1958). Such findings purport to offer a partial explanation for the hypovolemia of adrenocortical insufficiency.

ZWEIFACH et al. (1953) found that increased capillary permeability, as manifested by a greater flow of the dye T-1824 out of the capillaries, could be observed in the mesoappendix of the adrenalectomized rat only in the terminal state. RENKIN and ZAUN (1955) studied capillary permeability in adrenalectomized rats by perfusing the isolated lower trunk and hind legs. The animals were studied at various times ranging from the 3rd to the 13th postoperative day, and the results compared with those obtained in intact rats. They found that, within the post-operative time period studied, capillaries did not become more permeable to fluids, proteins or sucrose.

RAMEY and GOLDSTEIN (1957) have cited evidence to show that the susceptibility of small vessels to the toxic effects of epinephrine and norepinephrine is increased after adrenalectomy. They concluded that, in adrenocortical insufficiency, the organism responds to its cardiovascular inadequacies with the release of large quantities of epinephrine and norepinephrine, with the result that "the high titre of constrictor materials released ultimately acts to destroy the small blood vessels."

Capillary resistance to rupture, or fragility, is usually measured in terms of the amount of negative pressure (or vacuum) required to effect rupture of small vessels such as capillaries or venules. KRAMÁR and SIMAY-KRAMÁR (1953) found that the resistance of the capillaries in the abdominal skin of the rat was markedly reduced after adrenalectomy. Whereas negative pressures of from 35 to 70 cm of Hg were required to effect capillary rupture in the skins of intact animals, vacuums of less than 10 cm of Hg were effective in rats 4 to 5 days after adrenalectomy. These authors maintained their rats on saline postoperatively and found that decreased capillary resistance could be observed up to 9 months following adrenalectomy. WYMAN et al. (1953b) observed that the venules in the cheek pouch of the adrenalectomized hamster were abnormally susceptible to rupture during the first 3 postoperative days. Then, as generalized vasoconstriction became evident in this vascular bed, the resistance to rupture rose in parallel.

From a consideration of this evidence, it would seem that capillary permeability may increase in adrenocortical insufficiency. Thus the leakage of water and solutes through the capillary walls could contribute to hemoconcentration. However, this hemoconcentration probably reflects primarily the loss of osmotically active solutes from the blood to the urine. Furthermore, it is possible that increases in capillary permeability may not be of any great significance until the terminal stages of adrenocortical insufficiency. This must be regarded as a tentative conclusion, for the available evidence is scanty; moreover, as RENKIN and ZAUN (1955) have emphasized, reports of changes in capillary "permeability" must be interpreted with caution unless the investigators have been able to distinguish between true changes in capillary permeability and changes in the transcapillary filtration rate due to altered capillary hydrostatic pressure.

3. Blood

a) Volume and protein content

Plasma volume is usually subnormal in adrenocortical insufficiency; this has been found in human beings and other species (NOBLE 1950). The dog is a species in which hemoconcentration after adrenalectomy may be severe; the blood volume is reported to decrease by one-third (SWINGLE et al. 1934c) and the hematocrit to rise to 56 (G. HALL and CLEGHORN 1938). It is doubtful, however, that any accurate measure of blood volume can be obtained in terminal stages of adrenocortical insufficiency because the imminence of complete circulatory collapse makes it unlikely that any agent used to measure blood volume (such as a dye) will be uniformly distributed throughout the entire circulation (SWINGLE et al. 1934c).

The decrease in plasma volume appears to be a reflection of other changes that occur in adrenocortical insufficiency. If the adrenalectomized rat is supplied with saline, the plasma volume is normal (HECHTER 1945).

As plasma proteins are known to play a part in the regulation of plasma volume, it is of interest to note that their concentration in plasma is usually elevated in adrenocortical insufficiency (NOBLE 1950). This is not simply a manifestation of hemoconcentration, for the increase is somewhat greater than would be predicted on the basis of the increase in hematocrit (LEATHEM 1945). Additionally, the composition of the proteins in plasma is altered. In the saline-maintained, adrenalectomized rat, the albumin fraction increases while the globulin fraction decreases (LEATHEM 1945); thus the albumin/globulin ratio shifts from a normal value of 1.29 to 1.54. In adrenalectomized dogs, the opposite occurs: the concentration of albumin falls while that of the globulins rises (PAGE and LEWIS 1951). Table 5 lists the changes measured by these authors by electrophoretic analyses of the plasma of a dog before and after adrenalectomy.

It can be seen that the greatest decrease occurs in the albumin fraction whereas the greatest increase occurs in the $\phi + \gamma_1$ fraction of the globulins. The authors note that it is unusual to have a decrease in the γ_2 fraction at a time when all of the other globulin fractions are increased.

Table 5. *Alterations in plasma proteins after adrenalectomy in a dog* (PAGE and LEWIS 1951)

Status	Plasma proteins, g %	Albumin		Globulins									
		g %	% Total protein	α_1		α_2		β		$\phi + \gamma_1$		γ_2	
				g %	% TP[1]	g %	% TP	g %	% TP	g %	% TP	g %	% TP
Intact	6.05	2.67	44.6	0.37	6.1	0.74	12.3	0.97	16.0	0.65	10.7	0.65	10.7
Adrx.	6.40	1.92	30.0	0.51	8.0	0.98	15.3	1.09	17.0	1.28	20.0	0.62	9.7

[1] TP = total protein.

Except for the effect of hemoconcentration, it is not known why plasma protein concentrations increase after adrenalectomy. In rats fed N^{15}-labeled glycine, STOERK et al. (1947) found adrenalectomy to be without effect on the turnover of plasma proteins. These animals were maintained postoperatively on saline and deoxycorticosterone acetate.

b) The formed elements and hematopoiesis

α) Blood. It should be noted at the outset that values for the concentrations of formed elements in the blood in adrenocortical insufficiency may be deceptive because of the decreases in plasma volume that occur in this state. Such concentrations, in and of themselves, do not necessarily reflect alterations in the rate of hematopoiesis or in the longevity of formed elements in the blood.

Erythrocytes. In Addison's disease, mild normocytic and normochromic anemia exists. There is no evidence of increased destruction of blood, and the serum iron concentration is normal (BÁEZ-VILLASEÑOR et al. 1948). The concentrations of Na^+ and K^+ within erythrocytes are likewise normal (JONES 1955). The average hematocrit of the untreated patient is 41.4 (THORN et al. 1942).

The effect of adrenalectomy on blood erythrocyte concentrations in the rat is transient (A. GORDON 1954). The reversible nature of this response can be seen fully only in those animals maintained on saline in the postoperative period. The initial response is an anemic one, the erythrocyte count being lowest in the second to third week after adrenalectomy. At this time, the concentration of hemoglobin is reduced, mean corpuscular volume is increased somewhat and a reticulocytosis is present (CRAFTS 1941, 1953, A. GORDON et al. 1951). Erythrocyte fragility at this time has been variously described as being normal (CRAFTS 1941) or decreased (A. GORDON et al. 1951). MEGEL and GORDON (1951) found that the ability of erythrocytes to withstand lysis in hypotonic solutions increased after adrenalectomy. This was not true of washed cells, however, and the authors concluded that an "anti-lysis" factor is present in the plasma after adrenalectomy in the rat. By 60 days after adrenalectomy in the rat, the changes described above have been reversed and the erythrocyte concentrations in the blood are once again normal (CRAFTS 1941, 1953, A. GORDON et al. 1951). K. STEIN and MARTIN (1951) also noted a reticulocytosis in mice following adrenalectomy.

In rats maintained on saline in the postoperative period, HAMILTON et al. (1951) found that the plasma iron concentration falls from a normal value of 282 μg% to 155 by the third day after adrenalectomy. The authors did not believe that this decrease was a reflection of decreased food intake or an impairment of iron absorption from the gastrointestinal tract.

WATERMAN et al. (1939) found that the response to adrenalectomy in 5 dogs was variable with regard to the erythrocytes in blood. The hematocrit rose in only 2 animals. The concentration of erythrocytes fell in only 2 but the hemoglobin concentration fell in 3. The erythrocyte volume was increased in 4 of the 5 animals.

Leukocytes. The various reports concerning leukocyte concentrations in the blood in adrenocortical insufficiency vary not only among species but also within a given species.

In Addison's disease, although the total leukocyte concentration in the blood is within normal limits, a lymphocytosis and neutropenia are present (BÁEZ-VILLASEÑOR et al. 1948).

In adrenalectomized rats, BÜCHERL and SCHWAB (1952) found the total leukocyte count on the 6th postoperative day to be twice that of intact rats. However, CRAFTS (1941, 1953) found it to be essentially normal. ELMADJIAN et al. (1946) and CRAFTS (1941) found that the lymphocyte concentration did not change after adrenalectomy, but BÜCHERL and SCHWAB (1952) found it to be doubled. However, these latter authors found that the lymphocytes constitute 73% of the total leukocytes in normal rats but only 63% in adrenalectomized rats. This decrease is accounted for by the fact that the concentrations of neutrophils and eosinophils triple after adrenalectomy. HERNBERG (1953) reported that the diameter of lymphocytes increased from a normal value of 10.5 μ to 12–16 by the fourth day after adrenalectomy.

DOUGHERTY and WHITE (1944) found the total leukocyte concentration in the blood of mice to increase slightly after adrenalectomy, the increase reflecting a mild lymphocytosis. This finding was confirmed by ELMADJIAN and PINCUS (1945). In their study on adrenalectomized mice, K. STEIN and MARTIN (1951) found an increase in the mononuclear and eosinophilic cells in the blood but no change in the neutrophils. Since hematopoiesis was simultaneously depressed, they concluded that certain formed elements persist in the blood for a longer time after adrenalectomy than normally.

WATERMAN et al. (1939) found the following changes to occur in the bloods of the 5 dogs they studied before and after adrenalectomy: all of the animals showed an unchanged concentration of leukocytes in the blood. There were no consistent changes in the proportions of monocytes or eosinophils, and in only one animal did the lymphocytes constitute an abnormally large proportion of the total leukocytes. However, in all of the animals, the neutrophils constituted a subnormal proportion of the total leukocytes after adrenalectomy.

Platelets. E. ADAMS (1949) found that, on the fifth postoperative day, there were identical increases in the platelets in the blood of adrenalectomized and sham-adrenalectomized rats following a surgical stress. However, when adrenalectomized rats received a subcutaneous injection of saline, platelet concentrations in the blood fell to subnormal levels. That this response was not a reflection of hemodilution was evidenced by the fact that there was no similar decrease in the concentration of hemoglobin, and no decrease in platelet concentration resulted from the injection of water.

β) Bone marrow. In the bone marrow of adrenalectomized rats, the proportion of young granulocytes is increased while the proportion of developing erythrocytes is decreased (A. GORDON et al. 1951). MARINONE et al. (1956) found no change in the mitotic rate of erythroblasts on the third day after adrenalectomy of rats, and PILIERO (1955) found that when adrenalectomized rats maintained on saline were exposed to ambient pressures equivalent to an altitude of 20,000 feet, the erythroid response in blood and bone marrow was essentially normal. However, FISHER (1958) found that no polycythemia developed when cobalt was administered to adrenalectomized rats.

BERMAN et al. (1956) studied the phosphorous metabolism in the bone marrow of the adrenalectomized rat, by injecting $Na_2HP^{32}O_4$. They observed a transient and reversible depression of activity in animals maintained on saline postoperatively paralleling the transient anemia that develops in this species after adrenalectomy (see above). By the fifth postoperative day, the incorporation of P^{32} into deoxyribonucleic acid, ribonucleic acid and phospholipides of bone marrow was somewhat subnormal. However, by two weeks after adrenalectomy, all of these rates of incorporation were essentially normal.

K. STEIN and MARTIN (1951) noted that the number of mitotic figures in the bone marrow of the mouse was reduced after adrenalectomy. They observed also that the ratio of myeloid elements to erythroid elements was abnormally high. They concluded that the formation and maturation of red blood cells in the bone marrow are depressed after adrenalectomy of the mouse.

BÁEZ-VILLASEÑOR et al. (1948) reported that the incorporation of radioactive iron into hemoglobin was only slightly subnormal in patients with Addison's disease.

III. Skeletal muscle

Muscular weakness and easy fatigability are commonly encountered in adrenocortical insufficiency. Muscular asthenia is so characteristic of adrenocortical insufficiency that, unless it is present to some degree, the organism is probably not in a state of severe adrenocortical insufficiency.

In subsequent paragraphs it will be shown that attempts to attribute the alterations in muscular function to a single causative factor have not met with success. As is common in adrenocortical insufficiency, a multiplicity of disturbances appears to be operative; it has not yet been possible to demonstrate conclusively the prepotency of any one of them in the etiology of the muscular dysfunction. Any attempts to evaluate the relative significance of the various possible causes of muscular dysfunction, particularly the easy fatigability, are impeded by the fact that, all too often, the physiological and biochemical observations have been made in *resting* muscle. In such muscle, one can often see derangements after adrenalectomy that, *per se*, probably would not seriously hamper muscular function but that might increase upon exercise and make for a truly profound disturbance. Consequently, in attempting to find the causes of the muscular dysfunction of adrenocortical insufficiency, one is obliged to place only limited reliance upon the observations made in resting muscle and to speculate as to what may happen during prolonged exercise.

1. Manifestations of disturbances in muscular function

a) Addison's disease

Muscular asthenia figured in ADDISON's first description (1849) of the disease that bears his name. In the century that followed, this finding was confirmed by innumerable clinical investigators. Muscular weakness and easy fatigability are currently believed to be requisite features of the syndrome of adrenocortical insufficiency. "It is well to remember . . . that they represent cardinal manifestations of the disease, and one would hesitate to make the diagnosis in their absence" (THORN 1951).

The Addisonian "languor" has been described (SORKIN 1949) in the following way: "The asthenia is often obvious; the patient looks and acts fatigued, often even while lying in bed. The speech is frequently slow and languid, and the patient may state that even thinking requires effort." The degree of muscular weakness depends not only upon the severity of the disease but also upon the muscular status of the patient prior to the onset of the disease. A fair degree of muscular strength, for example, may still be present in those previously accustomed to a life of strenuous physical labor; even so, such patients are unable to perform physical feats once possible or may accomplish them only at the expense of marked exhaustion. In Addison's disease, muscular strength is characteristically greatest in the morning and reduced as the day progresses (THORN 1951).

b) Adrenalectomized animals

Muscular dysfunction has likewise been observed in animals following adre-al-ectomy. Investigators of this phenomenon have devised a number of "work tests" so that a quantitative estimate of the disability can be obtained. Such tests do not usually measure the strength of muscular contractions but rather the amount of work that can be done before exhaustion appears. They are commonly used also to evaluate the efficacy of various forms of replacement therapy.

In one such test, rats are made to swim and the time of the appearance of exhaustion is noted (GAARENSTROOM et al. 1937).

In another test, developed by INGLE and various collaborators (HERON et al. 1934), the gastrocnemius muscle is directly stimulated *in situ*. Upon contraction, the muscle lifts a weight a measured distance.

Test:

Although minor variations in stimulus strength and frequency are described in the various studies of INGLE, the test situation is essentially as follows (INGLE 1944): within one hour after adrenalectomy (and also nephrectomy, on occasion), the rat, anesthetized with a barbiturate, is prepared for the test. A stimulating electrode provides faradic stimuli directly to the gastrocnemius muscle for 20—25 msec, 3 or 5 times each second. The tendon of the gastrocnemius is dissected free and attached to a 100 g weight that the muscle lifts upon contraction. For each 400 g-cm of work, a recorder performs one revolution. The gastrocnemius muscle of a normal rat will readily work for 120 hours in this situation; the adrenalectomized rat usually dies after 6—32 hours (HALES et al. 1935).

Expressed quantitatively, a normal muscle, in one test situation, performed an amount of work recorded as approximately 17,000 revolutions within a test period (usually 24 hours); the muscle of the adrenalectomized rat performed only 4,000 revolutions before it would no longer contract (INGLE 1940). At the time of exhaustion, more than a state of simple fatigue in the working muscle is present, for, if the electrode is now shifted to the gastrocnemius muscle on the other leg, this will perform for only a few minutes and then it too becomes unresponsive to the stimulus. At this time, the rat is cyanotic and its body temperature subnormal.

A similar test was devised by EVERSE and DE FREMERY (1932).

Test:

Rats are used on the 4th day after adrenalectomy. The gastrocnemius muscle is subjected to short, intermittent stimulation by tetanizing currents. An ergograph is obtained by recording the tetanic contractions of the muscle. One unit of biological activity in this test is that amount of steroid hormone that will effect restoration of a normal ergograph in an adrenalectomized rat.

Despite the superficial similarities in the tests of Ingle and of Everse and de Fremery, the potencies of the various adrenocortical hormones are quite different in the two assays. In the Ingle test, compounds with predominantly "glucocorticoid" activity, such as cortisone, are the most potent forms of replacement therapy; compounds with predominantly "mineralocorticoid" activity, such as deoxycorticosterone, have little or no effect upon the capacity to perform muscular work (INGLE 1940). Directly opposite results are obtained in the test of Everse and de Fremery; in this test, deoxycorticosterone, on a weight basis, is approximately 30 times more potent than is cortisone (NOBLE 1950). The test of Everse and de Fremery probably poses a more severe stress for the rat than does the test of Ingle, for the animals are used 4 days after adrenalectomy rather than immediately after operation; hence, the severity of adrenocortical insufficiency is greater. The stress of surgery seems to be of no consequence in the Ingle test; the work performance of rats from which one adrenal and the contralateral kidney have been removed immediately prior to the test is not significantly different from that of intact rats (HALES et al. 1935). Additionally, it would seem that the tetanic stimulus employed by EVERSE and DE FREMERY would more rapidly lead to fatigue. INGLE (1940) found that the work period in the adrenalectomized rat was shortened when the intensity of stimulation was increased. The divergent findings reported above with regard to the relative potencies of the types of adrenocortical hormones in these two tests suggest that *the muscular dysfunction of adrenocortical insufficiency does not stem from any single physiological or biochemical disturbance.*

The measurement of certain electrical and mechanical aspects of muscular contraction in adrenalectomized animals has yielded information of great interest. In frogs, following adrenalectomy, the chronaxie of theg astrocnemius muscle becomes shorter (HOPPE and VOGEL 1951), indicating increased excitability. The studies of WALKER (1947, 1948) in the adrenalectomized rat likewise indicate disturbances in muscular function. Recording the isometric contractions of the triceps surae, he employed direct and indirect (via the motor nerve) stimuli, singly and repetitively. He found the threshold for response to single, indirect stimuli to be unaffected by adrenalectomy but the threshold for the production of supramaximal muscle responses by this route to be subnormal (WALKER 1947). Using a direct stimulus of a strength 6 times that required for *maximal* contractile response in the intact rat, he found that, after adrenalectomy, a greater tension developed and the sequence of events in time was disturbed. The rising time of contractile response was abnormally long but the time required subsequently for the maximal response to decrease by 50% ("half-falling time") was shorter than usual. As the heightened response in the triceps of the adrenalectomized rat could be antagonized by curare, he speculated that repetitive contractions were responsible. He also found that the maximal tetanic tension developed upon direct stimulation of the muscle was only slightly subnormal following adrenalectomy. Post-tetanic potentiation was likewise essentially normal in magnitude in the rested muscle but greatly reduced or absent following exercise of the triceps. Upon prolonged stimulation at a frequency of 2 per second, twitch tension gradually declined. However, the rate of decline was identical with the use of direct or indirect stimuli, suggesting that adrenalectomy had not seriously interfered with the transmission of impulses at the neuromuscular junction.

In a subsequent paper, WALKER (1948) investigated the possibility of repetitive contractions occurring in the muscle of the adrenalectomized rat. In this study, the test object was the gastrocnemius muscle and stimulation was indirect via the sciatic nerve. The twitch tension could be potentiated by adrenalectomy or KCl administration or by the production of tetanus or treppe. Although the duration of contraction was abnormally long in all of these instances, there was no evidence of repetitive contractions in the muscle. The duration of the muscle action potential (MAP) was also abnormally long following adrenalectomy. MILTON (1953) likewise found the duration of the MAP in the quadriceps femoris to be abnormally long in adrenalectomized rats. He found that this prolongation of the MAP could be prevented by saline or adrenocortical extract therapy; it could be observed in the muscles of normal rats 48 hours after bilateral nephrectomy or following the intraperitoneal injection of KCl. Anoxia was without effect on the duration of the MAP.

TORDA and WOLFF (1952a) found that, following adrenalectomy of rats, repetitive stimulation of a muscle via its motor nerve caused an abnormal decline in the amplitude and area of the MAP. Since the response of the muscle to direct stimulation declined less than that to indirect stimulation, they concluded that the disturbance in the nerve following adrenalectomy was greater than that in the muscle. They did not comment upon the matter of impulse transmission across the neuromuscular junction.

Under the experimental conditions in the studies reported above, two general types of disturbances seem to occur in skeletal muscle after adrenalectomy. *One* may be characterized as an initial hyper-responsiveness to stimulation, direct or indirect. Excitability is increased (decreased chronaxie). The electrical changes occasioned by stimulation persist for an undue period of time (prolongation of the MAP). Under certain circumstances, the muscles develop a greater force of contraction that persists for a longer period of time than is the case in the intact animal. However these electrical and mechanical phenomena are regulated in the normal animal, some derangement of this regulation is evident after adrenalectomy.

The *second* type of muscular disturbance is demonstrable only upon prolonged stimulation of the muscle. It is characterized by abnormally rapid onset of the inability of muscle to respond to a stimulus with a normal contraction during a period of repeated stimulation. The easy fatigability of skeletal muscle in adrenocortical insufficiency probably reflects not only derangements within the muscle itself but also systemic disturbances, such as cardiovascular inadequacies.

In the succeeding paragraphs some of the possible causes for the disturbances in muscular function will be examined.

2. Possible causes of muscular dysfunction

a) Alterations in energy metabolism

The well-known disturbances in intermediary metabolism that occur in adrenocortical insufficiency have been invoked often in an attempt to explain the origin of the muscular dysfunction that occurs in this condition. In particular, hypoglycemia and an inadequate supply of energy sources within the muscle have been considered contributory to the easy fatigability of muscle after adrenalectomy (NOBLE 1950). The relevant evidence does not always support this view, however. It has likewise proven difficult to assess the significance of such disturbances, since, following adrenalectomy, many upsets occur that may lead to the development of muscular dysfunction.

In the resting muscles of rats, fed or fasted for 24 hours, glycogen levels are essentially the same in intact and adrenalectomized animals (RUSSELL and BLOOM 1956, BOWMAN 1956). The glycogen content of the diaphragm may actually be significantly increased (RUSSELL and BLOOM 1956); however, such an increase may occur after the removal of only the adrenal medulla (BOWMAN 1956). The phosphorylation of glucose has likewise been found to occur at a normal rate in the muscles of adrenalectomized animals (THER and MUELLER 1950). Adrenalectomy of the rat does not cause any significant change in the content of adenosine triphosphate or phosphocreatine in muscle (ALBAUM et al. 1951).

From an examination of the literature it is difficult to determine to what extent muscular performance in states of adrenocortical insufficiency will be benefited by the administration of glucose. On the one hand, SORKIN (1949) reported that the muscular weakness of some patients with Addison's disease had been relieved temporarily by the use of glucose. INGLE and LUKENS (1941) reported that injections of Adrenalin and hypertonic glucose solutions (up to 50% glucose) enabled adrenalectomized rats to work as well as intact rats. They found glucose to be clearly superior to solutions of NaCl or sucrose. Also, the greater the total quantity of glucose given or the more hypertonic the glucose solution, the greater was the work output. However, INGLE and NEZAMIS (1948) found that constant intravenous infusions of glucose (as high as 200 mg/100 g of body weight/hour) were without effect on the work output of the adrenalectomized rat. At any given rate of infusion, adrenalectomized rats had lower blood glucose levels than their intact counterparts. When the muscle would no longer contract, however, blood glucose levels in the adrenalectomized rat rose considerably above those present in intact rats at the same time in the experiment. The authors were unable to reconcile these findings with the earlier ones of INGLE and LUKENS (1941). A year later INGLE and NEZAMIS (1949) reported that "work was greatly improved" by the administration of glucose to adrenalectomized rats already receiving adrenocortical extract.

These apparently contradictory findings make it difficult to determine to what extent the easy fatigability of skeletal muscle in adrenocortical insufficiency can be attributed to a lack of glucose. An additional complication is present in the findings (INGLE and KUIZENGA 1945) that the hormones having the greatest potency in the Ingle work test are those with predominantly glucocorticoid activity, for such compounds inhibit glucose utilization (NOBLE 1950). Tentatively, it may be concluded that, while it is possible under some experimental conditions to demonstrate a beneficial effect of glucose upon the capacity to perform muscular work, hypoglycemia is clearly not the only cause of easy fatigability in adrenocortical insufficiency. Adrenocortical steroids may also make energy sources available to muscle by the promotion of gluconeogenesis (INGLE and NEZAMIS 1948). However, with the evidence presently at hand, it is impossible to determine to what extent muscular function is impaired in adrenocortical insufficiency through disturbances in intermediary metabolism.

b) Alterations in electrolyte content and distribution

Alterations in the content and distribution of water and electrolytes within the body that occur in adrenocortical insufficiency have also been thought by some to contribute to the development of muscular dysfunction. Attention has been particularly directed towards the changes that may occur in the extracellular and intracellular concentrations of K^+.

The muscular asthenia of adrenocortical insufficiency resembles that produced by toxic doses of K^+ (ZWEMER and TRUSZKOWSKI 1937). HOPPE and VOGEL (1951) found that chronaxie in the gastrocnemius muscle of the frog was decreased by KCl administration or adrenalectomy, whereas NaCl administration exerted the opposite effect. The decrease in chronaxie that followed adrenalectomy could be accentuated by KCl and partially reversed by NaCl. MILTON (1953) found that the prolongation of the MAP that occurred in rats after adrenalectomy could be mimicked by the intraperitoneal injection of KCl into intact rats but not by the injection of NaCl or $CaCl_2$. He attributed the prolongation of the MAP to K^+ retention rather than Na^+ depletion, because it also occurred after bilateral nephrectomy, in which Na^+ retention takes place. WALKER (1948) found that the twitch tension in the gastrocnemius muscle in the rat was potentiated by KCl or adrenalectomy. He speculated that the increased concentration of K^+ in extracellular fluid might prolong the duration of the "breakdown" in the potential of the muscle membrane following stimulation and thereby increase the strength and duration of the contractile response. He also suggested that changes in the intracellular concentration of K^+ might affect the contractile process. The finding that compounds with predominantly mineralocorticoid activity, such as deoxycorticosterone, have the greatest potency of all of the steroid hormones in the work test of Everse and de Fremery (NOBLE 1950) provides indirect evidence that electrolyte disturbances are related causally to some aspect of the muscular dysfunction of adrenocortical insufficiency.

Other findings tend to minimize any possible role of electrolyte disturbances in the etiology of this muscular dysfunction. In the Ingle test compounds with predominantly mineralocorticoid activity have little or no potency (INGLE and KUIZENGA 1945). INGLE and LUKENS (1941) found that the beneficial effects of injections of hypertonic glucose solutions into adrenalectomized rats subjected to this test were not significantly modified by the simultaneous administration of KCl. In this context, the findings of FLÜCKINGER and VERZÁR (1954) are of some interest. They measured an increased uptake of K^{42} *in vitro* by the rat diaphragm and a decreased uptake of Na^{24} when tissues from adrenalectomized animals were compared with those of intact ones. They found that the addition of glucose to the incubation medium decreased the uptake of Na^{24} by diaphragms obtained from intact and adrenalectomized rats but was without significant effect upon the uptake of K^{42}. The findings of H. MILLER and DARROW (1941) would tend to suggest that a rat's capacity to perform muscular work is not limited by small increases in the intracellular concentration of K^+ in skeletal muscle. Using intact rats, they found that "within wide limits the amount of potassium in the muscle cells does not limit the capacity of rats to swim continuously for sixty minutes." In some of their animals, the concentration of K^+ in muscle was within the range encountered in adrenocortical insufficiency.

It seems pertinent at this juncture to consider some of the changes in muscle K^+ that have been reported to occur following adrenalectomy. The data in Table 6 (except for the ratios) are from the study by H. WHITE and ROLF (1955) on rats.

It is evident from these data that, regardless of the basis upon which the concentration of K^+ in muscle is computed, the increases after adrenalectomy in the rat are not very great. There is, however, a progressive decrease evident in the ratio of the concentrations of K^+ in intracellular and extracellular fluids.

Table 6. *Potassium in intact and adrenalectomized rats*
(Data of H. White and Rolf 1955)

	Status	Meq of K^+ ± S. D.	Reference unit	Intracellular concn. / Plasma concn.
Plasma	Intact	3.75 ± 0.9	per liter	—
	5—7 d adrx.	6.3 ± 0.8		—
	12—22 d adrx.	8.3 ± 1.9		—
Skeletal muscle	Intact	97.3 ± 4.4	per kg of fresh tissue	25.9
	5—7 d adrx.	102 ± 4.0		16.2
	12—22 d adrx.	102 ± 4.5		12.3
	Intact	172 ± 2.9	per kg of intra-cellular water	45.9
	5—7 d adrx.	177 ± 7.2		28.1
	12—22 d adrx.	177 ± 7.4		21.3
	Intact	45.9 ± 2.4	per 100 g of fat-free solids	12.2
	5—7 d adrx.	44.7 ± 0.9		7.1
	12—22 d adrx.	46.1 ± 1.6		5.6

The role of Na^+ in the changes in membrane potential that occur upon stimulation of nerve or muscle has received increasing recognition in recent years (Curtis 1956). One would expect that the Na^+ diffusion potential might be altered somewhat by the changes in the content and distribution of Na^+ in the organism that occur in adrenocortical insufficiency. The data in Table 7 (except for the ratios) were again taken from those of H. White and Rolf (1955).

Table 7. *Sodium in intact and adrenalectomized rats*
(Data of H. White and Rolf 1955)

	Status	Meq of Na^+ ± S. D.	Reference unit	Plasma concn. / Intracellular concn.
Plasma	Intact	143 ± 2.5	per liter	—
	5—7 d adrx.	135 ± 3.3		—
	12—22 d adrx.	133 ± 3.0		—
Skeletal muscle	Intact	30.4 ± 2.8	per kg of fresh tissue	4.7
	5—7 d adrx.	25.5 ± 1.7		5.3
	12—22 d adrx.	25.0 ± 1.1		5.3
	Intact	6.6 ± 2.8	per kg of intra-cellular water	21.7
	5—7 d adrx.	1.7 ± 1.5		79.4
	12—22 d adrx.	1 ± 1		133.0
	Intact	14.3 ± 1.5	per 100 g of fat-free solids	10.0
	5—7 d adrx.	11.7 ± 0.5		11.5
	12—22 d adrx.	11.3 ± 0.6		11.8

The data indicate that the changes in the concentrations of Na^+ in muscle and in the extracellular: intracellular ratios are not very great after adrenalectomy in the rat except when the Na^+ concentration is computed in terms of intracellular water. These latter data were computed on the assumption that Cl^- is entirely extracellular in location. If one calculates extracellular fluid on a Donnan distribution basis, the concentration of Na^+ in muscle (meq/kg of intracellular water) is normally 9.6 and drops to 5.8 at 5 to 7 days after adrenalectomy (H. White and Rolf 1955). The normal extracellular: intracellular fluid ratio for Na^+ would be 14.9 and it rises to 23.3 at 5 to 7 days after adrenalectomy. Whatever the initial premise employed for the computation of Na^+ concentrations in intracellular muscle water, the results portray the same trend: a decrease in the Na^+ concentration and an increase in the ionic gradient across the cellular membrane.

Recently, emphasis has been placed upon the concept that ionic gradients from extra- to intracellular fluids may be of as much physiological significance in nerve and muscle function as the absolute concentrations of Na^+ and K^+ in the two fluid compartments (GOODMAN and GILMAN 1955, their p. 797). The alterations in the ionic gradients depicted in the tables above may well account for the prolongation of the MAP and the potentiation of the twitch tension observed in rats after adrenalectomy (WALKER 1948, MILTON 1953). However, it does not seem likely that the increases in the concentration of K^+ *per se* in muscle can account for the easy fatigability of adrenocortical insufficiency. H. MILLER and DARROW (1941) observed unimpaired swimming activity in intact rats that had concentrations of K^+ in muscle, before and after exercise, that were actually slightly higher than those reported in Table 6 to occur after adrenalectomy. And yet, for lack of complete evidence, this must remain a tentative conclusion. Upon initial stimulation, the muscles of adrenalectomized rats are capable of an abnormally strong mechanical response (WALKER 1947, 1948), but unresponsiveness to stimulation ensues with unusual rapidity. The data reported above depict electrolyte disturbances after adrenalectomy in *resting* muscle; should the situation worsen with exercise, the muscle might well become totally unresponsive to external stimuli. Until similar analyses are made in the muscles of adrenalectomized animals *after exercise* we cannot know to what extent this possibility is realized during prolonged exercise and, consequently, to what extent electrolyte disturbances contribute to the easy fatigability of adrenocortical insufficiency.

c) *Alterations in neuromuscular transmission*

A muscle no longer responds to stimulation via its motor nerve if transmission of the impulse across the neuromuscular junction is disturbed. Such a disturbance could result from a failure of the motor nerve ending to liberate adequate quantities of acetylcholine, or it could arise from the accumulation of abnormal quantities of acetylcholinesterase at the junctional site. It could arise also from an altered response of the muscle end-plate to acetylcholine or an impediment in the spread of the wave of excitation from the end-plate to the muscle itself.

Investigators have paid scant attention to the possible role of a disturbance in neuromuscular transmission in the muscular dysfunction of adrenocortical insufficiency. As noted above, WALKER (1947) found that, upon prolonged stimulation at 2/sec, the twitch tension in the triceps surae of the adrenalectomized rat gradually declined. As the rate of decline was similar upon stimulation of the motor nerve or the muscle itself, he concluded that in the rat adrenalectomy produced no serious disturbance in neuromuscular transmission. While a work test, such as that of Ingle, in which the muscle is stimulated directly, does not rule out a contemporaneous disturbance in neuromuscular transmission, it does indicate that the potential for "fatigue" is resident within the muscle itself after adrenalectomy and does not require mediation via the neuromuscular junction.

TORDA and WOLFF (1952a) found that after adrenalectomy of rats, the amplitude and area of the MAP declined more upon stimulation of the motor nerve than upon stimulation of the muscle directly and so concluded that, in this situation, the disturbance in nerve was greater than that in muscle. Such a disturbance could result from an inability of the motor nerve to respond to stimuli, to conduct impulses or to liberate acetylcholine at its ending.

Subjecting rats to the work test of Everse and de Fremery, OVERBEEK (1949a) found that the work performance after adrenalectomy was improved by the prior injection of physostigmine or diisopropylfluorophosphate (DFP), but not by the prior injection of acetylcholine, carbachol, epinephrine or sodium citrate. Strangely enough, he found the prior injection of neostigmine to be without beneficial effect upon work performance. In a subsequent report OVERBEEK (1949b) sought additional evidence to support a thesis that work performance in adrenocortical insufficiency is subnormal because of a disturbance in neuromuscular transmission resulting from the presence of abnormally large quantities of acetylcholinesterase in the end-plate region. He found the content of cholinesterase in skeletal muscle was increased by 30% after adrenalectomy. In both intact and adrenalectomized rats, work caused a slight decrease in the content of cholinesterase in the muscles. Whereas his findings with physostigmine and DFP do indicate a possible disruption of normal neuromuscular transmission from an abnormal intervention of cholinesterase, the inability to demonstrate a similar beneficial effect with another cholinesterase inhibitor, neostigmine, weakens his thesis. Although the pharmacological effects exerted by physostigmine and neostigmine are not identical (GOODMAN and GILMAN 1955, pp. 444–458), the differences would lead one to predict that neostigmine would provide the greater restoration of work performance.

As is common in consideration of many aspects of adrenocortical insufficiency, no firm conclusion regarding the role of disturbances in neuromuscular transmission in producing the muscular dysfunction can be drawn because the available evidence is both inadequate in scope and internally conflicting.

d) Alterations in circulation

An inadequately perfused skeletal muscle obviously cannot be expected to work in a normal fashion. In adrenalectomized cats (SCHWEITZER 1945) and dogs (GOLDSTEIN et al. 1950), the onset of muscular fatigue is associated in time with the development of hypotension. Electrically elicited muscular contractions in the adrenalectomized dog cease when the systemic arterial blood pressure falls to 60 to 80 mm of Hg; they will resume, however, if the blood pressure can be raised above this apparently critical range (RAMEY et al. (1951).

The findings of HALES et al. (1935) indicate that, at the time of "fatigue" in the Ingle work test, the adrenalectomized rat is in a shock-like state, and that unresponsiveness can be observed in muscles that have been resting as well as in those that have been working. INGLE and NEZAMIS (1949) were unable, however, to achieve a total restoration of normal work performance in adrenalectomized rats by the administration of adrenocortical extract plus conventional doses of epinephrine. Large doses of epinephrine were found actually to depress work performance. As the cardiovascular responses to these various doses of epinephrine were not measured, their effect on the perfusion of muscle must remain a matter for conjecture.

It is of great interest that, under appropriate *in vitro* conditions, the muscles of adrenalectomized rats will perform work as well as those from intact animals, even if the muscles are taken from adrenalectomized rats previously exercised to the point of exhaustion (RAMEY et al. 1950). These findings suggest that distortions in the composition of the extracellular fluid may be of great importance in the development of the muscular unresponsiveness of adrenocortical insufficiency.

RAMEY and GOLDSTEIN (1957) believe that the easy fatigability of adrenocortical insufficiency is secondary to the circulatory collapse induced by the stress of muscular work. The studies in adrenalectomized cats and dogs lend strong support to this concept. The maintenance of systemic arterial blood pressure above an apparently critical level is associated with the capacity to do muscular work (RAMEY et al. 1951); muscular work ceases when the blood pressure falls below this level.

The situation obtaining in adrenalectomized rats is less clear. No one has measured blood pressure levels in the rat during the course of a work test; we only know that a shock-like state is present at the time of "fatigue" (HALES et al. 1935). In adrenalectomized rats, perhaps distortions in the composition of blood are of greater importance than decreases in the perfusion pressure. RAMEY et al. (1950) demonstrated that muscles taken from adrenalectomized rats previously exercised to the point of exhaustion will contract once again when stimulated in an appropriate *in vitro* environment. Under these conditions, the muscles are not being perfused, but must rely on the much less efficient process of diffusion for supplies of oxygen and nutrient materials and for the removal of the waste products of metabolism. Thus, the most significant factor in the *in vitro* situation would seem to be the "normal" extracellular fluid into which the muscles are placed.

It may be concluded tentatively that, while circulatory disturbances definitely could contribute to the easy fatigability of adrenocortical insufficiency, it is not yet known to what extent inadequate perfusion of skeletal muscle *per se* is related causally and directly. Moreover, one can make only cautious estimates of blood flow through any given vascular bed knowing just the perfusion pressure; the rate of blood flow is also markedly influenced by the vascular resistance in the bed. No one has ever measured directly the effect of exercise on the blood flow through the muscles of an adrenalectomized animal. Additionally, it seems reasonable to suppose that the composition of the blood with which the muscles are being perfused could be of great significance. We are essentially without information as to the chemical composition of the blood at times of muscular "fatigue" in states of adrenocortical insufficiency.

e) Absence of adrenal medullary secretions

The adrenal medullary secretions of mammals usually consist of a mixture of epinephrine and norepinephrine. These secretions probably constitute the only major source of epinephrine in the body, since adrenergic nerves are believed to secrete only norepinephrine (GOODMAN and GILMAN 1955, p. 400). Therefore, following adrenalectomy, circulating epinephrine is greatly reduced. Epinephrine can affect the function of skeletal muscle in at least two ways. It can promote glycogenolysis in muscle and can affect the diameter of some of the blood vessels in muscle. This latter effect appears largely in the form of arteriolar dilation (GOODMAN and GILMAN 1955, pp. 481–487). This dilation, coupled with its ability to increase cardiac output, means that epinephrine acts to increase blood flow through skeletal muscle. Thus epinephrine can be seen to assist the working muscle in two ways: it promotes blood flow through the muscle and it promotes the release of a stored source of energy in muscle by promoting glycogenolysis.

From the available evidence it is difficult to determine to what extent the reduction in epinephrine facilitates the early onset of "fatigue" in adrenalectomized animals. INGLE and LUKENS (1941) found that adrenal medullectomy in the rat resulted in some impairment of performance early in the period of the work test but that, after 10 hours of work, such animals seemed to have recovered and were now working in a normal fashion. They found, however, that the subcutaneous administration of Adrenalin into adrenalectomized rats early in the period of the test significantly increased the work performance of these animals. If Adrenalin is not given until the time of "fatigue" in the work test of Ingle, it causes only a transient enhancement of the work performance of adrenalectomized rats (INGLE 1936).

Overbeek (1949a) found that epinephrine was without effect upon the performance of adrenalectomized rats in the Everse and de Fremery test. To repeat, Ingle and Nezamis (1949) were unable to confirm the results of Ingle and Lukens (1941) reported above. They found epinephrine, in conventional doses, to be without effect upon the performance of adrenalectomized rats in the work test of Ingle; in fact, large doses of epinephrine were found to depress the work performance. Ther and Mueller (1950) found glycogenolysis to proceed at a normal rate in the *resting* muscle of adrenalectomized animals; a similar finding has also been made in the *working* muscle of adrenalectomized animals (Ramey et al. 1950).

Although the evidence reported above is contradictory, it does seem to suggest that the absence of the adrenal medullary secretions plays no major role in the etiology of the muscular dysfunction displayed by adrenalectomized animals.

IV. The nervous system

Although Addison's accounts of adrenocortical insufficiency contained descriptions of neuropsychiatric disturbances, in the century that followed only clinical observers systematically reported upon possible interrelationships between adrenocortical secretions and the activity of the nervous system (Cleghorn 1951, 1952a, b, 1957). These reports were perforce often simply descriptive; more detailed knowledge had to await experimentation with animals. Although many studies have been made upon the separate roles of nervous and endocrine mechanisms as effectors of homeostasis, experimental evidence bearing upon possible interrelations between these two important integrative systems (particularly in regard to the adrenal cortex) is largely a product of the last decade. An account of these recent developments can be found in the reviews of Hoagland (1954, 1955), Ramey and Goldstein (1957) and Woodbury (1958).

1. Peripheral nerve

a) Excitability

Both *in situ* and *in vitro* measurements have been made of the excitability of peripheral nerves after adrenalectomy. Most commonly the sciatic nerves of rats have been employed. The results have been calculated in terms of various values: chronaxie, the k of A. Hill (1936) or the k of Blair (1932). (The first two values vary inversely with excitability whereas the k of Blair varies directly.)

All investigators have found that the excitability of peripheral nerve is decreased in adrenocortical insufficiency (Slocombe et al. 1952, 1954, Chauchard 1952, Lecoq et al. 1952, Lecoq 1954, Hoagland et al. 1953a, b, Hoagland 1954, 1955, E. Wright and Lester 1959). That has been found to be true regardless of the species studied, the preparation employed or the aspects of neuronal function embodied in the value employed to calculate the results.

Slocombe et al. (1954) made observations in the rat under both *in situ* and *in vitro* conditions. They found that restoration of normal excitability after adrenalectomy could be partially achieved by the use of adrenocortical extracts, but simply assuring the nerve of an adequate oxygen supply and placing it in an environment normal with respect to electrolytes was without effect.

E. Wright and Lester (1959) found that adrenalectomy produced the following changes in the sciatic nerves of rats: decreased excitability, slightly decreased conduction velocity and decreased survival time of the nerve in the absence of oxygen. Similar alterations in excitability could be produced in the nerves of normal animals when glucose was substituted for NaCl in the medium bathing the nerve. These authors concluded, "This suggests that nerve tissue excitability change following adrenalectomy is due to the change in electrolyte concentration of plasma, especially the depletion of sodium from this medium."

Despite their findings of decreased excitability in peripheral nerve after adrenalectomy, E. WRIGHT and LESTER (1959) attempted to demonstrate that the reverse obtains so that they might reconcile the observations in peripheral nerve with the many reports (to be described below) of a contemporaneous increase in the excitability of central neurons. Their discussion of the problem is as follows. "In peripheral nerve increased oscillatory activity has been predicted mathematically (A. HILL 1936) and shown experimentally (E. WRIGHT and COLEMAN 1954, SOLANDT 1936, and E. WRIGHT and ADELMAN 1954) to be associated with a decreased k value (of Blair) or increased chronaxie. It has even been shown that the hyperexcitable state during recovery following an impulse is accompanied by increased chronaxie (BLAIR 1939). Consequently the effect of adrenalectomy to decrease k, or increase chronaxie, might be expected to enhance the tendency of the tissue to respond repetitively. A nerve fiber capable of repetitive firing is considered more excitable than a nonrepetitive one (E. WRIGHT and COLEMAN 1954); therefore the effect of adrenalectomy is to increase excitability rather than decrease it according to the classical definition of nerve excitability (SLOCOMBE et al. 1954). The contradiction between results obtained from brain tissue and nerve tissue is then resolved. Brain seizure threshold is determined by the tendency of brain cells to respond repetitively when stimulated and a lowered electroshock threshold, indicative of increased oscillatory activity in the CNS (WOODBURY 1958), might be expected then in the adrenalectomized rat on the basis of peripheral nerve results, assuming all nerve tissue to be affected similarly. It is concluded therefore that the effect of adrenalectomy on nerve and brain tissue is not contrary but alike in the sense that the tendency for oscillatory activity to occur in either tissue is enhanced by this treatment. In whole nerve experiments repetitive firing is not readily demonstrated and has not been observed in the present study."

It must be noted that increased oscillatory activity in peripheral nerve following adrenalectomy has never been demonstrated nor does the above explanation account for the numerous findings of decreased excitability as established by classical criteria. It is quite possible that different processes are being measured in the peripheral and central studies, and the "contradiction" discussed by WRIGHT and LESTER is only apparent and not real. Furthermore, corticosterone can exert a stabilizing influence upon some aspects of the activity of central neurons (WOODBURY 1958); therefore, in the early stages of adrenal insufficiency, at least, one might well observe increased excitability in the central nervous system in the absence of this hormone.

b) Neuromuscular junction

The status of the neuromuscular junction in adrenocortical insufficiency is discussed in the portion of this chapter devoted to skeletal muscle (p. 103ff.).

2. Central nervous system

a) Anatomical changes

In adrenocortical insufficiency, no anatomical lesions in the central nervous system are detectable consistently by standard methods (WOODBURY 1958).

b) Metabolic disturbances

α) Oxygen consumption. Unresolved discrepancies exist in the several reports concerning the effect of adrenocortical insufficiency upon cerebral oxygen consumption: data for man do not agree with those for experimental animals, and *in vivo* and *in vitro* findings with animal tissue are not in accord.

The hazards of making measurements in Addisonian patients receiving no therapeutic support can be readily appreciated and it is thus not surprising to find few such reports in the literature. SCHEINBERG (cited by F. ENGEL 1949) did however investigate this problem in two untreated Addisonian patients and found no change in cerebral oxygen consumption as measured by arteriovenous oxygen differences. HAFKENSCHIEL et al. (1954) reported similar findings in patients in whom 90 % of the adrenal tissue had been removed.

In adrenalectomized rats, however, the *in vivo* studies of HOAGLAND (1954) indicate an 18 % reduction in oxygen consumption.

Investigators measuring the Q_{O_2} of animal brain slices and homogenates have found no change occasioned by adrenalectomy (TIPTON 1939, CRISMON and FIELD 1940, and BERGEN et al. 1952). However, HIMWICH et al. (1934), employing minced brain obtained from adrenalectomized animals in severe states of insufficiency, reported a subnormal Q_{O_2}.

Because so few data are available upon which to base a conclusion, it is necessary at present to suspend judgment concerning the effects of adrenal insufficiency upon cerebral oxygen consumption.

β) Carbohydrate and protein metabolism. Although the effects of adrenocortical insufficiency on the intermediary metabolism of carbohydrate and protein have been studied intensively in liver and in skeletal muscle, few studies have been made upon the brain. SASS-KORTSAK (1944) and WOODBURY (1958) reported that there is no change in the glycogen content of the brain after adrenalectomy of the rat, but VACCARI and ROSSANDA (1951) found subnormal concentrations of glycogen in the brain and a slight decrease in the concentration of total carbohydrates.

Because other tissues show increased glucose utilization following adrenalectomy, the last report cited has been seized upon as evidence that a similar change occurs in brain tissue. It would, therefore, be pertinent to recall that increased glucose utilization in other tissues probably reflects, in a large measure, the action of insulin unopposed by the normal adrenocortical secretions, and that, in the case of brain tissue, no one has able to demonstrate an action of insulin upon its carbohydrate metabolism (STADIE 1958). (See also Chapter 7.)

VERNADAKIS and WOODBURY (cited by WOODBURY 1958) studied the effect of adrenalectomy upon the concentrations of various amino acids and related compounds in the brain, and found that the total concentration of these compounds (measured as free acids) was markedly decreased. More specifically, the concentrations of taurine, valine, cystine, glutamine, glutamic acid and γ-aminobutyric acid (GABA) were decreased. However, the concentration of aspartic acid was unchanged and the concentrations of glutathione and cysteic acid were actually increased. WOODBURY (1958) concluded that these findings are indicative of increased protein synthesis in the brain during adrenocortical insufficiency.

K. ELLIOTT and FLOREY (1956) and BAZEMORE et al. (1957) felt that GABA may function as an inhibitory mediator substance in the central nervous system. Its decrease in the brain following adrenalectomy affords a possible explanation for the contemporaneous increase in brain excitability (to be described below).

γ) Electrolyte content and distribution. WOODBURY (1954, 1958) has reviewed the numerous investigations of alterations in the content and distribution of brain electrolytes following adrenalectomy. The turnover of Na^+ in the brain is decreased postoperatively whereas the turnover of K^+ may possibly be increased. The extracellular: intracellular fluid gradient for Na^+ is decreased after adrenalectomy whereas this gradient for K^+ increases somewhat. The experimental evidence for these statements will be considered in the paragraphs to follow.

The K^+ in the brain is unaffected by adrenalectomy, both with respect to total content (V. DAVENPORT 1949, FLANAGAN et al. 1950, BERGEN and HOAGLAND 1951, STERN et al. 1951, TIMIRAS et al. 1954) and intracellular concentration (TIMIRAS et al. 1954). The intracellular concentration of Na^+ had been found to rise in the postoperative period (TIMIRAS et al. 1954), but the total content of Na^+ in the brain does not change (FLANAGAN et al. 1950, STERN et al. 1951, BERGEN and HOAGLAND 1951). The sum of the findings concerning the changes in Na^+ and K^+ would not lead one to expect that the intracellular concentration of Na^+ would rise. As the total brain content of Na^+ and K^+ remains unchanged after adrenalectomy, a change in the intracellular concentration of Na^+ should be accompanied, as a first approximation, by an equal change in the intracellular concentration of K^+. As the latter event does not occur, the need for further investigation of this matter is clearly indicated.

HOAGLAND and STONE (1948), HOAGLAND (1954, 1955) and BERGEN et al. (1955) reported accumulations of K^{42} by the brains of adrenalectomized rats to be 16 to 24% above normal. Since total brain K^+ remains unchanged, an increased K^+ turnover is suggested. A further analysis of all factors operative in the situation, however, prompted BERGEN et al. (1955) to point out that, as the normal rats were eating well while the adrenalectomized ones were eating poorly, if at all, dilution of K^{42} with dietary K^+ was occurring in the intact animals. The apparent increase in K^{42} uptake vanished when all of the animals were fasted prior to the observations.

In 1953, LEIDERMAN and KATZMAN reported some provocative findings concerning K^+ turnover in the brains of normal and adrenalectomized rats. Influx and outflux values were determined with the use of the following equation [developed as a general exchange equation by A. SOLOMON (1949)]:

$$\frac{d\,[K^{42}]_{\text{brain}}}{dt} = k_1\,[K^{42}]_{\text{plasma}} - k_2\,[K^{42}]_{\text{brain}}\,.$$

They found that in normal brain, only 80% of brain K^+ was exchangeable with plasma K^+. In adrenalectomized animals, however, the exchangeable K^+ of brain was 100%. However, numerous investigators (HOAGLAND 1954, 1955; BERGEN et al. 1955; WOODBURY 1958) have been unable to demonstrate the presence of a nonexchangeable K^+ fraction in the normal brain. It is to be hoped that future investigations will resolve the presently existing discrepancies.

δ) Acetylcholine and serotonin. The role played by acetylcholine in the function of the central nervous system remains obscure, but it is obvious that alterations in central neuronal activity after adrenalectomy cannot be ascribed to a lack of it. GREENBERG (1949) and TORDA and WOLFF (1952a) found acetylcholine synthesis to be essentially normal in the brains of adrenalectomized rats. GREENBERG employed acetone-dried powders while TORDA and WOLFF used tissue minces.

DE MAIO (1959) measured the serotonin (5-hydroxytryptamine) content of various areas of the brains of intact and adrenalectomized rats. The averages of 20 animals in each group showed that levels of serotonin in blood serum fell from 50 to 12 μg/100 ml, 24 hours after the operation. At the same time, the following increases occurred in the brain: hemispheres, from 85 to 195 μg/100 g of tissue; medulla oblongata, 97 to 115; and base of the brain, 37 to 140. The cerebellum contained no serotonin, and this was unaltered by adrenalectomy. Such rises could not be confirmed by TOWNE and SHERMAN (1960).

c) Cerebral circulation

In any assessment of central nervous system activity, the status of the cerebral circulation is clearly a matter of major concern. In adrenalectomized rats, cerebral blood flow has been found to be 39 to 43 % of normal (HOAGLAND et al. 1953a, 1953b, HOAGLAND 1954). These investigators concluded that the slowed electrocorticogram of adrenal insufficiency reflects the reduced oxygen supply to the brain resulting from an inadequate cerebral circulation.

The available data from man bearing upon this point do not permit of any firm conclusion. The findings of SCHEINBERG (cited by F. ENGEL 1949) and HAFKENSCHIEL et al. (1954) indicate no change or a slight increase in cerebral blood flow in adrenocortical insufficiency. SCHEINBERG, however, reported upon only two cases, and the patients of HAFKENSCHIEL et al. did not undergo total adrenalectomy and were suffering from severe essential hypertension before surgery. Additional data will probably not be forthcoming readily because of the hazards inherent in the making of such measurements in untreated adrenally-insufficient human beings.

d) Neurophysiological disturbances

α) Electrical activity of the brain. The electrical activity of the brain is modified by adrenocortical insufficiency. G. ENGEL and MARGOLIN (1941, 1942) were the first to describe these changes in man, and their findings have been confirmed and amplified (see the review of WOODBURY 1958). In sum, the following alterations have been observed: the electroencephalogram is characterized by many bursts of high-voltage (up to 75 μV), low-frequency (2 to 8/sec) activity. Such activity occurs most commonly in the frontal areas and is accompanied by a relative refractoriness to the events usually attendant upon opening the eyes. Concomitantly, there is a reduced incidence of low-voltage, fast-frequency beta waves. In addition, the response to voluntary hyperventilation is exaggerated. G. ENGEL and MARGOLIN reported that the response to hyperventilation can be restored to normal by the administration of glucose. They believed that a rough correlation exists between blood glucose concentrations and the presence of abnormalities in the electroencephalogram. However, W. HOFFMAN et al. (1942), while agreeing in the matter of hyperventilation, found that intravenous glucose did not restore the electroencephalogram to normal in the resting stage.

The electrocorticogram in the adrenalectomized rat is likewise dominated by low-frequency rhythms (BERGEN 1951). HOAGLAND (1954) felt that "the evidence supports the hypothesis that reduced brain oxygen consumption following adrenalectomy slows the electrocorticogram. The reduced oxygen consumption in turn appears to result from decreased cerebral circulation brought about by a general reduction in cardiovascular tone and responsivity." As comparable evidence of hypoxia is lacking in the human being, it is not known if such a state is contributory to the electroencephalographic abnormalities described for man.

Another expression of slowed activity in the central nervous system is the finding that in the rat impulses from the toe are conducted to the sensory cortex more slowly after adrenalectomy (HOAGLAND et al. 1953b). The decreased conduction time was attributed to the neurons of the central nervous system involved in this pathway.

β) Brain excitablity. Another aspect of the function of the central nervous system that has received much attention is brain excitability. As used today, the term "brain excitability" refers to the threshold for convulsions. The convulsions may occur spontaneously or be deliberately induced by an appropriate stimulus that can be graded in intensity. In general, it has been found that, in adrenalectomized animals, brain excitability is increased (i. e., the threshold for convulsions is decreased). One cannot say precisely what facets of neuronal function are being assessed in these studies. Certainly in this gross criterion one has an algebraic summation of the activities of a huge population of neurons. What appears to be increased excitability may be a manifestation of decreased activity of inhibitory neurons that normally prevent the widespread discharges that lead to convulsions.

Spontaneous convulsions. Clinical observers have known for almost a century that convulsions may occur in Addisonian crises (KLIPPEL 1899, G. ENGEL and MARGOLIN 1941, CLEGHORN 1951, STORRIE 1953). Such seizures often are of the grand mal type and may occur during periods characterized by other disturbances, such as stupor or paranoid delusions (STORRIE 1953). (See Chapter 6 in Part 1 of this Volume.)

Likewise, convulsions may occur spontaneously in adrenalectomized animals. In adrenalectomized rhesus monkeys, KNOBIL et al. (1953) associated convulsions and prostration with hypoglycemia. Providing the animals were not moribund, recovery from these episodes was produced in 5 to 60 minutes by the administration of glucose.

The findings of ARNETT et al. (1942) and STORRIE (1953) suggested that there is not necessarily a connection between hypoglycemia and convulsions since, in both Addisonian patients and in adrenalectomized rats, they observed no good correlation between blood glucose levels and the frequency of neurological symptoms. One of STORRIE's patients, for example, was found to have a blood glucose level of 119 mg % in the post-ictal state. Obviously the uptake and utilization of glucose by the brain must be the primary consideration in these situations; an adequate blood glucose level is only the first requirement for normal function.

Experimentally-induced convulsions. Appropriate electrical, auditory and chemical stimuli will provoke convulsions and they can be graded in intensity. Therefore such stimuli have been used to measure brain excitability in adrenalectomized animals.

The thresholds for electrically-induced seizures (EST) have been studied by V. DAVENPORT (1949) and TIMIRAS et al. (1954); a summary of the findings in this area is provided by WOODBURY (1954). In both adrenalectomized rats and mice, the EST was found to be approximately 25% below normal. This decline was accentuated by the administration of KCl or $MgCl_2$, but prevented by isotonic NaCl or $CaCl_2$. Not surprisingly, these data indicate the importance of Ca^{++}. But, in consideration of the Na^+ losses that accompany adrenocortical insufficiency and the findings with the use of NaCl, these investigators logically stressed the importance of the extracellular: intracellular Na^+ ratio in the maintenance of a normal EST.

The threshold for seizures induced by pentylenetetrazol (Metrazol) is unaffected by adrenalectomy (CICARDO 1945, 1946, TORDA and WOLFF 1952b). The sensitivity to insulin-induced convulsions is increased some 24 times by total adrenalectomy (SWANN and FITZGERALD 1938), but only 2-fold by the removal of the adrenal medulla alone.

Adrenalectomized laboratory rats show a lowered threshold for audiogenic seizures (GRIFFITHS 1949). If such animals are maintained on large doses of NaCl, however, they actually show a subnormal susceptibility to such stimuli. Using electrical stimuli, WOODBURY et al. (cited by WOODBURY 1958) likewise found that the EST of adrenalectomized animals could be elevated above normal by very large doses of NaCl. Such findings reveal the importance of Na^+ in protecting the organism from convulsions induced by certain stimuli, and point up the significance of the Na^+ shifts and losses that occur in adrenocortical insufficiency.

Interestingly enough, GRIFFITHS (1949) was unable to provoke audiogenic seizures (an air blast was used) in wild rats, intact or adrenalectomized.

As adrenalectomy increases the susceptibility to convulsions provoked by electrical and auditory stimuli but not to those induced by pentylenetetrazol, WOODBURY (1958) attempted the following reconciliation of these findings: "The problem of chemically induced convulsions in adrenalectomized animals is complicated by alterations in the absorption, fate, and excretion of the convulsant drug, and it is likely that adrenalectomy influences one or more of these factors; for example, if absorption of the convulsant is delayed, a larger dose would be required to produce a given effect." However, TORDA and WOLFF (1952b) administered the pentylenetetrazol intravenously; and it would appear that the effect of insulin was not seriously disturbed by the operation of the factors cited above. The matter is further complicated by the finding of SWINYARD et al. (1955) that, when hyponatremia was induced by peritoneal lavage in intact mice, their susceptibility to the convulsant effects of pentylenetetrazol was increased.

Convulsions provoked by "water loading" are discussed in Chapter 4, Part 1 of this Volume.

γ) Neurological disturbances. *Behavioral changes.* Adrenalectomy decreases the spontaneous running activity of rats, wild or domesticated (GRIFFITHS 1949, RICHTER 1952, GOLDMAN and ABOOD 1953). Restoration of the status of carbohydrate metabolism to normal (by appropriate hormonal therapy) is a potent means of repairing this deficiency, although NaCl supplements effect a partial repair. GOLDMAN and ABOOD (1953) found creatine phosphate stores to be lowered and inorganic phosphate levels elevated in the hypothalamus and pituitary of adrenalectomized rats exhibiting decreased spontaneous activity; no such changes could be detected in cerebral cortical tissue. These findings may or may not be relevant to decreased running activity.

Psychic disturbances. Investigators have noted that psychological derangements are frequently present in patients with Addison's disease. Reviews of this topic have been made by GORMAN and WORTIS (1947), CLEGHORN (1951, 1952a, b, 1953), CLEGHORN and PATTEE (1954), WOODBURY (1958) and STOLL in Chapter 6 of this Volume. The aberrations may range from the neurotic to the frankly psychotic.

The incidence of such disturbances is rather high. G. ENGEL and MARGOLIN (1941, 1942), in a review of 25 cases, found 16 to have neuropsychiatric disturbances, of which 3 were frankly psychotic. In another review of 25 patients, CLEGHORN (1951) reported that the commonest disturbances were apathy and negativism, encountered in 80 to 84 % of his patients. Less common findings were seclusiveness, depression and irritability. More severe upsets, such as suicidal tendencies and paranoid delusions, seem to occur relatively rarely.

e) The response to drugs

Adrenocortical insufficiency has been shown to modify the response to some drugs that act upon the central nervous system.

The *barbiturates*, in particular, have been studied by numerous investigators. The time required for the onset of the effects of these drugs after administration is shorter after adrenalectomy (KOMIYA and SHIBATA 1956) and the duration of the effects is prolonged (TUREMAN et al. 1951, 1952, SHIBATA and KOMIYA 1953, ROBILLARD et al. 1954). Animals are more susceptible to the toxic effects of the barbiturates after adrenalectomy (RICHARDS 1941, EICHOLTZ et al. 1949). ROBILLARD et al. (1954) demonstrated that hepatic tissue from adrenalectomized animals metabolizes pentobarbital at a subnormal rate. KOMIYA and SHIBATA (1956) found that the brains of adrenalectomized animals concentrate barbital to an abnormally high degree. It seems likely that in these observations there is a partial explanation for the altered responses of adrenalectomized animals to barbiturates. Additionally, as barbital is excreted wholly in an unchanged form, inadequate renal function may possibly contribute to the enhancement of its activity in states of adrenocortical insufficiency.

Susceptibility to *urethan* (SINDRAM 1935) and to *morphine* (J. LEWIS and TORINO 1926, N. TALBOT et al. 1952, p. 214, WHITAKER et al. 1956) is increased in states of adrenocortical insufficiency.

V. Gastrointestinal system

1. Food intake

a) Anorexia

A reduced intake of food is characteristic in adrenocortical insufficiency. A capricious appetite is often displayed by both man and animals, and, in the more severe stages of this condition, a refusal of all food is common.

In Addison's disease, anorexia may be of insidious onset and slow development (SORKIN 1949). It is encountered in 80—90% of all Addisonian patients (MARAÑON et al. 1934, SORKIN 1949, THORN 1951). Other gastrointestinal disturbances, such as nausea, vomiting, diarrhea, constipation and abdominal pain, occur less frequently than anorexia. However, these latter signs and symptoms can, on occasion, be so violent as to suggest diagnoses like acute peritonitis or poisoning (SORKIN 1949) and to lead mistakenly to surgery for "chronic appendicitis" or "chronic cholecystitis" (THORN 1951).

Until the careful studies of ROGOFF and STEWART (1926a) were initiated, the food intake of adrenalectomized dogs could not be measured, for the period of survival after operation was too short. Inexpert surgery and inadequate care in the postoperative period meant that "only animals in a state of acute collapse were available for experimental study" (GAUNT and EVERSOLE 1949). With the possible exception of the studies using rats, a similar condition prevailed in the instances of work with other species. The insistence of ROGOFF and STEWART on high standards of surgical technique and animal care (G. STEWART 1924) enabled their wholly-adrenalectomized dogs to live for approximately 10 days and permitted them to observe the reduction in food intake that occurred in the postoperative period. While the rate at which food intake declined was not predictable, once all food was refused death would occur within 3 days. Reports of reduced food intake following adrenalectomy have subsequently been made for many species (HARTMAN and BROWNELL 1949, NOBLE 1950).

b) "Salt craving"

The experiments of RICHTER and others (summarized by KENDALL 1948) indicate that the adrenalectomized rat, when offered a choice of several diets, will voluntarily select one that provides a high daily intake of sodium chloride. RICHTER (1939) has postulated that an increased sensitivity of the sense of taste to salt may be the basis for these findings.

In these selection experiments, the implication is present that they constitute an example of the "wisdom of the body" at work: the Na^+ losses attendant upon adrenocortical insufficiency are somehow recognized and an attempt is made to compensate for them. If true, it must be noted that such "wisdom" is not evident in all species, nor indeed is it even evident in the rat under all experimental conditions. An increased appetite for salt is not a consistent finding in Addison's disease (SORKIN 1949) and occurred in only 19 % of one group of patients (THORN 1951). The adrenalectomized rhesus monkey likewise displays no unusual avidity for salt (HOFMANN et al. 1954). HARRIMAN (1955) offered a group of intact rats the choice of an 8 % sugar solution or a 1.2 % NaCl solution to drink; most of the animals quickly showed a preference for the sugar solution. These animals constituted Group I. Another group of rats (Group II) was maintained on distilled water as a drinking fluid. After 18—20 days on these regimens, half of the animals in each group were adrenalectomized and the other half subjected to sham adrenalectomy. All animals were then offered the choice of sugar and salt solutions. Only the adrenalectomized rats in Group II showed a preference for the salt solution. The adrenalectomized rats in Group I that had established a preoperative preference for the sugar solution continued to display this preference after operation. They drank no more of the salt solution than did the sham-operated animals in Groups I and II.

"Salt craving" would appear to be a possible, though far from inevitable, occurrence in adrenocortical insufficiency.

2. Stomach

a) *Anatomical changes*

Gastric ulcers have been observed in about 2 % of patients with Addison's disease (MARAÑON et al. 1934), and gastric ulcers or hemorrhages in 39 % of adrenalectomized cats (HARTMAN and BROWNELL 1949). Lesions in the gastric mucosa of adrenalectomized mice, rats, cats and dogs were described by MANN (1916). The incidence of gastric ulcers seems to be a function of species. They occur commonly in adrenalectomized cats and dogs (HARTMAN and BROWNELL 1949) but only rarely in adrenalectomized rhesus monkeys (GREEP et al. 1952) or rats (MADDEN and RAMSBURG 1951a, WELBOURN and CODE 1953, HIRSCHOWITZ et al. 1956).

BAKER and BRIDGMAN (1954) made a histological study of the influence of adrenalectomy on the gastric mucosa in the rat. Most noticeably affected were the zymogenic cells. They were subnormal in size, had a reduced granular content and contained little basophilic (ribonucleoprotein) material. The parietal cells were but little affected, showing a slight reduction in size and some increase in fuchsinophilic bodies (mitochondria). The content of mucus within the mucous neck cells was reduced. No consistent changes in the argentophile cells were noted.

b) *Motor activity*

Very few studies have been made of the effects of adrenocortical insufficiency upon gastric motility. TUERKISCHER and WERTHEIMER (1945) reported that no abnormalities in gastric motility could be observed roentgenologically following the administration of barium sulfate to adrenalectomized rats.

c) *Secretory activity*

In many but not all species, gastric secretions are depressed in adrenocortical insufficiency. The pertinent reports can be summarized under several headings.

Volume of secretions. Studies of the volume of gastric secretions have been carried out largely on adrenalectomized rats. The techniques employed have entailed pyloric ligation (MADDEN and RAMSBURG 1951a, b, WELBOURN and CODE 1953) or the injection of the parasympathomimetic drug carbachol (TUERKISCHER and WERTHEIMER 1945). In all instances, the volume of gastric secretions has been found to be subnormal. That this decrease is evident immediately following adrenalectomy is indicated in the report of MADDEN and RAMSBURG (1951b). After a 24-hour fast, pyloric ligation and, immediately thereafter, adrenalectomy were performed in a group of rats. Another group was subjected to pyloric ligation alone. Stomachs were removed at the 6th postoperative hour and the volumes of their contents measured. Stomachs from control animals were found to contain 8.7 ± 0.17 ml of fluid on the average and those from adrenalectomized animals 5.7 ± 0.19 ml.

Acidity of secretions. The pH of the stomach contents rises after adrenalectomy of the rat. TUERKISCHER and WERTHEIMER (1945) observed a rise from a pH of about 1.5 in intact rats to a pH of about 6 after adrenalectomy. A smaller decrease in the acidity of the gastric contents (to a pH of about 3.8) was observed by WELBOURN and CODE (1953) in adrenalectomized rats. In mice, on the other hand, H. DAVENPORT and CHAVRE (1950) reported that the stomachs of adrenalectomized mice, male and female, secreted acid *in vitro* at a rate identical to those of intact or sham-operated, pair-fed controls. Their observations were made at various times from the 4th hour to the 7th day after operation.

TUERKISCHER and WERTHEIMER (1945) measured, in addition, the free and total acid content of the gastric secretions after the injection of carbachol, and found them to be subnormal at 2—5 days after adrenalectomy. Furthermore, this drug was unable to evoke the gastric secretion of injected neutral red in adrenalectomized rats, although it induced such secretion readily in intact animals. In addition, the gastric secretory response to histamine and to insulin was depressed following adrenalectomy. Gastric secretory activities were essentially normal in rats subjected to adrenal medullectomy.

In Addison's disease, the gastric secretion of HCl is often reduced (THORN 1951). In his series of 50 patients, SORKIN (1949) reported a 50% incidence of achlorhydria. He noted, in addition, that "no definite relationship could be established in the present series between the occurrence of gastrointestinal symptoms and the presence of achlorhydria. The administration of dilute hydrochloric acid did not correct the symptoms in all of the patients who had achlorhydria."

STEMPIEN and DAGRADI (1954) reported on a patient with Addison's disease in whom the secretion of HCl following the injection of histamine was markedly subnormal. By gastroscopy, the gastric mucosa of this individual appeared to be normal. J. KYLE et al. (1956) studied the gastric secretion of free HCl in patients with Cushing's disease before and after adrenalectomy, using the gruel fractional test-meal technique. They found that the hyperresponsiveness of Cushing's patients was reduced by adrenalectomy.

Enzymic content of secretions. The enzymic content of gastric secretions likewise is lowered by adrenocortical insufficiency. TUERKISCHER and WERTHEIMER (1945) found that the rennin and pepsin content of gastric secretions is lowered by adrenalectomy of rats and that the peptic activity of the secretions is also diminished. HIRSCHOWITZ et al. (1956) studied the effect of adrenalectomy upon gastric pepsinogen in the rat. By the 7th postoperative day, the total pepsinogen content of the stomachs of saline-maintained adrenalectomized rats had fallen from a control value of 17,720 units of pepsin equivalents per stomach to 10,280 U per stomach.

No further depletion of pepsinogen occurred on a low sodium diet, but an additional significant decrease was observed after starvation.

The excretion of uropepsin, a urinary proteolytic enzyme believed to be secreted by the gastric mucosa (SPIRO et al. 1950), has been found to be subnormal in Addison's disease. GRAY and RAMSEY (1957) reported that, in 265 normal adults, the average uropepsin excretion was 3,670 U per day; 12 Addisonian patients were found to have an average excretion of 542 U per day. SPIRO et al. (1950) found an excretory rate of less than 500 U per day in Addison's disease.

TUERKISCHER and WERTHEIMER (1945) reported that the gastric secretions of adrenalectomized rats are relatively rich in mucin. However, the content of mucus within the mucous neck cells is reduced (BAKER and BRIDGMAN 1954).

3. Intestines

a) Anatomical changes

VOGT (1943a) reported that abnormal spastic contractions can be observed in the small intestines of moribund adrenalectomized rabbits, and that, in general, the entire gastrointestinal tract is "atrophic" at this time. In these rabbits, death occurred, on the average, on the 10th postoperative day. SATAKE et al. (1931), by means of a Thiry-Vella fistula, observed congestion and diffuse hemorrhage in the intestines of adrenalectomized dogs and rabbits. ROGOFF and STEWART (1928b) made similar findings at autopsy in many adrenalectomized dogs but noted that, on occasion, the gastrointestinal tract was wholly devoid of congestion and that there was no blood in the lumen. GUNN (1927) reported that the intestinal mucosa of the adrenalectomized cat is anemic in appearance.

In their histological and histochemical study of the gastrointestinal tract of the adrenalectomized rat, BAKER and BRIDGMAN (1954) observed that the duodenal villi became relatively low and broad. The goblet cells in the villous epithelium became smaller and contained less mucus. The alkaline phosphatase activity of the absorptive cells was slightly reduced after adrenal extirpation. The most striking alteration occurred in the duodenal Paneth cells; they underwent several-fold enlargement and their content of granules increased markedly. No major changes in the histology of the colon occurred.

SORKIN (1949) observed that hyperplasia of the lymphatic elements of the gastrointestinal tract has been found in the postmortem examination of individuals with Addison's disease (see p. 126).

b) Motor activity

VOGT (1943a) made a detailed physiological and pharmacological study of the small intestine in adrenalectomized rabbits. Portions of the jejeunum were removed from adrenalectomized rabbits when death seemed imminent and were suspended in a bath containing oxygenated, glucose-free Tyrode's solution. The muscles in the longitudinal coat showed normal spontaneous activity and a normal response to a variety of agents (physostigmine, nicotine, histamine, acetylcholine, epinephrine, ephedrine and hypertonic NaCl solutions). The muscles in the circular coat frequently displayed powerful spontaneous spasmodic rhythmic contractions; such activity is not seen in strips of jejeunum from intact rabbits. The response of the circular muscles to agents that presumably stimulate them via an effect on the nerve plexus was unpredictable. Nicotine and hypertonic NaCl solutions provoked normal responses in the circular muscles but, in some strips, the response to physostigmine was weak or wholly absent. The responses to agents that presumably affect the circular muscles directly (muscarine and KCl) were normal. The alterations in intestinal function resulting from adrenalectomy (by the criteria of this study) seem to be confined to the muscles of the circular coat and seem to consist of abnormal spontaneous activity and a reduced responsiveness to some drugs that are presumed to act via the plexus of Auerbach. VOGT found no evidence of a disturbed carbohydrate metabolism in the intestine of the adrenalectomized rabbit, and, since the addition of an adrenocortical extract (Eucortone) to the bath did not cause restoration of normal function, the abnormalities reported above were interpreted as "secondary effects of suprarenalectomy caused by primary deficiencies in other organs."

WYMAN (1953) studied the propulsive motility of the small intestine in adrenalectomized rats. Between the 4th and 7th postoperative day, adrenalectomized rats were given a suspension of powdered charcoal by stomach tube after a fast of 24 hours. Control observations were made in similarly treated intact and sham-operated animals. The animals were sacrificed 30 minutes after the administration of the charcoal and the findings expressed in terms of the fraction of the total length of the small intestine traversed by the test "meal." Two-thirds of the adrenalectomized animals displayed increased propulsive motility while one-tenth displayed intestinal activity less than that found in the control animals. Propulsive motility could not be correlated with severity of adrenocortical insufficiency, since signs of intense adrenocortical insufficiency were present in some rats with increased motility and in some with subnormal motility. WYMAN suggested that the K^+ retained after adrenalectomy acts as a stimulant to smooth muscle and thus accounts for the finding of enhanced propulsive motility.

The findings of VOGT (1943a) and WYMAN (1953) indicate that increases in the motor activity of the intestine occur after adrenalectomy in the rabbit and rat but that this finding is not an invariable one.

c) *Absorptive and secretory activities*

Many studies have shown clearly that the ability of the intestinal mucosa to absorb inorganic electrolytes, carbohydrates and fats is impaired in adrenocortical insufficiency. These findings have been summarized in earlier reviews (SWINGLE and REMINGTON 1944, KENDALL 1948, NOBLE 1950). VERZÁR and his associates have attempted to determine an enzymic basis for these absorptive disturbances. They reported that alkaline phosphatase activity in the intestine decreased after adrenalectomy (VERZÁR and SAILER 1952). The activities of certain enzymes in the intestinal mucosa of the rat before and after adrenalectomy were also determined in *in vitro* assays (LÜTHY and VERZÁR 1954). They found that hexokinase (glucohexokinase) activity was unaffected between the 6th and 10th postoperative day, but that adenosinetriphosphatase activity had fallen to two-thirds of control values. VERZÁR has long maintained that impaired phosphorylation plays an important role in the retarded absorption of certain foodstuffs in adrenocortical insufficiency, but the findings of other workers do not support this concept. This situation has been reviewed by KENDALL (1948) and NOBLE (1950), and is discussed by LEVINE and GOLDSTEIN in Chapter 7 of this Volume. Also of interest in this area are the observations of MOOG (1959) that demonstrate a direct relationship between adrenocortical hormones and the phosphatase activity of the duodenum.

No studies of the effects of adrenocortical insufficiency upon intestinal secretions have come to the attention of the authors.

4. Histamine content

ROSE and BROWNE (1941) reported that the histamine content of the stomach and intestine of the rat was over 300 % of control values by the 12th day after adrenalectomy. P. MARSHALL (1943) found that adrenalectomy resulted in increases in the histamine content of many organs in the rat, but that the stomach was one organ in which an increase did not occur. We cannot, therefore, state what effect adrenalectomy has on the histamine content of the rat stomach. ROSE and BROWNE (1941), P. MARSHALL (1943) and BARTLET and LOCKETT (1959) all agreed that the histamine content of the rat intestine rises after adrenalectomy. Such unanimity lends credence to this observation. However, it makes interpretation of the report of VALETTE and HUIDOBRO (1952) more difficult. These authors found that the histaminase content of the rat intestine also increases after adrenalectomy. Further investigations may resolve the puzzle of an increase in a tissue's content of histamine contemporaneous with an increase in the activity of an enzyme capable of accelerating the detoxification of this amine. In the cat, however, HAEGER et al. (1952) reported that the histaminase activity of the intestine is decreased after adrenalectomy.

VI. Reproductive system

1. Ovarian function

a) Sexual maturation

The age at vaginal opening in adrenalectomized albino rats was found to be the same as in intact animals (WADE and HAZELWOOD 1941).

b) Estrous cycle

Data on alteration of the estrous cycle following adrenalectomy are not consistent. CARR and CONNOR (1933) found that in all of 80 rats, which had been cycling normally, estrous cycles ceased following adrenalectomy. S. MARTIN (1932) noted that cycles were suppressed or prolonged in 88 % of 121 rats that survived adrenalectomy by 7 to 27 days. In a later paper (S. MARTIN and FAZEKAS 1937), abnormalities of the estrous cycle were reported in only 45 % of the animals; in this instance physiological saline was made available after adrenalectomy. AGATE and ZWEMER (1935) found no effect of adrenalectomy on estrous cycles in untreated animals; many of these animals had accessory adrenal cortical tissue and were regarded by the authors as having only moderate adrenocortical insufficiency.

From the results obtained by different experimenters, it appears that maintenance of the estrous cycle depends on the general health of the adrenalectomized animal and that it may be normal if an adequate intake of sodium salts is provided (PARKES 1945).

c) Ovulation

Coitus-induced ovulation in rabbits is not interfered with by adrenalectomy (FRIEDGOOD 1939, PARKES 1945). Impairment of ovulation was demonstrated in cats, but also followed a variety of other surgical procedures (FRIEDGOOD 1939). Since many animals do become pregnant after adrenalectomy (see below), it is evident that ovulation may occur.

d) Menstrual cycle

In women with primary adrenal insufficiency, given adequate replacement therapy, the menstrual cycle can be expected to be undisturbed (SAMUELS et al. 1943, THORN 1951), although SORKIN (1949) states that amenorrhea and sterility are common.

2. Testicular function

Brolin and Lindbäch (1951) found no degeneration of the seminiferous epithelium in adrenalectomized rats. In chickens, however, adrenalectomy is followed by testicular atrophy even when saline and adrenocortical extract are given (Herrick and Torstveit 1938, Parkes 1945). In male cats sexual behavior is not impaired by prepuberal adrenalectomy (M. Levy et al. 1956); this finding was interpreted as an indication of normal testicular function, since animals castrated before puberty failed to mature in this respect. Adrenalectomized rats are able to fertilize intact females (Britton and Kline 1936).

3. Gonadotrophic activity

The occasional interference with the estrous cycle in rats after adrenalectomy has led to investigation of the possibility of impaired response to gonadotrophin or of reduced gonadotrophin production.

a) Response to gonadotrophin

Parkes (1945) stated that adrenalectomized rats show normal responses to injections of gonadotrophin(s). Brolin and Lindbäch (1951) found that ovaries and testes of adrenalectomized rats responded to gonadotrophin (LH or mixed FSH and LH preparations) as did those of intact animals. Zuckerman (1953) summarized work that indicates that responsiveness to either FSH or LH is reduced by adrenalectomy; the rats studied were immature. Mandl (1954) also found that the ovaries of immature adrenalectomized rats respond less than those of intact animals to pregnant mare serum, while adrenalectomy of adult animals does not affect the response. Thus there seems to be an age differential in responsiveness of the gonads to adrenalectomy.

b) Gonadotrophin production

Bioassay of the gonadotrophin content of the pituitaries of adrenalectomized rats was made in immature animals by S. Martin (1932). The increase in the ovarian weight of the test animals was 27 mg, as compared to 42 mg when the donor animals had intact adrenals, indicating less gonadotrophin in the pituitaries of adrenalectomized animals. S. Martin and Fazekas (1937) reported that the gonadotrophin content of the pituitaries of adrenalectomized female rats was reduced in only 40 % of the animals studied.

These findings suggest that disturbance of the estrous cycle may be the result of reduced gonadotrophin production, which in turn may reflect the general condition of the animal, rather than being a specific consequence of adrenalectomy.

4. Pregnancy

a) Influence of adrenal insufficiency on pregnancy in animals

α) Adrenalectomy before pregnancy. The incidence of pregnancy in adrenalectomized rats is variable. Britton and Kline (1936) found that only 5 of 19 animals became pregnant. These pregnancies were carried through successfully; however, accessory adrenal tissue was found in all 5 animals. In untreated adrenalectomized rats, 14 of 38 that had mated became pregnant; only 3 litters were delivered (Tobin 1941). Parkes (1945) concluded that, if adequate salt and water are available to the animal, adrenalectomy does not interfere with reproduction.

CUPPS (1955) found that untreated adrenalectomized rats did not become pregnant; treatment with 0.25 mg cortisone per day allowed pregnancy to occur but the number of young per litter was reduced; daily treatment with 1.25 mg cortisone and 0.5 mg deoxycorticosterone restored reproductive capacity and litter size to normal.

β) Adrenalectomy during pregnancy. The effect of adrenalectomy depends on the stage of pregnancy at which the operation is performed. BRITTON and KLINE (1936) did not mention the stage of pregnancy in the rats they adrenalectomized; in all abortion or fetal death resulted. CARR and CONNOR (1933) found that abortion occurred if rats were adrenalectomized before the 16th day of gestation; those operated upon later went to term but died in labor or a few days after parturition. TOBIN (1941) reported that 16 of 40 rats remained pregnant when adrenalectomy was performed on the 2nd day, and that gestation was undisturbed in 23 of 25 operated upon between the 17th and 19th days.

In cats, adrenalectomy consistently interrupts pregnancy (BRITTON and KLINE 1936, PARKES 1945).

DEWAR (1957) adrenalectomized mice on the 13th to 15th day of pregnancy without ill effect. He found that the animals gained weight as did normal ones, and attributed this effect to the presence of the placenta, as it was not affected by destruction of the fetuses.

b) Influence of pregnancy on adrenal insufficiency in animals

Prolongation of life in adrenalectomized animals by pregnancy was observed many years ago by ROGOFF and STEWART (1926/27), who suggested that the adrenals of the fetuses were not responsible for the benefits observed. Since pseudopregnancy has the same effect as actual pregnancy in cats and dogs (reviewed by PARKES 1945), it appears that this early suggestion was correct. The careful work of CHRISTIANSON and CHESTER JONES (1957) makes it clear that the fetal adrenals play no role in the physiology of the adrenalectomized mother rat. As DEWAR (1957) demonstrated in the mouse, weight loss does not become evident until after the delivery of the placentas, even when the fetuses have been destroyed prior to the termination of pregnancy.

c) Pregnancy in Addison's disease

See p. 144 for discussion of this subject.

5. The mammary glands and lactation

a) Mammary growth and development

Studies of the effects of adrenocortical insufficiency upon mammary growth and development have been carried out largely in rats. Conflicting findings have been reported.

BUTCHER (1939), REEDER and LEONARD (1944) and R. JOHNSTON and SMITHCORS (1948) all reported that adrenalectomy acts as a stimulus to the growth and development of the mammary glands of immature rats. BUTCHER adrenalectomized female rats at 30 days of age, permitted them access to a 3 % NaCl solution thereafter and sacrificed them on the 10th to 15th postoperative day. At this time, he found the adrenalectomized rats to have a greater area of mammary tissue than did unoperated littermate controls, and that many more lateral and end buds were visible in these enlarged glands. The same results were obtained in animals from whom the ovaries were removed at the time of adrenalectomy. REEDER and

LEONARD adrenalectomized male rats at 22 to 35 days of age, maintained them on saline in the postoperative period and sacrificed them on the 10th to 12th postoperative day. They too observed more lateral and end buds on the ducts of the mammary glands than could be found in the glands of intact or castrated rats, but they did not find that adrenalectomy was followed by any increase in the area of the mammary glands. The combination of castration and adrenalectomy gave results identical to those obtained with adrenalectomy alone. R. JOHNSTON and SMITHCORS (1948) found that elongation of the mammary ducts occurred after adrenalectomy of immature, castrated rats.

COWIE (1949b) adrenalectomized male and female hooded Norway rats at 22 days of age. Despite maintenance on saline, most of the animals died at various intervals up to the 40th postoperative day. By relating total mammary-gland area to body surface at various ages in intact animals, a relative growth rate for the glands could be determined. COWIE found that adrenalectomy produced no alterations in the relative growth rate of the mammary glands. RICHARDSON (1955), studying immature female mice, likewise found adrenalectomy to be without effect on mammary growth. In her study, gonadectomized animals were compared with gonadectomized-adrenalectomized animals. No changes could be found in the doubly-operated animals that were not present in those that had undergone gonadectomy alone. TRENTIN and TURNER (1947) employed male rats weighing 200—300 grams. The animals were castrated 41—103 days before sacrifice. In addition, some of the animals were adrenalectomized 10 days prior to sacrifice. These authors found that castration resulted in some reduction of the alveolar development normally found in the mammary glands of adult male rats. Adrenalectomy of castrated males was followed by complete regression of alveolar development and the ducts had a thin, atrophic appearance. COWIE and FOLLEY (1947) also found, frequently but not invariably, that adrenalectomy of gonadectomized immature male and female hooded Norway rats was followed by some slight regressive changes in the mammary glands. They felt the changes to be so slight, however, that they concluded that their findings provided "little evidence that the adrenal cortex plays an outstanding and essential role in the normal development of the mammary gland."

It does not seem possible to reconcile these discrepant findings at this time. It is of interest that observations of adrenalectomy-induced growth and development of mammary glands were made in immature rats (BUTCHER 1939, REEDER and LEONARD 1944, R. JOHNSTON and SMITHCORS 1948), while pronounced regressive changes occurred after adrenalectomy of adult animals (TRENTIN and TURNER 1947). However, this does not materially assist in evaluating findings that adrenalectomy is without effect upon the mammary glands of immature rats and mice (COWIE 1949b, RICHARDSON 1955).

b) The response of the mammary glands to estrogens

Adrenalectomy of the immature male rat seems to potentiate the response of the mammary glands to estrogenic compounds. REEDER and LEONARD (1944) administered 5 μg of estradiol dipropionate per day for 10 to 12 days to animals that had been adrenalectomized at 22 to 35 days of age. They observed a greater dilation of the lateral buds in these animals than they did in intact rats receiving the same quantities of estradiol. R. JOHNSTON and SMITHCORS (1948) likewise found a greater increase in the area of mammary tissue and more pronounced lobular and alveolar development with diethylstilbestrol in adrenalectomized-castrated rats than in animals that had been subjected to gonadectomy alone.

In the adult male rat, however, adrenalectomy results in a reduced response on the part of the mammary glands to estrogens. TRENTIN and TURNER (1947) injected 5 μg of estradiol in oil per day for 10 days to castrated and adrenalectomized-castrated rats. The administration of estradiol to castrated rats resulted in a marked stimulation of duct and alveolar growth, while in the doubly-operated animals, duct growth alone was produced.

c) *Lactation*

The quantities of milk secreted after parturition by mammals with adrenocortical insufficiency are ordinarily too scanty to meet the needs of the newborn. Such observations have been made in all mammals studied thus far and have been well summarized in previous reviews (PETERSEN 1944, HARTMAN and BROWNELL 1949, FOLLEY 1953, 1956, LYONS et al. 1958).

VII. The thyroid gland

Many studies, too numerous to summarize here, have been made on the effect of adrenocortical hormones on thyroid function. Such studies have particularly concerned themselves with the impact of exogenous adrenal steroids upon some aspect of thyroid function, such as the uptake of iodide by the gland. Relatively few studies have been made of the function of the thyroid gland (and of the biological effects of its hormones) in states of adrenocortical insufficiency. These few studies leave many questions unanswered, for numerous aspects of the problem remain unexplored, and, even among themselves, not all the findings are in agreement.

1. Thyroid function in adrenocortical insufficiency

a) *The secretion of thyrotrophin*

From their studies on thyroid-adrenocortical interrelationships, S. HILL et al. (1950) concluded that adrenal steroids are capable of inhibiting the secretion of thyrotrophin (TSH) by the anterior pituitary gland. In support of this conclusion, they cited instances of patients in whom thyroid function, as measured by I^{131} uptake, was depressed by the administration of ACTH. If TSH was given to such patients, I^{131} uptake and serum protein-bound iodine (PBI) levels returned to normal (and even above normal), despite the continued administration of ACTH. This report (and others to be cited below) suggest that adrenocortical hormones (in particular, those with biological effects like cortisol) normally exert some inhibitory effect (direct or indirect) upon the thyroid gland. However, there is not invariable evidence, in states of adrenocortical insufficiency, of a diminution of any restraint of thyroid function. With respect to TSH-release, however, one piece of evidence is available. G. HARRIS and WOODS (1956) studied the effects of stimulating rabbits in the area of the tuber cinereum. Electrodes implanted in this area were designed to respond to remote signals, and thus observations could be made in unanesthetized, unrestrained animals. Thyroid function was assessed by administering I^{131} to the rabbits and measuring the glandular content of this isotope and the PBI concentrations in blood before and after stimulation of the tuber cinereum. Of 22 animals, only one gave evidence of increased thyroid activity following electrical stimulation of this area. Fourteen of these animals were subsequently adrenalectomized and, of this group, 10 now gave evidence of marked increases in thyroid activity when the tuber cinereum was stimulated.

b) The secretion of thyroid hormones

Thyroid function in states of adrenocortical insufficiency has been variously found to be increased, decreased or essentially normal.

In 1921, MARINE and BAUMANN reported that adrenalectomy or damage to the adrenal glands by freezing resulted in an increased heat production and CO_2 output in approximately three-quarters of the rabbits treated. They also observed thyroid hypertrophy following adrenalectomy of rabbits. The following year, these authors (MARINE and BAUMANN 1922) reported that prior thyroidectomy prevented or greatly diminished the increased metabolic activity found commonly in rabbits after adrenalectomy. They concluded from these studies that adrenocortical secretions normally act as a restraint upon thyroid function, as evidenced by the increased thyroid activity that followed adrenalectomy. This concept was supported by the findings of DAVIS and HASTINGS (1933), who studied the aerobic metabolism of excised abdominal muscles obtained from intact and adrenalectomized mice. The oxygen consumption *in vitro* of abdominal muscle increased after adrenalectomy, and such an increase did not follow prior thyroidectomy. DAVIS and HASTINGS suggested that the increased oxygen consumption reflects increased thyroid activity which, in turn, may reflect a physiological attempt to compensate for the metabolic disturbances occasioned by adrenalectomy. VIDOVIC and POPOVIC (1954) reported that adrenalectomy of the hibernating ground squirrel resulted in a rapid return of body temperature to normal (within 1 to 2 hours). The thyroid gland of this animal, when hibernating, takes up no iodide. Within 3 days after adrenalectomy, thyroidal I^{131} uptake was equivalent to that found in non-hibernating animals. One might interpret these findings as indicating activation of the thyroid gland by adrenalectomy. However, as animals subjected to sham adrenalectomy or prior thyroidectomy were not employed in this study, judgment must be reserved.

In a study on patients with Addison's disease, S. HILL et al. (1950) reported that thyroidal I^{131} accumulations, PBI concentrations and basal metabolic rates (BMR) fell within normal limits, but that the mean values for this group of subjects were slightly below the corresponding mean values for a group of normal individuals.

In castrated rats, the uptake of I^{131} by the thyroid can be doubled by the prior administration of estrogens (FELDMAN 1956). This estrogenic effect is in no way influenced by adrenalectomy.

AUB et al. (1921) reported that the basal metabolism of the cat is reduced 25 % by adrenalectomy; this finding was confirmed by SWINGLE et al. (1931) and by WEBSTER et al. (1931). A similar decline in basal metabolism in the dog following adrenalectomy was made by HARROP et al. (1931a).

In a consideration of these differences between species, the following facts must be borne in mind. The reports of apparent decreases in thyroid activity following adrenalectomy were made in cats and dogs, animals that usually deteriorate rapidly after the removal of the adrenals and die within one to two weeks (NOBLE 1950). It is difficult to know how to interpret data obtained from animals for whom death may be very near; in the preterminal state, one would expect a body-wide decline in cellular activity. On the other hand, it is equally difficult to interpret data obtained from animals that may have less than total adrenocortical insufficiency. MARINE and BAUMANN (1921, 1922) employed animals (rabbits) in whom accessory adrenocortical tissue occurs frequently (KOJIMA 1929, FIROR and GROLLMAN 1933). Although mice are reported to die, on the average, 5 days after adrenalectomy (FIROR and GROLLMAN 1933), the observations of DAVIS and

HASTINGS (1933) reported above were made in mice 40 to 70 days after adrenalectomy. Here, too, there is certainly the possibility of accessory adrenal tissue or incomplete adrenalectomy. These differences could be reconciled, after a fashion, by invoking the possibility of species differences in response to adrenalectomy. In general, however, the investigations cited above seem inadequate to answer the question regarding the status of thyroid function in states of adrenocortical insufficiency.

2. Tissue responses to thyroid hormones in adrenocortical insufficiency

In general, it has been found that the biological effects of thyroid hormones are reduced in adrenocortical insufficiency. Tissues seem to require adrenocortical hormones if they are to respond normally to thyroid hormones.

HOFFMANN et al. (1948) reported that the usual increase in oxygen consumption of the whole animal that follows administration of thyroxine was much less evident in the adrenalectomized rat. If oxygen consumption was first increased by the administration of thyroxine, it declined following adrenalectomy despite continued administration of the thyroid hormone. The normal response to thyroxine could be restored in adrenalectomized rats by giving them adrenocortical extract (Eschatin). In 1951, HOFFMANN et al. extended these studies. Immature male and female rats were given 10 μg of thyroxine per day throughout the entire period of the experiment. Urinary nitrogen excretion rose from 88 to 125 mg per day. At this point, adrenalectomy was performed in some of the animals and they were given access to a 1 % NaCl solution. The remainder of the animals were subjected to sham operation and, in the postoperative period, pair-fed with the adrenalectomized rats. Despite pair-feeding, by the 7th postoperative day, nitrogen excretion in the sham-operated animals had further increased to 143 mg per day and the O_2 consumption of the whole animal at this time was 33 % greater than it had been prior to the daily administration of thyroxine. In the adrenalectomized rats, nitrogen excretion on the 7th postoperative day had decreased to the pretreatment level or lower and O_2 consumption was actually 10 % less than it had been in the intact rats prior to thyroxine administration.

TIPTON et al. (1946) studied the effects of feeding desiccated thyroid substance upon succinoxidase and cytochrome oxidase activities in rat liver. Increases in these activities could first be detected on the 4th experimental day. From that point on, the activities of these enzymes increased steadily. If at any time in the experimental period adrenalectomy was performed, no further increases were noted.

These findings of *reduced tissue responses to thyroid hormones in adrenocortical insufficiency* have a definite bearing on the problem discussed in the previous section. In that section it was reported that many investigators measured thyroid function in adrenocortical insufficiency by various indirect criteria such as heat production or O_2 consumption. Implicit in such measurements is the assumption that the tissues' capacity to respond to thyroid hormones is not significantly modified by adrenocortical insufficiency. In the case of the rat, at least, such an assumption would appear to be invalid. If tissue responses to thyroid hormones are also reduced following adrenalectomy in rabbits and mice, and if changes in heat production, O_2 consumption and CO_2 excretion are valid, albeit indirect, indices of thyroid activity, then the positions of MARINE and BAUMANN (1921, 1922) and DAVIS and HASTINGS (1933) that thyroid activity increases after adrenalectomy gain support. With reduced responsiveness on the part of tissues, a greater quantity of thyroid hormones would be required to achieve any given intensity of biological effect. Again, however, any position in this matter must be tentative until more facts become available.

3. The role of thyroid hormones in adrenocortical insufficiency

Despite the fact that the capacity of the tissues to respond to thyroid hormones is reduced following adrenalectomy of at least one species, the rat, thyroid hormones are capable of intensifying the signs and symptoms of adrenocortical insufficiency and thereby threatening the survival of the organism. Sensitivity to the potentially deleterious effects of an excess of thyroid hormones is increased in adrenocortical insufficiency.

Perhaps the earliest intimation of this fact is to be found in the report of ZWEMER (1926/27b). In his laboratory at that time, cats survived, on the average, for 53 hours following completion of two-stage adrenalectomy. Animals previously thyroidectomized survived for an average of 200 hours after adrenalectomy, while those fed desiccated thyroid substance survived for only 18 hours. Subsequently, numerous reports of crises induced in states of adrenocortical insufficiency by the administration of thyroid hormones were made from studies of animals (CARR and CONNOR 1933, KOELSCHE and KENDALL 1935, KENDALL 1940) and human beings (HARROP et al. 1933b, I. ANDERSON and LYALL 1937, MEANS et al. 1940).

There seems little doubt that an excess of thyroid hormones creates stresses within the organism that increase its need for adrenocortical hormones. THORN (1951) has suggested that thyrotoxicosis of long standing can be an important factor in the initiation of Addison's disease.

Thyrotoxicosis can significantly modify the signs and symptoms of adrenocortical insufficiency. Body temperature, BMR and cardiac size and activity may be greater than those ordinarily encountered in uncomplicated Addison's disease, and hypoglycemic episodes may be more frequent and muscular weakness more pronounced (THORN 1951). While coexistent thyrotoxicosis threatens the well-being of the patient with Addison's disease and makes maintenance therapy more difficult, the signs and symptoms of thyrotoxicosis and its response to appropriate therapeutic measures appear to be unaffected in states of adrenocortical insufficiency (RUPP and PASCHKIS 1957).

In patients suffering from both Addison's disease and hypothyroidism, the disturbances in carbohydrate metabolism so characteristic of adrenocortical insufficiency are less intense. Attempts to treat the hypothyroidism with exogenous hormones must be undertaken cautiously because of the increased sensitivity to thyroid hormones (THORN 1951).

4. The incidence of thyroid dysfunction in Addison's disease

In a study of 180 cases of Addison's disease, FREDERICKSON (1951) reported that 8 of the patients gave evidence of having *hyperthyroidism*. Such an incidence is 10 times greater than one encounters in the general population. RUPP and PASCHKIS (1957), studying coexistence of Addison's disease and thyrotoxicosis, concluded that only 8 authentic cases of this combination of diseases exist in the literature.

BLOODWORTH et al. (1954), in a study of 35 cases of Addison's disease, reported that 13 of their patients gave clinical or laboratory evidence of *hypothyroidism*. Pituitary failure in these patients was ruled out on the basis of laboratory, clinical and autopsy findings. The most striking finding upon autopsy of hypothyroid patients with Addison's disease was a lymphocytic infiltration of the thyroid gland. The authors hypothesized that the hypertrophy of lymphatic tissue so characteristic of adrenocortical insufficiency can manifest itself in the thyroid gland, producing secondary destruction of the thyroid acini.

On the basis of the presently available evidence, it must be concluded that signs of some degree of thyroid dysfunction are not uncommon in Addison's disease.

D. The status of some other tissues

I. Lymphatic tissue

Increases in the mass of the thymus and of the lymph nodes are common sequels to adrenalectomy (NOBLE 1950). The weight gains are considerably greater in the systemic (cervical, epitrochlear, etc.) lymph nodes of the rat than they are in the mesenteric nodes (REINHARDT and HOLMES 1940). The uptake of O_2 and of P^{32} by lymphatic tissue is also increased after adrenalectomy of the rat (SHIBATA 1954). The composition of the soluble proteins in these tissues, however, is little affected; in analyses of the electrophoretic patterns of soluble lymphatic-tissue proteins, PELLEGRINO and TORCIGLIANI (1956) found that slight decreases in two small, fast components were the only changes that could be observed after adrenalectomy.

Attempts have been made to determine if the lymphatic hyperplasia is related causally to the lymphocytosis observed in some species after adrenalectomy (see p. 95). On the 20th day after adrenalectomy, in rats maintained postoperatively on saline, HUNGERFORD et al. (1952) measured an increase in the total cellular concentration in thoracic duct lymph that was possibly of statistical significance ($p < 0.05$). An extension of these observations to the 40th postoperative day was made by SHREWSBURY and REINHARDT (1955); the rate of flow and the cellular content of thoracic duct lymph were found to be normal at this time, although the thymus and the lymph nodes were still enlarged. These authors concluded that the lymphocytosis of adrenocortical insufficiency probably is a reflection of an abnormally long period of survival of lymphocytes in blood.

Contrary to the general view, STOERK (1944) has doubted that thymus weight (expressed in terms of body weight) increases after adrenalectomy. He concluded that optimal thymus weights are rarely achieved in laboratory rats because of the intervention of such growth-inhibiting influences as undernourishment in large litters before weaning, the use of nutritionally inadequate diets subsequent to weaning, and the transient but significant effects of respiratory infections. It was his contention that the apparent hyperplasia of lymphatic tissue in adrenocortical insufficiency is, in reality, simply a manifestation of the normal growth potential of this tissue. In his view, adrenalectomy merely makes the lymphatic tissue less sensitive to the "suboptimal conditions" under which he holds the laboratory rat usually lives. This unique viewpoint has garnered little support from other investigators.

II. The reticuloendothelial system

A. GORDON and KATSH (1949) made a study of the effect of adrenalectomy on the reticuloendothelial system in the rat. They found the concentration of Kupffer cells in the liver to be normal but the concentration of macrophages in the spleen to be reduced.

The phagocytic cells, in general, were slightly subnormal in size; the density of cytoplasm was increased, condensation of the nuclear chromatin occurred and there was reduced evidence of phagocytosis. These changes were seen most prominently in the Kupffer cells in the liver and were less pronounced in the histiocytic elements in the spleen, thymus, lymph nodes and bone marrow.

To test the phagocytic activity of the reticuloendothelial system after adrenalectomy, these authors administered intravenously a colloidal solution of thorium dioxide. They found that the uptake of thorium by the spleen was subnormal, as did REICHARD and GORDON (1956). In addition, they found that thorium produced severe degenerative changes in the macrophages of the spleen that were not seen in intact rats. These changes consisted of vacuolation of the cytoplasm, nuclear condensation and pyknosis, and the coalescence of cells. The injection of thorium likewise produced a decrease in the total number and frequency of Kupffer cells in the liver.

III. Calcified tissues

1. Bone

The influence of adrenocortical insufficiency upon bone is discussed in a previous section (p. 68).

2. Teeth

SCHOUR and ROGOFF (1936) observed that calcification of the dentin in incisors of rats was disturbed after adrenalectomy. Globules could be seen throughout the predentin in the middle third of the incisors. The labial dentin stained more intensely than normally with hematoxylin, and a prominent stratification could be seen in the lingual dentin. No supportive therapy was furnished to these animals after adrenalectomy.

PERLA and SANDBERG (1937) likewise observed abnormalities in the predentin of rats' teeth after adrenalectomy. They noted that the predentin zone either disappeared or there was a marked condensation of the calcified portion. Vacuolation and pyknosis of nuclei were observed in the odontoblasts. Metaplasia and islands of osteoid tissue were observed in the dentin. These animals also had no supportive therapy in the postoperative period.

These findings of changes in the predentin and dentin after adrenalectomy of the rat were not confirmed by APPLEBAUM and SEELIG (1955); the authors attributed this discrepancy to the fact that their animals were supplied with 0.85% NaCl in the postoperative period. They confined their studies to fully erupted, first lower molars. They observed a loss of supporting bone in the lower jaw in the vicinity of the molars, and an unusually large amount of marrow in the septal bone. Within the molars themselves, the pulpal blood vessels were seen to be dilated and there was evidence of hemorrhage.

DOMM and WELLBAND (1960) observed that, 6 weeks after the adrenalectomy of 39-day old female rats, there was a significant reduction in the rate of growth of incisors.

IV. Pigmentation of the skin

1. Man

ADDISON (1855) reported that there is usually a generalized increase in skin pigmentation in adrenal insufficiency, but that vitiligo may sometimes develop. Although vitiligo is relatively rare, it has occurred in well-documented instances of Addison's disease (MOEHLIG 1947, SORKIN 1949, THORN 1951) and may be the only manifestation of a change in skin pigmentation. T. HALL et al. (1953) described the use of reflectance spectrophotometry for the evaluation of intensity of skin pigmentation. JEGHERS (1944) reviewed pigmentation of the skin in man, and emphasized the yellowish pallor of patients with hypopituitarism as a feature distinguishing them from individuals with primary adrenal insufficiency. Abnormalities of pigmentation of the skin and mucosa may be especially difficult to detect in Negroes (BERGNER and EISENSTEIN 1951). Administration of ACTH to individuals with normal adrenal function may lead to considerable darkening of the skin (T. HALL et al. 1953, GILBERT-DREYFUS et al. 1953).

LERNER et al. (1954) and BEERMAN and COLBURN (1956) discussed the relationship of melanocyte stimulating activity to the skin pigmentation of Addison's disease and its reversal by treatment with cortisone or cortisol. JOHNSSON and HÖGBERG (1952, 1953) and SULMAN (1953) found increased concentrations of a melanocyte stimulating substance in the blood of patients with adrenal insufficiency. SHIZUME and LERNER (1954) found an increased rate of urinary excretion of such a substance in patients with adrenal insufficiency, and that it was reduced by cortisone administration. HALKIER (1953) was unable to confi m the observations of the previous authors, perhaps because a different species of frog was used in the bioassay. SULMAN (1956) investigated the melanocyte stimulating activity of the blood of 1200 individuals; he found it always increased in patients with Addison's disease and also under other circumstances in which ACTH output could be expected to be elevated.

The relationship between ACTH and the melanocyte stimulating hormone is discussed by JOHNSSON and HÖGBERG (1952, 1953), SULMAN (1953), LERNER et al. (1954), LERNER and TAKAHASHI (1956), LI (1956) and DEUTSCH and MESCON (1957). The question of their chemical interrelationship is not yet completely resolved (STEELMAN and GUILLEMIN 1959, B. HUDSON 1960, SCHALLY and GUILLEMIN 1960). An amino acid sequence is common to melanocyte stimulating hormones and ACTH and apparently accounts for the melanocyte stimulating activity of ACTH (GESCHWIND 1959).

2. Animals

The changes in pigmentation of the skin and mucous membranes commonly seen in human beings with adrenal insufficiency have never been described in animals.

SULMAN (1953) studied a variety of species (chickens, dogs, mice, rats, golden hamsters), and found no change in melanocyte stimulating activity in the plasma, following adrenalectomy, in any of them. JOHNSSON and HÖGBERG (1953) reported a 4-fold increase in melanocyte stimulating activity in the plasma of adrenalectomized rats — which, however, was not reduced by treatment of the animals with cortisone.

V. Growth of hair and its pigmentation

Adrenalectomy has been found to be a potent stimulus to the growth of hair in rats (BUTCHER 1937, DIEKE 1948, CHASE 1954). Investigators of this phenomenon usually remove the hair on the back of the animal by shaving or with the aid of a depilatory at the time of adrenalectomy; the rate of regrowth is then determined with the aid of photographs or drawings.

In black or brown rats maintained on saline after adrenalectomy, RALLI and GRAEF (1943) noted a marked but transitory hyperplasia of the hair bulbs and follicles; it was most prominent on the 16th postoperative day and began to subside after the 30th day. ZECKWER (1953a) observed a period of latency following adrenalectomy that was 12 to 23 days in duration. After this time, hair grew at an accelerated rate, but diffusely instead of in normal cycles. This investigator also observed (ZECKWER 1953b) that the rate of hair growth of adrenalectomized rats joined in parabiosis to intact rats was somewhat delayed in comparison to that of a single adrenalectomized animal, suggesting that the intact member of the pair had provided a hormone that inhibited hair growth.

The fact that a tissue will grow more rapidly after adrenalectomy is a notable finding, because retardation or cessation of growth processes is the common sequel of this operation (see p. 68ff.). Hair growth occurs in the adrenalectomized animal at a time when the rate of bodily growth is subnormal, and the animal appears overtly sick (BAKER 1951). The impetus to hair growth supplied by adrenalectomy prevails even in the presence of normally inhibitory factors such as underfeeding (BUTCHER and RICHARDS 1939), or the administration of estradiol to ovariectomized rats (BAKER and WHITAKER 1949) and to male rats (INGLE and BAKER 1951).

In view of the many reports of accelerated regrowth of hair in adrenalectomized rats, the findings of L. STEIN and WERTHEIMER (1941) are of particular interest. They sought to determine if adrenalectomy affected the rate of hair loss in the rat. Studying male and female rats maintained on the Rubin-Krick solution postoperatively, they measured the daily molting of hair gravimetrically. They found that a 10-fold increase in hair loss occurred in male rats after adrenalectomy and a 50% increase occurred in female rats. (Disturbances in the skins of these animals ranged from the presence of abnormal amounts of dandruff to extensive areas of ulceration.) All of these changes were transitory and persisted for only 3 to 4 weeks. Changes similar in type and magnitude occurred after removal of only the adrenal medulla, and they could be prevented by the administration of Adrenalin.

Adrenocortical insufficiency in human beings, especially in women, leads to a reduction in the rate of growth of pubic and axillary hair (ALBRIGHT et al. 1942a, KEPLER et al. 1943, KINSELL et al. 1954). HUGGINS and SCOTT (1945) noted that pubic and axillary hair became sparse 2 months after adrenalectomy of a man who had been previously gonadectomized because of prostatic cancer.

The pigmentation of the hair is usually not affected by the development of adrenocortical insufficiency. However, it has been reported that adrenalectomy can influence changes in hair coloration that are the result of certain dietary deficiencies. RALLI and GRAEF (1943) found that in black rats which had become gray while on a pantothenic acid-deficient diet, pigmentation of the hair was restored by adrenalectomy, appearing a few days after acceleration in the rate of growth of the hair became evident. The pigment was limited to the hair buds and follicles. By chemical studies of the skin, SPOOR and RALLI (1944) demonstrated that the decrease in melanin content resulting from a diet deficient in pantothenic acid was partially restored by adrenalectomy. A diet deficient in copper also leads to graying of the hair of black rats, and this can be prevented by either adrenalectomy or hypophysectomy, according to HUNDLEY and ING (1951). Since either of these two maneuvers for reducing adrenal activity restores the hair pigment, the mechanism for the pigment changes could not be the same as that observed in man.

E. Susceptibility to noxious influences

I. Increased susceptibility

It has been long known that, in states of adrenocortical insufficiency, the organism has a reduced capacity for withstanding stress. It shows an increased susceptibility to the noxious effects of chemical agents such as histamine and bacterial toxins; it is likewise more susceptible to changes in its environment such as extremes of heat and cold, or reduction in the O_2 content of the inspired air. It succumbs more easily to the effects of ionizing radiations. All of these findings have been treated in previous presentations (HARTMAN and BROWNELL 1949, NOBLE 1950, 1955, GERMUTH 1956, SAUNDERS 1956).

The cellular basis for this enhanced susceptibility to noxious influences in adrenocortical insufficiency is not well understood. In the opinion of one investigator (GERMUTH 1956), many stresses impose demands on the cardiovascular system.

In states of adrenocortical insufficiency, the cardiovascular system is unable to cope with these demands and, hence, the organism's ability to withstand stresses is reduced.

II. Unchanged or decreased susceptibility

In view of the many findings of enhanced susceptibility to a wide variety of noxious influences in states of adrenocortical insufficiency, it is of particular interest to note that the ability of the organism to cope with certain stresses may be unchanged or even increased after adrenalectomy.

Viral infections. Several investigators have observed that *the course of certain viral infections in mice is not altered by adrenalectomy.* SOUTHAM and BABCOCK (1951) noted that the susceptibility of adrenalectomized mice to lethal infection by West Nile, Ilheus and Bunyamwera viruses was no different than that of intact animals. KUDO et al. (1954) made similar observations in adrenalectomized mice infected with the Japanese encephalitis virus. J. SMITH et al. (1951) found that adrenalectomized mice were slightly less susceptible to the effects of the pneumonia virus of mice than were intact animals; adrenalectomy seemed to inhibit viral growth somewhat. All of the adrenalectomized animals employed in these three studies were maintained on saline in the postoperative period.

The findings that adrenalectomy does not increase the susceptibility of mice to viral infections stand in sharp contrast to the findings of a subnormal resistance to *bacterial* infections in adrenalectomized rats (STEINBACH 1929, H. ROBINSON et al. 1953).

Wound healing. One response of the intact organism to certain stresses, such as severe burns, is an increased urinary output of nitrogen; after adrenalectomy of the rat, this response is abolished or nitrogen retention develops (SELLERS et al. 1950). In general, the failure of the adrenalectomized animal to demonstrate certain usual responses to stress, coupled with the retardation of many growth processes (see p. 68ff.) might lead one to suspect that wound healing would not proceed normally in states of adrenocortical insufficiency. This has not been found to be true in the case of the rat.

ALRICH et al. (1951) performed experimental laparotomies in rats that had been adrenalectomized for one week. The tensile strength of the wound in the adrenalectomized animals at various times after laparotomy was comparable to that in the intact rats. CHASSIN et al. (1954) performed similar experiments in rats that had been adrenalectomized for 15 days. On the 7th day after laparotomy, the tensile strength of the healing incision in adrenalectomized rats was actually significantly greater than it was in intact rats. The adrenalectomized rats used in these two studies were furnished with saline in the postoperative period.

Oxygen poisoning. The inhalation of oxygen at pressures greater than one atmosphere for an appreciable period of time produces tracheobronchitis and disturbances in the function of the central nervous system that may lead to convulsions (GOODMAN and GILMAN 1955, p. 912). Although adrenalectomized animals have a subnormal tolerance for hypoxia (NOBLE 1950), they have an unusual resistance to the toxic effects of O_2. Adrenalectomized rats survive in an atmosphere of O_2 at a pressure of 6 atmospheres for a greater period of time than do intact animals (GERSCHMAN et al. 1954, 1955). In the rat, adrenalectomy confers greater protection against the effects of O_2 at high pressures on the central nervous system than it does against the effects on the lungs (D. TAYLOR 1958). The rats used in all these studies were maintained on saline after adrenalectomy, although D. TAYLOR (1958) found no significant difference in the responses of adrenalectomized rats supplied with water.

Diagnosis of adrenal insufficiency

A. General symptoms and signs

The possibility of chronic adrenal insufficiency should be considered in any patient who complains of *weakness, weight loss, anorexia and vomiting*, especially when increased pigmentation of the skin also occurs. [The interpretation of pigment changes in Negro patients may be difficult and misleading (CONN and MATTHEWS 1946, BERGNER and EISENSTEIN 1951).] According to THORN (1951) abdominal pain, constipation or diarrhea are relatively infrequent symptoms, occurring in only 20—30 % of his patients. An increased appetite for salty foods is also ground for suspicion, although habit or renal disease may produce salt-craving.

In some instances the neurological component of adrenal insufficiency may be so prominent that the presence of a brain tumor is suspected. Although such symptoms may be due to hypoglycemia and dehydration (THORN 1951) they may occur when the concentration of glucose in the blood is normal (see also brain excitability, p. 111). *Papilledema and increased intracranial pressure* may be prominent findings in a patient with unsuspected Addison's disease (BOUDIN et al. 1950a, JEFFERSON 1956); their occurrence may be the consequence of water intoxication (BOUDIN et al. 1950b). An erroneous diagnosis of brain tumor, pseudotumor or thrombosis of the intracranial dural sinuses may have disastrous consequences for the patient (JEFFERSON 1956). Some of the reports of papilledema in Addison's disease describe patients who were being treated with DOC and sodium chloride; confusion in these instances would arise only from failure to realize that the neurological signs could be the result of excessive therapy (THORN 1951, WALSH 1952; see also treatment, p. 141). Since there is an association of adrenal insufficiency with certain *organic neurological disorders* (see also etiology, p. 52), the possibility of Addison's disease must be considered in the evaluation of patients whose complaints are primarily neurological.

Hypotension and a small diameter of the heart are to be expected except in patients in whom hypertension or heart disease were present before the onset of adrenal insufficiency. A *small pulse pressure* is also to be expected. A *moderate decrease in the basal metabolic rate* is a common finding, and THORN (1951) has emphasized that a normal basal metabolic rate in a patient with adrenal insufficiency is suggestive of coexisting thyrotoxicosis. Adrenal insufficiency should always be suspected (1) in an infant with virilism or pseudohermaphroditism (to be discussed in Part 3 of this Volume), congenital adrenal hyperplasia with electrolyte imbalance, and (2) in patients with moniliasis involving the nails and corners of the mouth.

Laboratory findings of *elevated BUN or NPN, and of decreased* Na^+ *and* Cl^- *and elevated* K^+ *concentrations in the plasma* provide support for the diagnosis. However, the concentrations of these substances may be so slightly altered as not to be detectable in patients with chronic adrenal insufficiency in reasonably good balance. The existence of moderate anemia is not especially helpful for diagnosis. *Relative lymphocytosis with a low total white blood cell count* is commonly found in adrenal insufficiency and in panhypopituitarism (DE LA BALZE et al. 1946, LABHART 1957, p. 312); while the number of eosinophils is usually increased. BÁEZ-VILLASEÑOR et al. (1948) stated that lymphocytosis and neutropenia occur, but that the total white blood cell concentration is within normal limits.

In most instances more extensive diagnostic efforts must be undertaken, as indicated below. Detailed outlines for the investigation of adrenal status are available (THORN 1951, THORN et al. 1951, N. TALBOT et al. 1952, VENNING et al. 1954, FORSHAM and THORN 1955, SOFFER 1956, LABHART 1957, WILKINS 1957).

B. Evaluation of adrenal function

I. Water and electrolyte metabolism

1. Response to ingestion of water

The Robinson-Kepler-Power water test is based on the inability of patients with adrenal insufficiency to respond to a water load with prompt diuresis. Details of its performance are given by F. ROBINSON et al. (1941), THORN (1951), N. TALBOT et al. (1952).

Test: Ingestion of extra NaCl should be eliminated for 3 days preceding the test. Fluids are withheld after 6 P. M. of the day immediately before the test. The patient voids at 10 P. M. and all urine passed until 7 A. M. is measured. Between 7 and 8 A. M. the patient drinks 20 ml of water per kg, and voids at hourly intervals thereafter until 11 A.M. If the volume of a single hourly voiding exceeds that of the urine passed during the night, adrenal insufficiency may be considered ruled out. If diuresis does not occur, a sample of blood is drawn for blood urea nitrogen and Cl^- determinations, and the same determinations are made on the urine collected during the night. The following ratios are calculated:

$$\frac{\text{urinary urea nitrogen}}{\text{blood urea nitrogen}}, \quad \frac{\text{plasma Cl}^-}{\text{urinary Cl}^-}, \text{ and } \frac{\text{volume largest hourly urine}}{\text{volume urine 10 P. M. — 7 A. M.}}.$$

"A", the product of multiplication of these three ratios is less than 25 in indviduals with adrenal insufficiency.

This test is relatively free of danger to the patient, although it may be unpleasant. It cannot be considered highly significant, since failure of water diuresis may occur in patients with hypothyroidism, intestinal insufficiency, cirrhosis or nephritis. In renal disease, especially of the salt-losing type, "A" may also be low (M. S. LEVY et al. 1946, CRISPELL et al. 1954).

2. Concentrations of sodium and chloride in plasma

The concentrations of Na^+ and Cl^- are usually low in adrenal insufficiency, but they may also be abnormal in other conditions.

Low plasma Na^+ or Cl^- may be found in patients with tuberculosis (THORN et al. 1940a, SIMS et al. 1950), acute porphyria (PRUNTY 1949), liver disease (BONGIOVANNI and EISENMENGER 1951), chronic renal disease (THORN et al. 1944, JOHNSSON et al. 1951) and in a variety of other disorders (WINKLER and CRANKSHAW 1938, DANOWSKI et al. 1955). Water intoxication may be mistaken for adrenal insufficiency if reliance is placed entirely on plasma Na^+ concentration (WYNN and ROB 1954).

PERLMUTTER (1956) suggested that simultaneous determination of Na^+ concentrations in plasma and urine might facilitate the diagnosis of adrenal insufficiency; in most non-adrenal causes of low plasma Na^+, urinary Na^+ is also low, although excess urinary excretion of Na^+ is found in patients with salt-losing renal disease.

3. Inability to conserve sodium

a) Restriction of sodium intake

Limitation of Na^+ intake to 200 mg or less per day may reveal adrenal insufficiency (HARROP et al. 1933c, NETTROUR and RYNEARSON 1934), when plasma electrolytes are normal. The individual with normal adrenal function is able to maintain a normal concentration of Na^+ in the plasma for 6 days, while reducing the urinary excretion of Na^+ to a very small amount. This test is not without danger, and may lead to death (GARVIN and REICHLE 1940). The patient's weight and blood pressure should be determined daily and *the test should be terminated promptly if any adverse signs or symptoms develop.*

Although patients with pulmonary tuberculosis may have low plasma Na^+ and Cl^- concentrations, they are able to conserve Na^+ on a restricted diet (SIMS et al. 1950) and respond in a normal manner to the Cutler-Power-Wilder test (THORN et al. 1940a).

b) Use of carbonic anhydrase inhibitor

HAMWI and BROWN (1956) proposed the use of a carbonic anhydrase inhibitor (Diamox), which promotes excretion of Na^+ and water, as a diagnostic test for adrenal insufficiency as an alternative to the restriction of Na^+ intake. They applied this test to 5 patients with Addison's disease and induced crises in 4 of them. This result suggests that *the dangers of the test are too great to justify its use.*

4. Potassium tolerance

The Cutler-Power-Wilder test (CUTLER et al. 1938b) requires a shorter period of time than does Na^+ restriction but is more *hazardous for the patient.* A low Na^+ diet is supplemented with potassium citrate [42 mg per pound (454 g) of body weight]. On the third day of such a program, urine is collected for a 4-hour period in the morning. In healthy individuals, Cl^- excretion is less than 60 meq/l; in Addisonians, the excretion is greater than 60 meq/l. Details of the diet for this test are given by WILLSON et al. (1942) and THORN (1951), and a program modified for children is outlined by N. TALBOT et al. (1952). WILLSON et al. (1942) reported one false-positive response to this test when applied to 44 non-Addisonian patients, and one atypical response among 16 individuals with Addison's disease; in 10 of these 16 patients the test had to be discontinued because a crisis was induced.

I. JAFFE and BYRON (1942) described a simplified test for K^+ tolerance, which depends on a rise in plasma K^+ concentration following administration of a single oral dose of K^+ [10 mg/lb (22 mg per kg)]. They concluded that the test was of limited usefulness since abnormal elevations of plasma K^+ were found in patients who did not have adrenal insufficiency by other criteria.

5. Composition of sweat

CONN (1949) reported elevated Cl^- concentration in the thermal sweat of untreated patients with Addison's disease — 110 meq/l, as compared to 60 meq, the greatest concentration found for a healthy person. LOCKE et al. (1951) found that Cl^- concentration increased with the rate of sweating, while K^+ decreased. There was considerable overlap between the Na^+ and K^+ concentrations found in normals and those of patients with poorly-controlled Addison's disease. They found the best correlation with adrenal status was provided by an index calculated from the rate of sweating and Cl^- concentration of the sweat. KLEEMAN et al. (1954) further defined some of the variables that must be controlled for confidence in interpretation of the results of sweat tests. GRØNBAEK (1955) collected sweat under carefully controlled conditions and found that the ratio of Na^+ to K^+ ranged from 0.8 to 2.4 (mean 1.5) in 74 normal children; the ratio was high in their 2 patients with adrenal insufficiency as well as in one with pancreatic fibrosis.

6. Composition of saliva

FRAWLEY and THORN (1951) studied the composition of saliva produced by chewing on paraffin. In patients with adrenal insufficiency the concentration of Na^+ averaged 60 meq/l and of K^+ 15 meq/l, compared to 22 meq/l of Na^+ and 20 meq/l of K^+ found in healthy individuals. The overlap was so large that they proposed the use of the ratio of Na^+ to K^+. This ratio for normal persons was

0.33 to 2.1 (mean 1.3) and in 4 untreated patients with Addison's disease 2.2 to 8.1 (mean 5.0). It is noteworthy that when Na^+ intake was restricted, salivary Na^+ tended to fall as plasma Na^+ fell.

Shannon and Prigmore (1958) reported that the normal value for salivary Cl^- is 23 $\pm$ 4 meq/l for men and 24 $\pm$ 4 for women when no stimulus is used; in both sexes stimulation of flow by chewing on rubber bands yields a higher Cl^- concentration. A. G. White et al. (1955) found the Na^+ to average 26.4 meq/l, Cl^- 29.0 and K^+ 19.7 in 73 normal people.

II. Carbohydrate metabolism

1. Response to fasting

Fasting for a 24-hour period is probably the least dangerous procedure for evaluating carbohydrate metabolism; in normal individuals the blood sugar is well-maintained and may be somewhat higher toward the end of the 24-hour period than it was at the beginning of the fast (Craig et al. 1958). In patients with adrenal insufficiency the blood sugar *falls* during this period, and may become low enough to produce symptoms of hypoglycemia or convulsions. It is wise to withhold food, but not water, following an evening meal so that, should hypoglycemia ensue, it would be during the daytime hours when it is easiest to keep a patient under close observation (Thorn 1951, Thorn et al. 1951, N. Talbot et al. 1952).

2. Glucose tolerance tests

Glucose tolerance tests are a source of danger to the patient with adrenal insufficiency, since the most characteristic abnormality is the development of hypoglycemia 4 or 5 hours after the ingestion of glucose, 1.75 mg/kg (Thorn et al. 1940b), whereas in the normal individual the blood sugar will have returned to approximately the fasting level by the 3rd hour. The intravenous glucose tolerance test has the additional hazard of a febrile reaction, which probably represents water intoxication (Thorn 1951).

3. Insulin tolerance tests

The intravenous insulin tolerance test, while far from innocuous, is especially useful since its short duration allows the patient to be under constant observation. In the normal individual, mild symptoms of hypoglycemia may develop in 20 to 30 minutes, but the blood sugar will be normal 60—90 minutes after insulin administration. However, *no more than 0.03 to 0.05 U of insulin per kg should be given when adrenal insufficiency is suspected,* even though Perlmutter and Mufson (1951) found that 2 of their patients could tolerate the full dose of 0.1 U per kg. Should symptoms of hypoglycemia develop during the 90 minute test it can sometimes be treated most safely by the administration of epinephrine. If intravenous administration of glucose is needed, a later episode of hypoglycemia can be averted by giving adrenal extract or cortisone at the same time. *It is sometimes extremely difficult to restore a patient in whom hypoglycemic shock has occurred.*

F. Engel and Scott (1950) proposed an insulin-glucose tolerance test, which has the advantage of providing glucose (0.8 mg/kg by mouth) 30 minutes after the intravenous administration of 0.1 U of insulin/kg. In normal individuals, the blood sugar rises above the fasting concentration by 30—60 minutes after the ingestion of glucose while in none of those with adrenal insufficiency does it return to the fasting concentration by 60 minutes, although hypoglycemia does not occur (Thorn et al. 1940b, Fraser et al. 1941b, Thorn 1951, N. Talbot et al. 1952).

In patients with hypopituitarism, the fall in blood sugar in response to insulin is more rapid than in normal individuals or in patients with Addison's disease, and they demonstrate the same failure of response to hypoglycemia as do patients with adrenal insufficiency (FRASER et al. 1941b).

III. Steroid excretion

1. 17-Ketosteroids

Excretion of 17-ketosteroids is almost always low in patients with adrenal insufficiency, although occasionally it may be normal (H. MASON 1955). The excretion of these substances is also decreased in gout (WOLFSON et al. 1949), hypothyroidism (G. WILLIAMS et al. 1957), malnutrition, chronic illness, etc., so that this determination is useful especially in eliminating the possibility of Addison's disease if normal values are found (FRASER et al. 1941b, N. TALBOT and BUTLER 1942, MCCULLAGH et al. 1948, AMPERS et al. 1953, HUDSON et al. 1953). The values obtained for an individual may have to be compared to those for his own population group, determined in the same laboratory. Among West Africans the range was 4—14 mg per 24 hours (mean 8) compared to 10—20 (mean 16) for Europeans living in the same environment (BARNICOT and WOLFFSON 1952, BARNICOT and LADELL 1954). Indians excreted 7—18 mg of 17-ketosteroids daily (mean 11) while Europeans living in the same area excreted 12—31 (mean 21) (H. FRIEDMANN 1954).

Since there is substantial diurnal variation in the excretion of 17-ketosteroids, the use of only overnight collections of urine can give unduly low values for the 24-hour excretion (PINCUS 1943).

Pre-adolescent children excrete such small amounts of 17-ketosteroids that this measurement is not useful when one is seeking evidence of underactivity of the adrenal cortex (KENIGSBERG et al. 1949, N. TALBOT et al. 1952, WILKINS 1957).

2. 17-Hydroxycorticosteroids

Except for the neonatal period, excretion of 17-hydroxycorticosteroids is related more to body size than to age (KING and MASON 1950, N. TALBOT et al. 1951, N. TALBOT et al. 1952), and is a more direct index of adrenal function than is the rate of excretion of 17-ketosteroids (BIRKE et al. 1958). Corticosteroid excretion is distinctly lower during the first 10 days of age (VENNING et al. 1949, MATSON and LONGWELL 1949, READ et al. 1950). Diurnal variation in the excretion of corticosteroids precludes use of less than a 24-hour urine collection as the basis for calculating the daily output (MIGEON et al. 1956).

The excretion of corticosteroids is reduced in patients with either primary or secondary adrenal insufficiency (N. TALBOT et al. 1945, HEARD et al. 1946, N. TALBOT et al. 1947, VENNING and BROWNE 1947, DAUGHADAY et al. 1948, FORSHAM et al. 1948, FORBES et al. 1950, THORN 1951, N. TALBOT et al. 1952, LAIDLAW et al. 1955b, WILKINS 1957). However, there is some overlap in the values obtained for normal and abnormal individuals (H. MASON 1955), and the resting rate of excretion may be within the normal range in patients who actually do have adrenal insufficiency (ABU HAYDAR et al. 1958). COPE and BLACK (1959) proposed that the excretion of 17-ketogenic steroids may be very useful in detecting adrenal hypofunction. The actual values found depend on which of several methods is used for the determination.

3. Aldosterone

LUETSCHER and AXELRAD (1954) found no sodium-retaining corticoid (aldosterone) in the urine of 3 patients with adrenal insufficiency. Patients who have been adrenalectomized excrete no aldosterone (DUDLEY et al. 1957, LIEBERMAN and LUETSCHER 1960); the amount excreted by individuals with hypopituitarism is somewhat reduced from normal and increases only moderately when Na^+ intake is restricted (MULLER et al. 1958, LIEBERMAN and LUETSCHER 1960). The values for normal individuals are 0.8 to 23.5 μg per 24 hours (NEHER and WETTSTEIN 1956, AYRES et al. 1957, LIEBERMAN and LUETSCHER 1960).

Determination of the aldosterone content of urine is still difficult, so that this measurement is not yet practical as a routine procedure although potentially of great value.

IV. Plasma steroids

In the absence of stress, the concentration of 17-hydroxycorticosteroids in plasma is probably the most direct index of resting adrenal function and is quite constant throughout life, except for the very low concentration characteristic of the neonatal period (KLEIN et al. 1954). The findings for healthy individuals, in μg/100 ml of plasma, are given as 4 to 10 (NELSON et al. 1951, NELSON and SAMUELS 1952), 6 to 20 (SILBER and PORTER 1954), and 4 to 32 (WALLACE et al. 1955). The range of variation found by others is somewhat larger, 0 to 28, so that finding a low value in the fasting state is not in itself adequate to establish the diagnosis of adrenal insufficiency (ELY et al. 1953, LANMAN 1953, PERKOFF et al. 1954, BERGSTRAND and GEMZELL 1955, SILBER and BUSCH 1956). SWEAT (1955) and EIK-NES et al. (1955) described plasma concentrations well within the normal range in some patients with adrenal insufficiency (see p. 138).

Since plasma concentrations of corticosteroids are highest in the early morning hours (MIGEON et al. 1956, REDDY et al. 1956), it is evident that blood samples should be obtained in the morning, in order that meaningful interpretation of the findings can be made.

V. Response to epinephrine

The epinephrine test at one time was thought to measure the release of ACTH, which in turn stimulated the output of adrenal steroids, leading to a decrease in total eosinophil count (RECANT et al. 1948, 1950, N. TALBOT et al. 1952). This test can no longer be considered reliable, for BEST et al. (1952) found that many normal persons did not respond in this fashion to epinephrine and that a marked eosinophil decline was induced in 5 adrenalectomized patients who were receiving cortisone therapy.

VI. Response to ACTH

The response to administration of ACTH is helpful in establishing a definitive diagnosis of Addison's disease, distinguishing it from adrenal insufficiency secondary to decreased pituitary activity. The tests applied, discussed in increasing order of specificity are changes in (1) circulating eosinophils, (2) urinary excretion of uric acid, (3) urinary excretion of steroids and (4) plasma corticosteroid levels.

1. Circulating eosinophils

The percentage decrease in total eosinophil count was one of the early criteria used in evaluating the response to ACTH. FORSHAM et al. (1948) and THORN et al. (1948) proposed that *a decrease of more than 50 % 4 hours following intramuscular administration of ACTH be considered a normal response.* A decrease of 90 % was observed 12 hours after the initiation of an 8-hour infusion containing 20 mg (approximately 20 U) of ACTH (Renold et al. 1951). Intravenous administration of ACTH on 2 successive days provoked an 80—100 % decrease in eosinophils in normal individuals, but no significant fall in patients with adrenal insufficiency (Laidlaw et al. 1955b).

There are several precautions to be observed in performing eosinophil counts (Randolph 1949, Best et al. 1952, Bonner 1952). The most important criticism of the test is the observation that a decline in eosinophil count may occur in individuals who do have adrenal insufficiency (Thorn et al. 1951, Best et al. 1953, K. West 1953). Bliss et al. (1954) found that the eosinophil count might decrease even when there was no rise in plasma corticosteroids. De Mowbray and Bishop (1953) reported one patient in whom a 75 % fall in circulating eosinophils was observed 4 hours after 25 mg of ACTH were given intramuscularly; at autopsy 14 days later no histological evidence of adrenal tissue could be found.

Errors in the other direction may also occur, with no decrease in eosinophils in persons with normal adrenal function (Best et al. 1953); most of these misleading responses apparently can be attributed to local inactivation of the intramuscularly-administered ACTH (Renold et al. 1951, D. Jenkins et al. 1955).

The eosinophil count is an unreliable index of adrenocortical function.

2. Excretion of uric acid

A 50 % increase in the ratio of urinary uric acid to creatinine has also been found to be an unreliable index of adrenal reaction to ACTH. Early work with this procedure was promising (Forsham et al. 1948, Thorn et al. 1948). However, normal responses have been elicited in patients with Addison's disease, and failure of increase of uric acid excretion has been found in normal persons (de Mowbray and Bishop 1953). Taussky et al. (1951) pointed out that the response depends in part on the rate of uric acid excretion before ACTH is given; if this is high, little increase is found. D. Gordon et al. (1954) found that the ratio of uric acid to creatinine in the urine correlated poorly with other indices of adrenal function.

3. Excretion of steroids

An increase in the urinary excretion of 17-ketosteroids and of corticosteroids in response to ACTH is a useful indication of integrity of the adrenal cortex (Browne 1943, H. Mason et al. 1947, Sandberg et al. 1953, D. Gordon et al. 1954, Labhart 1957, Hartenbach 1958). This procedure originally involved giving ACTH intramuscularly at 6-hour intervals for 2 days (Forsham et al. 1948, Thorn 1951).

It is now more usual to administer the ACTH by infusion during a 4- to 8-hour period (Laidlaw et al. 1955b, D. Jenkins et al. 1955) or to take advantage of the prolonged action of ACTH gel or zinc-ACTH given intramuscularly (Laidlaw et al. 1955b, Jeanneret et al. 1954, de Fillipis and Young 1957, Gold and Starr 1959). Infants increase their corticosteroid excretion (Read et al. 1950) when given ACTH, even though their resting levels are very low. In older children the relative increase in 17-ketosteroid excretion is less than that found in adults (Gottfried et al. 1957).

Abu Haydar et al. (1958) reported that 7 patients with Addison's disease showed no response to ACTH, although their unstimulated excretion of 17-hydroxycorticosteroids was normal. *In patients with hypopituitarism, several days of ACTH treatment may be required before a response occurs* (D. Gordon et al. 1954, Laidlaw et al. 1955b).

4. Plasma corticosteroids

The change in concentration of corticosteroids in plasma 4 hours after the initiation of an infusion of ACTH is a definite measure of adrenal function. [EIK-NES et al. (1958) have compared the responses to different routes of administration and different preparations of ACTH.] In patients with adrenal insufficiency no rise at all may be found, or the increase may be substantially less than the 40 μg expected for normal individuals (CHRISTY et al. 1955, EIK-NES et al. 1955, 1958, SWEAT 1955, PETERSON 1957, SANDBERG et al. 1957).

The usefulness of this test can be judged by the fact that in patients with adrenal insufficiency who have normal resting plasma concentrations of corticosteroids, no increase is found when ACTH is given intravenously (EIK-NES et al. 1955). The test is also valuable for evaluating patients who are receiving corticosteroid therapy. If pituitary function is suppressed but the adrenal gland itself is potentially responsive, administration of ACTH for several days will promote an increase in plasma corticosteroid concentration (LAIDLAW et al. 1955b, SANDBERG et al. 1957).

C. Evaluation of ACTH production

I. Indirect

Among the compounds that inhibit adrenocortical enzyme systems, CHART et al. (1958) described 1,2-bis(3-pyridyl)-2-methyl-1-propanone (Su 4885; Metopirone, Ciba), which leads to a decrease in the production of neutral reducing steroids and an increase in the adrenal weight of dogs. J. JENKINS et al. (1958), extending these observations to man, found a decrease in the excretion of urocortisone and urocortisol and an increase in the excretion of tetrahydro-11-deoxycortisol when Su 4885 was given to a patient with Cushing's syndrome.

LIDDLE et al. (1958) suggested that Su 4885 might act by inhibition of 11β-hydroxylation. They confirmed this hypothesis by analysis of the steroids produced by bovine adrenal glands when the compound was added to an *in vitro* system, and proposed its potential usefulness as a test for ACTH reserve. They found no increase in steroid excretion in patients with known pituitary hypofunction. LIDDLE et al. (1959) extended their observations and confirmed that either oral or intravenous administration of Su 4885 could be used to evaluate the ability of the pituitary gland to release ACTH in response to a decreased adrenal output of cortisol as a consequence of the inhibition of 11β-hydroxylation. They used an increase in the excretion of Porter-Silber chromogens (17-hydroxysteroids) as an indication of ACTH release; subsequent response to exogenous ACTH was used to distinguish pituitary from adrenal insufficiency in those patients in whom no change in corticosteroid excretion had occurred. HOLUB et al. (1959) have applied this test to patients in whom long-term steroid therapy might have suppressed pituitary function; in 3 of 4 patients the response was normal. GOLD et al. (1960a) found that the excretion of 17-ketogenic steroids increases about twice as much as the Porter-Silber chromogens, and therefore suggested that 17-ketogenic steroids be measured in the performance of this test. An increase in the plasma concentration of non-conjugated 17-hydroxycorticosteroids was found by GOLD et al. (1960b) when they used a method modified so as to improve the estimation of 11-deoxycortisol, which is the major compound secreted after giving this inhibitor.

From the work of the investigators cited above, *it appears that Su 4885 is a useful substance for evaluation of pituitary function.* Some unpleasant reactions to the compound have been noted, especially dizziness, nausea and epigastric discomfort, when large amounts were given either intravenously or by mouth (GOLD et al. 1960a).

II. Direct measurement of ACTH in blood

In patients with primary adrenal insufficiency who are not being treated, the presence of ACTH can be demonstrated by suitable bioassay of blood (A. TAYLOR et al. 1949, SYDNOR et al. 1953, PARIS et al. 1954). This procedure is not of practical diagnostic value, since the amount of blood required is large and the technique of bioassay is complicated.

D. Therapeutic trial

A trial of therapy for adrenal insufficiency may be of assistance in those patients in whom the diagnosis cannot be established clearly. For this purpose *the careful use of deoxycorticosterone, given intramuscularly, is probably the most satisfactory and specific.* Cortisone and its analogues may induce subjective improvement in patients with a wide variety of disorders, and additional Na^+ intake will benefit individuals who are losing Na^+ for reasons other than adrenal insufficiency (THORN 1951).

E. Conclusion

The diagnosis of adrenal insufficiency can be established most clearly if there is no increase in concentration of plasma corticosteroids in response to ACTH administration. The next most useful procedure is the measurement of urinary corticosteroids after ACTH is given. If facilities for these studies are not available, reliance must be placed on the other tests described, which evaluate normality of carbohydrate metabolism and the distribution of water and electrolytes. Drop in blood eosinophil count and increase of relative concentration of urinary uric acid in response to ACTH have not proved to be reliable tests.

Treatment of adrenocortical insufficiency in man

A. Treatment of chronic Addison's disease

Treatment of *chronic Addison's disease* involves two major aspects: replacement of deficiencies of both aldosterone and cortisol. The requirements of individual patients are related to the degree of adrenal insufficiency present. The function of residual adrenal tissue is depressed by therapy with corticoids.

I. Hormonal treatment

1. Replacement of aldosterone deficiency

a) Deoxycorticosterone

Deoxycorticosterone is available in a variety of forms. It may be administered intramuscularly in oil, the usual requirement being 1 to 5 mg daily (THORN 1951). [In adrenalectomized dogs receiving a moderate supplement of sodium, 0.03 to 0.17 mg/kg is effective (CLEGHORN et al. 1939).] Since the extent of adrenal insufficiency is variable in human beings, the dose needed must be arrived at individually for each patient, and cannot be related directly to body weight. When DOC is administered by the sublingual route (E. ANDERSON et al. 1940b), the daily requirement is 3 to 4 times larger than it is by intramuscular administration

(THORN et al. 1941). Buccal tablets are somewhat more convenient than solutions and may prove adequate for patients whose requirements for deoxycorticosterone are not great (E. ANDERSON et al. 1949).

Pellets of crystalline deoxycorticosterone acetate may be implanted subcutaneously after the patient's requirements have been established by daily injections for a period of approximately 2 months (DEANESLY and PARKES 1938, LEVY SIMPSON 1939, THORN et al. 1939a, b, HENI 1941, McCULLAGH et al. 1943, SHIPLEY 1944, THORN 1951). More recently it has become possible to achieve similar results with *deoxycorticosterone trimethylacetate;* a single intramuscular injection provides hormone for about 4 weeks (THORN et al. 1953b, LABHART 1957). The patient's needs must first be established with daily injections of deoxycorticosterone acetate in oil, just as before implantation of pellets. One 125 mg pellet of deoxycorticosterone acetate or 15 mg of deoxycorticosterone trimethylacetate is approximately equivalent to 0.5 mg deoxycorticosterone acetate in oil given daily (McMILLAN et al. 1955). When treatment with one of these long-acting preparations is planned, it is wise to establish the dose while the patient is taking about 3 gm of NaCl daily so that signs and symptoms of overdosage of the hormone can be counteracted by reducing the salt intake. With proper calculation of the dose, it is rarely necessary to remove any of the pellets. The dose must be re-established every 10 to 12 months before fresh pellets are implanted.

Deoxycorticosterone alone does not provide complete substitution therapy.

b) 9α-Fluorocortisol

9α-Fluorocortisol is effective when given by mouth (LIDDLE et al. 1954, FRIED 1955). Since fluoride substitution on the cortisol molecule enhances its electrolyte activity more than its carbohydrate effect, this compound also cannot be used alone as replacement therapy in patients with total adrenal insufficiency (HENI and MAST 1957). The compound has about 5 times the effect of cortisone in glycogen deposition and 10 times the activity of deoxycorticosterone in Na^+ retention (FRIED 1955, SCHAFER 1957, GROSS and LICHTLEN 1958).

The daily dose of 9α-fluorocortisol is given as 0.10 to 0.5 mg (RENOLD et al. 1955, EMERSON 1956, SCHAFER 1957, LABHART 1957, LEITH and BECK 1957). Even when the dose seems appropriate for the individual, as judged by stability of weight, plasma electrolyte concentrations and blood pressure, hypertension may develop after some months of treatment (OWEN et al. 1957).

c) Aldosterone

Aldosterone, which is not available commercially, has approximately the same potency in Na^+ retention as 9α-fluorocortisol, and 250 μg of aldosterone given intramuscularly every 6 hours was about equivalent to a single oral dose of 1 mg of 9α-fluorocortisol. Thus, MACH et al. (1954) found 150 to 200 μg of aldosterone in oil, injected daily, was adequate for maintenance of electrolyte balance in short-term treatment. REICHSTEIN (1954) found the required dose to be about 4% that of deoxycorticosterone and suggested that excess water retention or hypertension need not be anticipated. KEKWICK and PAWAN (1954) were able to restore Na^+ balance in one patient with a daily oral dose of 100 μg, whereas 40 μg had been inadequate; this patient required 3 mg deoxycorticosterone acetate in oil daily. PRUNTY et al. (1954) noted an effect 2 hours following the intramuscular administration of 100 μg in oil; the effect began to dissipate 8 hours after the injection. SALASSA et al. (1957) reported the development of moderate edema in a patient given 800 μg of aldosterone intramuscularly; 400 μg proved excessive in two other patients with adrenal insufficiency. PRUNTY et al. (1955) found 200 μg per day to be the effective replacement dose, given by the parenteral route. MACH et al. (1954) and KEKWICK and PAWAN (1954) found that aldosterone restored glucose tolerance tests towards normal; there was no effect on response to a water load, nor other indications of a substantial influence on carbohydrate metabolism. The preparation is active when given intravenously (THORN et al. 1955), as well as intramuscularly or by the oral route. It is interesting that aldosterone did not induce hypertension in adrenalectomized rats, although a comparable dose of deoxycorticosterone did (GROSS et al. 1955).

2. Replacement of cortisol deficiency

Since none of the preparations most active in influencing electrolyte distribution is able to avert hypoglycemic reactions in doses that do not lead to excess Na^+ retention (LOEB et al. 1937, LUKENS 1942, THORN 1951, SMART 1953, THORN et al. 1958), administration of a carbohydrate-active steroid is necessary in the treatment of adrenal insufficiency.

a) Cortisone acetate

Cortisone acetate was the first of the adrenal steroids available in quantities large enough to permit its use in the treatment of chronic adrenal insufficiency (PERERA et al. 1949). Prior to this time, aqueous adrenal extract or lipo-adrenal extract were used for those patients who could not be maintained in reasonably satisfactory condition on deoxycorticosterone and salt alone.

A basis for the calculation of average daily requirement is available from investigation of the rate of production of corticosteroids by the adrenal glands. BONDY and ALTROCK (1953), by analysis of blood collected from the renal vein, estimated the rate of secretion of 17-hydroxycorticosteroids to be 17.4 to 28.5 mg per 24 hours. By the use of C^{14}-labeled cortisol, PETERSON and WYNGAARDEN (1956) found the turnover rate to be 17—29 mg daily. COPE and BLACK (1958a, b), using a similar approach, calculated the production rate of cortisol in healthy subjects at rest; the range was 4.2 to 27.9 mg per 24 hours. These figures correlate well with the effective replacement doses for patients with adrenal insufficiency, discussed below.

When given intramuscularly as a microcrystalline suspension, cortisone acetate is effective for 24 hours or more, although it does not reach therapeutic concentrations in the blood until 8 to 12 hours after a single injection (THORN et al. 1953a). Given by this route, 10 to 25 mg daily is adequate in Addison's disease (THORN et al. 1949, THORN 1951). The oral requirement is 12.5 to 25 mg a day (THORN 1951, THORN et al. 1958), although some patients are given as much as 50 mg daily (NABARRO and WALKER 1957, THORN et al. 1958). BECK and MONTGOMERY (1956) found that a patient who had seemed well when receiving only cortisone therapy had difficulty in hot weather, when Na^+ loss was increased. Since patients with Addison's disease are quite sensitive to overstimulation of the central nervous system by cortisone (PERERA et al. 1949, THORN 1951, THORN et al. 1951, EMERSON 1956), it is probably wise to use the smallest amount possible. LIPSETT and PEARSON (1956) advocated replacement therapy with cortisone alone, but their own data demonstrate negative Na^+ balance in patients so treated, and the experience of others with similar patients indicates that the addition of deoxycorticosterone or 9α-fluorocortisol is useful in their management (HILLS et al. 1956, NABARRO and WALKER 1957, HART 1957).

The cortisone needs of totally adrenalectomized patients seem to be somewhat greater than the needs of those with spontaneous Addison's disease. The oral doses used range from 37.5 to 50 mg daily (HILLS et al. 1956, NABARRO and WALKER 1957, HART 1957).

Subcutaneous implants of cortisone pellets are usually not used, since they require renewal at approximately 3-month intervals (THORN et al. 1949, 1953a) and do not simplify the therapeutic program.

b) Cortisol

The free alcohol, cortisol, differs from cortisone acetate chiefly in that it has about 20% more activity, so that the doses given should be correspondingly smaller (EMERSON 1956). Since crystalline cortisol is absorbed more slowly from intramuscular sites than is cortisone acetate, somewhat larger doses are needed in initiating therapy by this route (LAIDLAW et al. 1955a). Peak concentrations of cortisol in the blood are reached one hour after ingestion (THORN et al. 1953a), which is earlier than the time of peak concentration of cortisone.

c) Unsaturated analogues

Unsaturated analogues of cortisone and cortisol (Δ^1-cortisone and Δ^1-cortisol, or prednisone and prednisolone) are about 4 times more active than the parent compounds (BUNIM et al. 1955, TOLKSDORF et al. 1956, THORN et al. 1958). They differ from cortisone and cortisol in that they apparently induce relatively less Na^+ retention and K^+ excretion. In patients whose requirements are small, it might be difficult to avoid overdosage when using one of these potent compounds.

d) Corticosterone and 11-dehydrocorticosterone

Corticosterone and 11-dehydrocorticosterone, while theoretically useful for replacement therapy in adrenal insufficiency, offer no major advantages and have not come into common usage (FORSHAM et al. 1946, PERERA et al. 1946, SPRAGUE et al. 1948, CONN et al. 1951, THORN 1951, THORN et al. 1953a).

3. Testosterone

Testosterone was at one time given to Addisonian patients for its effect on muscle function. However, corticosteroids serve this purpose better (R. WILLIAMS et al. 1945). Since there is no evidence that androgens of the adrenal cortex are important to health (see Chapter 5, Part 1), this form of therapy is probably neither wise nor necessary.

4. Adrenal transplants

Adrenal transplants have generally proved unsuccessful in man (BEER and OPPENHEIMER 1934). However, it is possible that with the use of human fetal adrenals this procedure may prove rewarding (BAILEY and KEELE 1939).

Improvement in the incidence of successful transplants of human tissues was demonstrated by GAILLARD (1954) who cultivated fetal parathyroid glands *in vitro* before placing them in the perivascular tissues of hypoparathyroid patients. MARTINOVITCH (1955) found that *in vitro* cultivation of rat adrenals facilitated successful transplantation to hosts. POOR (1958) succeeded in transplanting fetal adrenals in golden hamsters; the transplants maintained life when the hosts were subsequently adrenalectomized.

II. Non-hormonal therapy

1. Licorice root and glycyrrhizic acid

Licorice root and glycyrrhizic acid have a deoxycorticosterone-like action on electrolyte distribution and may serve as partial replacement therapy in Addison's disease. GROEN et al. (1951, 1952) gave an oral dose of 15 g a day. MOLHUYSEN et al. (1950) found an effect on electrolytes only in individuals with normal adrenals. BORST et al. (1953) proposed that they might act by slowing the rate of breakdown of adrenal steroids, since glycyrrhizic acid was effective in Addison's disease when cortisone was given as well. HUDSON et al. (1954) suggested combining 5 mg of cortisone with 4 to 6 g of glycyrrhizin. ELMADJIAN et al. (1956) reported the use of 10 mg of cortisol with 4 g of mono-ammonium glycerrhizinate; further reduction of the dose of cortisol led to edema and symptoms of adrenal insufficiency. HART (1957) also found that patients so treated might develop edema while the Na^+ concentration in the plasma remained reduced. These results seem difficult to understand.

2. Sodium chloride

Sodium chloride supplements probably should be provided for all patients having adrenal insufficiency, although not as the only maneuver for maintaining electrolyte equilibrium. By this means, a margin of safety is provided for individuals treated with long-acting deoxycorticosterone preparations, since Na^+ intake can be reduced readily should evidence of over-dosage develop (THORN 1951, WALSH 1952); variations in Na^+ need, depending on environmental temperature, minor gastrointestinal upsets, or other stresses, can be met. However, when 9α-fluorocortisol is used, salt supplements should not be added to the diet (THORN et al. 1958).

3. Potassium

Potassium supplements may be necessary for some patients treated with deoxycorticosterone or 9α-fluorocortisol, should muscle weakness or other signs of K^+ depletion develop (MCGAVACK 1941, LUKENS 1942, THORN 1951). Potassium deficiency may be a factor in sudden death of patients with apparently well-regulated Addison's disease (HAMPTON and KEPLER 1941). The painful tendon contractures that sometimes occur following prolonged treatment are possibly due to Na^+ retention or K^+ loss and can be handled by providing extra K^+ (THORN 1951, WARMING-LARSEN and SPRECHLER 1953).

4. General measures

General measures, such as avoidance of excess fatigue, scrupulous avoidance of cathartics or morphine, provision of a high carbohydrate diet with high levels of ascorbic acid and B vitamins, facilitate the rehabilitation of patients with adrenal insufficiency.

III. Treatment of complications

1. Tuberculosis or other infections

Tuberculosis, naturally, requires attention; with the availability of specific anti-tubercular therapy, cortisone need not be regarded as having a deleterious effect on the infection (BROWNE et al. 1954). FRENKEL (1960b) has made some suggestions about controlling infections of the adrenal cortex proper (see footnote, p. 53).

2. Associated endocrine disorders

The management of diabetes mellitus or thyrotoxicosis complicating Addison's disease has been outlined by KNOWLTON and KRITZLER (1949), THORN (1951) and STANTON et al. (1954). THORN and CLINTON (1943) pointed out that long-acting insulin preparations should not be used for treating diabetes in patients with adrenal insufficiency.

3. Pregnancy

With the availability of effective replacement therapy it has become evident that women with adrenal insufficiency can safely have children. The increased rates of excretion of 17-ketosteroids and corticosteroids during pregnancy have been attributed chiefly to the placenta[1] (SAMUELS et al. 1943, HILLS et al. 1954). Hormone requirements are not reduced (HILLS et al. 1954) and pregnancy must be regarded as a time of increased hazard to the patient. The nausea of early pregnancy may be misinterpreted as a sign of impending crisis (GURLING and RACKOW 1954); or crisis may actually occur, particularly if an infection develops or an attempt is made to reduce the amount of substitution therapy (HILLS et al. 1954). Additional cortisone should be provided during labor and in the immediate postpartum period (KNOWLTON et al. 1949, SCHWARTZ et al. 1952, HUNT and MCCONAHEY 1953, KAISER 1956).

4. Surgery

Special consideration must be given to the use of drugs and anesthetic agents in Addisonian patients who require surgical procedures. Essentially, the use of reduced amounts of barbiturates, atropine, or other drugs used before anesthesia is induced, and an increase in replacement therapy suffice for the successful management of surgical problems. It is important to adjust the amount of hormone given and the salt in intravenous infusions, since either hypotension or pulmonary edema may develop (PAPPER and CAHILL 1952, SCHWARTZ et al. 1952, L. KYLE et al. 1953).

5. Other stresses

The occurrence of any stress, as from infection or injury, is an indication to increase temporarily the hormonal therapy, without waiting for signs of inadequate treatment to develop.

B. Treatment of adrenocortical crisis

Adrenal crisis requires heroic therapy, with the intravenous administration of fluids and steroids. Since cortisol has become available for i. v. use, 200 to 300 mg of this compound are administered by infusion in the first 24 hours of crisis in adults; doses for children are proportionately smaller. Estimations of adrenal secretion of corticosteroids under stress are 100 to 260 mg daily (BONDY and ALTROCK 1953, PETERSON and WYNGAARDEN 1956, COPE and BLACK 1958b). DIRAIMONDO and FORSHAM (1958) calculated that in human beings under stress the adrenals produce about 10 mg of cortisol per hour, and recommended the use of 250 mg/24 hours. EMERSON (1956) proposed that most patients be given 200 mg of cortisol i. v. in the first 24 hours of acute adrenal insufficiency, plus 5 mg of deoxycorticosterone intramuscularly. [Deoxycorticosterone glucoside is soluble enough to be given intravenously, thus obviating the delay of absorption. A dose of 15 mg is approximately equivalent to 1 mg of deoxycorticosterone acetate in oil (THORN et al. 1953a, LABHART 1957); this compound is no longer readily available.] HAMWI and GOLDBERG (1955) used 9α-fluorocortisol in 10 to 15 mg doses i. v. for patients undergoing bilateral adrenalectomy; the infusion fluid was free of Na^+. HART (1957) suggested that in crisis 9α-fluorocortisol also be given, in about one-fifth the dosage of cortisol. There is a substantial danger of producing

[1] There is still, in 1964, no unequivocal evidence for *placental* synthesis of corticosteroids. The stimulus to *adrenal* corticoid production in pregnancy is a response to the high levels of circulating estrogens. See discussion, p. 106, Part 1 of this Volume. — Sub-eds.

pulmonary edema by such means, and it might be safer under such circumstances to rely on compounds with less profound effects on Na^+ retention, and to provide i. v. Na^+ as needed (THORN et al. 1958).

Before the discovery of aldosterone, the presence of unknown life-sustaining substances in whole adrenocortical extract was indicated by a potency that was substantially greater than could be accounted for by the known constituents (KUIZENGA et al. 1945, TAIT et al. 1952, THORN et al. 1953a). However, it now seems (SIMPSON and TAIT 1955) that aldosterone is the active principle of the amorphous fraction. LIPSETT and PEARSON (1956) maintain that there is no evidence that adrenocortical extract is as effective as cortisol in crisis, or that there are important unknown steroids in extracts. Nevertheless evidence of overdosage with adrenocortical extract given in crisis has not been reported, whereas one may easily get into difficulty when using crystalline compounds, especially deoxycorticosterone. Therefore it is possibly premature to discard whole extract as a useful drug.

VOGT (1943b) demonstrated that the adrenals of a variety of species put out a large amount of hormone under stressful circumstances. From her calculations it appears that the adrenals of a stressed 10 kg dog may secrete the equivalent of 230 ml of commercial adrenocortical extract in 24 hours. However, approximately 200 ml per day, given as a constant infusion, may suffice for an adult human being in adrenal crisis; doses based on these calculations have proved very satisfactory in children undergoing removal of an adrenal tumor (SOBEL et al. 1953). The extract contains 10% alcohol, which may cause drowsiness and tachycardia (SOBEL et al. 1953).

THORN (1951) and LABHART (1957) have outlined details for the management of adrenal crisis. It is important that all intravenous solutions contain glucose, and it seems reasonable that Na^+ requirements [calculated on the basis of maintenance needs and repair of prior losses (N. TALBOT et al. 1952)] be met by giving Na^+ throughout the 24 hour period. Additional therapy depends on the cause of the crisis, for example, antibiotics for infection or attention to a surgical problem.

In view of the scarcity of data on the occurrence of adrenal insufficiency in overwhelming infection or in surgical shock (KOSTER and KASMAN 1942, RUKES et al. 1955) it seems self-evident that appropriate therapy for the primary disturbance must not be neglected while instituting replacement therapy for an adrenal insufficiency that may or may not occur. Nevertheless, administration of cortisol or one of its analogues may be useful as additional therapy. The finding that large doses of ACTH can induce "tubular degeneration" of the adrenals in rats (WILBUR and RICH 1953) suggests that ACTH should not be used under such circumstances (see Chapter 1, Part 1).

C. Treatment of secondary adrenal insufficiency

There is some disagreement about the desirability of providing sodium-retaining steroids for patients with hypopituitarism. BECK and MONTGOMERY (1957) found deoxycorticosterone and salt unnecessary in four women who had postpartum pituitary failure; THORN (1951) pointed out that only small doses of deoxycorticosterone are tolerated without the development of edema. It is clear, however, that some form of therapy for the avoidance of hypoglycemia is important for these patients, and BECK and MONTGOMERY (1957) found that they could withdraw all substitution therapy except cortisone from 3 of their 4 patients. Since the institution of thyroid therapy in pituitary myxedema can lead to profound hypoglycemia (CASTLEMAN and HERTZ 1939; THORN 1944) it is wise to begin with cortisone replacement, adding thyroid hormone only after cortisone therapy is well under way. The effect of cortisone in inhibiting growth (BLODGETT et al. 1956, SOBEL 1956) is especially important in the treatment of children with panhypopituitarism, since very small amounts of cortisone may bring their already slow growth to a complete halt.

Patients with hypopituitarism require additional supportive therapy during infection, surgery, etc., as do those with primary adrenal insufficiency.

Literature

ABELOUS, J. E., et P. LANGLOIS: Note sur les fonctions des capsules surrénales chez la grenouille. C. R. Soc. Biol. (Paris) **3**, 792—798 (1891).
— — Sur les fonctions des capsules surrénales. Arch. Physiol. **4**, 465—472 (1892a).
— — Recherches expérimentales sur les fonctions des capsules surrénales de la grenouille. Arch. Physiol. **5**, 269—278 (1892b).
ABU HAYDAR, N., J. R. ST. MARC, W. J. REDDY, J. C. LAIDLAW and G. W. THORN: Adrenocortical insufficiency with normal basal levels of urinary 17-hydroxycorticoids: diagnostic implications. J. clin. Endocr. **18**, 121—133 (1958).
ADAMS, E.: Observations on the influence of the hypophysis and the adrenal cortex on blood platelet levels. Blood **4**, 936—946 (1949).
ADAMS, E. W.: The result of organotherapy in Addison's disease. Practitioner **71**, 472—502 (1903).
ADDIS, T., J. MARMORSTON, H. C. GOODMAN, A. L. SELLERS and M. SMITH: Effect of adrenalectomy on spontaneous and induced proteinuria in the rat. Proc. Soc. exp. Biol. (N. Y.) **74**, 43—46 (1950).
ADDISON, T.: Anaemia-disease of the supra-renal capsules. Lond. med. Gazette **8**, 517—518 (1849).
Discussion of ADDISON, T.: Lond. med. Gazette **8**, 562 (1849).
— On the Constitutional and Local Effects of Disease of the Supra-renal Capsules. London 1855. In: Med. Classics **2**, 244—280 (1937); and In: ROWNTREE, L. G., and A. M. SNELL: A Clinical Study of Addison's Disease. Philadelphia: W. B. Saunders Co. 1931.
AEGERTER, E. E.: The Waterhouse-Friderichsen syndrome. J. Amer. med. Ass. **106**, 1715—1719 (1936).
AGATE, F. J., JR., and R. L. ZWEMER: Some factors influencing adrenal insufficiency in the rat. Amer. J. Physiol. **111**, 1—8 (1935).
ALBANESE, M.: La fatigue chez les animaux privés des capsules surrénales. Arch. ital. Biol. **17**, 239—247 (1892a).
— Recherches sur la fonction des capsules surrénales. Arch. ital. Biol. **18**, 49—53 (1892b).
ALBAUM, H. G., A. I. HIRSHFELD, N. E. TONHAZY and W. W. UMBREIT: Phosphocreatine and adenosine triphosphate content of rat tissues after adrenalectomy and cortisone treatment. Proc. Soc. exp. Biol. (N. Y.) **76**, 546—548 (1951).
ALBRIGHT, F., P. H. SMITH and R. FRASER: A syndrome characterized by primary ovarian insufficiency and decreased stature. Report of 11 cases with a digression on hormonal control of pubic and axillary hair. Amer. J. med. Sci. **204**, 625—648 (1942a).
— C. H. BURNETT, P. H. SMITH and W. PARSON: Pseudo-hypoparathyroidism — an example of Seabright-Bantam Syndrome; report of three cases. Endocrinology **30**, 922—932 (1942b).
ALLERS, W. D.: The influence of diet and mineral metabolism in dogs after suprarenalectomy. Proc. Mayo Clin. **10**, 406—409 (1935).
— and L. A. CRANDALL, JR.: Growth in adrenalectomized puppies on diet low in potassium, high in sodium chloride, sodium citrate. Proc. Soc. exp. Biol. (N. Y.) **34**, 878—879 (1936).
— and E. C. KENDALL: Maintenance of adrenalectomized dogs without cortin through control of the mineral constituents of the diet. Amer. J. Physiol. **118**, 87—94 (1937).
— H. W. NILSON and E. C. KENDALL: Studies on adrenalectomized dogs: the toxic action of potassium. Proc. Mayo Clin. **11**, 283—288 (1936).
ALRICH, E. M., J. P. CARTER and E. P. LEHMAN: The effect of ACTH and cortisone on wound healing. Ann. Surg. **133**, 783—789 (1951).
AMES, R. G., and H. B. VAN DYKE: Antidiuretic hormone in the serum or plasma of rats. Endocrinology **50**, 350—360 (1952).
AMPERS, M. E., M. ECHAVE and F. GOMEZ MONT: Determination of urinary 17-ketosteroids. Normal values and excretion in various pathological states (in Spanish). Rev. Invest. clin. **5**, 171—183 (1953).

ANDERSON, E., and V. V. HERRING: Effect of NaCl and desoxycorticosterone on body weight and carbohydrate stores of adrenalectomized rats. Proc. Soc. exp. Biol. (N. Y.) **43**, 363—366 (1940).

— M. JOSEPH and V. V. HERRING: Salt after adrenalectomy. I. Growth and survival of adrenalectomized rats given various levels of NaCl. Proc. Soc. exp. Biol. (N. Y.) **44**, 477—481 (1940a).

— W. HAYMAKER and E. HENDERSON: Successful sublingual therapy in Addison's disease. J. Amer. med. Ass. **115**, 2167—2168 (1940b).

— L. W. KINSELL, T. C. DANIELS and E. HENDERSON: Intraoral DCA tablets as treatment of Addison's disease. J. clin. Endocr. **9**, 1324—1332 (1949).

ANDERSON, I. A., and A. LYALL: Addison's disease due to suprarenal atrophy. Lancet **1937 i**, 1039—1043.

ANDREWES, F. W.: Haemorrhage into the supra-renal capsules. Trans. path. Soc. Lond. **49**, 259—260 (1898).

ANGEVINE, D. M.: Pathologic anatomy of hypophysis and adrenal in anencephaly. Arch. Path. **26**, 507—518 (1938).

ANNES-DIAS, C.: L'insuffisance surrénale. Presse méd. **48**, 139—142 (1940).

APPLEBAUM, E., and A. SEELIG: Histologic changes in jaws and teeth of rats following nephritis, adrenalectomy and cortisone treatment. Oral Surg. 8, 881—891 (1955).

ARNETT, V., M. KESSLER and E. GELLHORN: The role of the adrenal cortex in preventing hypoglycemic convulsions. Amer. J. Physiol. **137**, 653—657 (1942).

ARNRICH, L., L. S. HURLEY, B. R. FORKER and A. F. MORGAN: Response to stress by riboflavin-deficient and pantothenic acid-deficient dogs. Amer. J. Physiol. **184**, 515—520 (1956a).

— M. R. NELSON, M. R. GRAM and A. F. MORGAN: Effect of adrenal hormones on carbohydrate metabolism in riboflavin and pantothenic acid-deficient dogs. Amer. J. Physiol. **186**, 427—434 (1956b).

ASCOLI, G., u. T. LEGNANI: Die Folgen der Extirpation der Hypophyse. Münch. med. Wschr. **59**, 518—521 (1912).

ATRIA, A., R. SANZ and S. DONOSO: Necropsy study of a case of Turner's syndrome. J. clin. Endocr. 8, 397—405 (1948).

ATWELL, W. J.: Effects of administering adrenotropic extract to hypophysectomized and thyroidectomized tadpoles. Amer. J. Physiol. **118**, 452—456 (1937).

AUB, J. C., J. FORMAN and E. M. BRIGHT: The effect of adrenalectomy upon the total metabolism of the cat. Amer. J. Physiol. **55**, 293 (1921).

AYRES, P. J., O. GARROD, S. A. SIMPSON and J. F. TAIT: A method for the determination of aldosterone, cortisol and corticosterone in biological extracts, particularly applied to human urine. Biochem. J. **65**, 639—646 (1957).

BACCHUS, H., and M. H. HEIFFER: Urinary corticosteroids and 17-ketosteroids in ascorbic acid-deficient female guinea pigs. Amer. J. Physiol. **174**, 243—246 (1953).

BÁEZ-VILLASEÑOR, J., C. E. RATH and C. A. FINCH: The blood picture in Addison's disease. Blood **3**, 769—773 (1948).

BAILEY, H., and K. D. KEELE: Addison's disease treated by adrenal grafting. Proc. roy. Soc. Lond. B **32**, 1824—1835 (1939).

BAIRD, P. C., JR., E. CLONEY and F. ALBRIGHT: Effect of cortical hormones in preventing extreme drop in colonic temperature displayed by hypophysectomized rats upon exposure to cold with preliminary observations upon the effect of hypophyseal and other hormones. Amer. J. Physiol. **104**, 489—501 (1933).

BAKER, B. L.: The relationship of the adrenal, thyroid and pituitary glands to the growth of hair. Ann. N. Y. Acad. Sci. **53**, 690—707 (1951).

— and R. M. BRIDGMAN: The histology of the gastro-intestinal mucosa (rat) after adrenalectomy or administration of adrenocortical hormones. Amer. J. Anat. **94**, 363—397 (1954).

— and W. L. WHITAKER: Relationship of the adrenal cortex to inhibition of hair growth by estrogen. Amer. J. Physiol. **159**, 118—123 (1949).

DE LA BALZE, F. A., E. C. REIFENSTEIN, JR., and F. ALBRIGHT: Differential blood counts in certain adrenal cortical disorders (Cushing's syndrome, Addison's disease and panhypopituitarism). J. clin. Endocr. **6**, 312—319 (1946).

BANERJEE, S., and C. DEB: Effect of scurvy on cholesterol and ascorbic acid in guinea pig adrenals. J. biol. Chem. **190**, 177—180 (1951).

— — Adrenal cortical activity in scurvy. Endocrinology **51**, 572—576 (1952).

BANTING, F. G., and S. GAIRNS: Suprarenal insufficiency. Amer. J. Physiol. **77**, 100—113 (1926).

BARKER, N. W.: The pathologic anatomy in twenty-eight cases of Addison's disease. Arch. Path. 8, 432—450 (1929).

BARLOW, O. W., and M. M. ELLIS: Effects of double adrenalectomy on the blood coagulation time in cats. Amer. J. Physiol. **70**, 58—67 (1924).

BARNICOT, N. A., and W. S. S. LADELL: Physiological effects of ACTH in some West African subjects. J. trop. Med. Hyg. **57**, 180—191, 210—220 (1954).
— and D. WOLFFSON: Daily urinary 17-ketosteroid output of African Negroes. Lancet **1952 i**, 893—895.
BARTLET, A. L., and M. F. LOCKETT: Effect of adrenalectomy on tissue histamine in rats. J. Physiol. (Lond.) **147**, 51—57 (1959).
BAUER, J.: Der Einfluß der Nebennieren und Hypophyse auf die Blutregulation und Umstimmung der Geschlechtscharaktere beim Menschen. Klin. Wschr. **14**, 361—367 (1935).
BAUMANN, E. J., and S. KURLAND: Changes in the inorganic constituents of the blood in suprarenalectomized cats and rabbits. J. biol. Chem. **71**, 281—302 (1927).
BAZEMORE, A. W., K. A. C. ELLIOTT and E. FLOREY: Isolation of factor I. J. Neurochem. **1**, 334—339 (1957).
BEAN, W. B., and R. E. HODGES: Pantothenic acid deficiency induced in human subjects. Proc. Soc. exp. Biol. (N. Y.) **86**, 693—698 (1954).
— — and K. DAUM: Pantothenic acid deficiency induced in human subjects. J. clin. Invest. **34**, 1073—1084 (1955).
BECK, R. N., and D. A. D. MONTGOMERY: Treatment of Addison's disease. Brit. med. J. **1956 ii**, 921—923.
— — Treatment of hypopituitarism. Brit. med. J. **1957 i**, 441—444.
BEER, E., and B. S. OPPENHEIMER: Transplantation of the adrenal cortex for Addison's disease. Ann. Surg. **100**, 689—703 (1934).
BEERMAN, H., and H. L. COLBURN: Some aspects of pigmentation of the skin. Amer. J. med. Sci. **231**, 451—475 (1956).
BELLVILLE, J. W., O. SCHWEIZER and W. S. HOWLAND: Anesthesia for bilateral adrenalectomy. A statistical and clinical analysis. Anesthesiology **19**, 217—230 (1958).
BENDER, S., et A. LERI: De l'atrophie des capsules surrénales chez le foetus anencéphales. C. R. Soc. Biol. (Paris) **55**, 1137—1139 (1903).
BENIRSCHKE, K., E. BLOCH and A. T. HERTIG: Concerning the function of the fetal zone of the human adrenal gland. Endocrinology **58**, 598—625 (1956).
BERGEN, J. R.: Rat electrocorticogram in relation to adrenal cortical function. Amer. J. Physiol. **164**, 16—22 (1951).
— and H. HOAGLAND: Brain and muscle potassium in normal and in adrenalectomized rats. Amer. J. Physiol. **164**, 23—25 (1951).
— C. A. HUNT and H. HOAGLAND: In vitro oxygen consumption of tissues from normal and adrenalectomized rats. Amer. J. Physiol. **171**, 624—629 (1952).
— D. STONE and H. HOAGLAND: Studies on brain potassium in relation to the adrenal cortex. Internat. Conf. on the Peaceful Uses of Atomic Energy (1955).
BERGNER, G. E., and A. B. EISENSTEIN: Addison's disease in the Negro. J. clin. Endocr. **11**, 322—329 (1951).
BERGSTRAND, C. G., and C. A. GEMZELL: Plasma level of 17-hydroxycorticosteroids and urinary excretion of 17-ketosteroids in normal children. Acta paediat. (Stockh.) **44**, 318—321 (1955).
BERMAN, I., A. EDELMANN and A. S. GORDON: The relation of the adrenal to phosphorus metabolism of bone marrow. Endocrinology **59**, 656—665 (1956).
BERNARD, L., et M. BIGART: Sur les réactions histologiques générales des surrénales à certaines influences pathogènes expérimentales. C. R. Soc. Biol. (Paris) **54**, 1219—1221 (1902).
— — et H. LABBÉ: Sur la sécrétion de lécithine dans les capsules surrénales. C. R. Soc. Biol. (Paris) **55**, 120—122 (1903).
BERNSTEIN, D. E.: Diabetes mellitus followed by Addison's disease and hypothyroidism, simulating panhypopituitarism. J. clin. Endocr. **8**, 687—693 (1948).
BEST, W. R., R. C. MUEHRCKE and R. M. KARK: Studies on adrenocortical eosinopenia: a clinical and statistical evaluation of four-hour eosinophil response tests. J. clin. Invest. **31**, 733—742 (1952).
— R. M. KARK, R. C. MUEHRCKE and M. SAMTER: Clinical value of eosinophil counts and eosinophil response tests. J. Amer. med. Ass. **151**, 702—706 (1953).
BIEDL, A.: The Internal Secretory Organs. New York: W. Wood and Co. 1913.
BILLMAN, F., u. F. D. GRATHWOHL: Über die Wirksamkeit der subcutanen Einpflanzung von Tabletten krystallisierten Dehydrocortisonacetats an nebennierenlosen Hunden. Klin. Wschr. **19**, 1030—1033 (1940).
BIRKE, G., B. BOEMEIJ, B. OLHAGEN and L. O. PLANTIN: Measures for the avoidance of adrenal atrophy in prolonged corticosteroid therapy. Acta med. scand. **155**, 245—252 (1956).
— E. DICZFALUSY and L. O. PLANTIN: Assessment of the functional capacity of the adrenal cortex. I. Establishment of normal values. J. clin. Endocr. **18**, 736—754 (1958).
BLACK, D. R.: Addison's disease. J. Kans. med. Soc. **26**, 190—192 (1926).
BLAIR, H. A.: On the measure of excitability. J. gen. Physiol. **15**, 709 (1932).
— Time constant of excitation and velocity in supernormal phase of nerve. J. Neurophysiol. **2**, 249 (1939).

BLISS, E. L., D. H. NELSON and L. T. SAMUELS: Effects of intravenous ACTH on blood levels of 17-hydroxycorticosteroids and circulating leucocytes. J. clin. Endocr. **14**, 423—439 (1954).

BLIZZARD, R. M., and M. ALBERTS: Hypopituitarism, hypoadrenalism and hypogonadism in the newborn infant. J. Pediat. **48**, 782—792 (1956).

BLODGETT, F. M., L. BURGIN, D. IEZZONI, D. GRIBETZ and N. B. TALBOT: Effects of prolonged cortisone therapy on the statural growth, skeletal maturation and metabolic status of children. New Engl. J. Med. **254**, 636—641 (1956).

BLOODWORTH, J. M. B., JR., W. M. KIRKKENDALL and T. L. CARR: Addison's disease associated with thyroid insufficiency and atrophy (Schmidt syndrome). J. clin. Endocr. **14**, 540—553 (1954).

BØGGILD, D. H.: Experimentelle Untersuchungen über die Bedeutung der Nebennieren für die Blutzuckerregulierung. Acta path. microbiol. scand. **2**, 68—91 (1925).

BOINET, E.: Action antitoxique des capsules surrénales sur la neurine. C. R. Soc. Biol. (Paris) **48**, 364—367 (1896).

— Dangers de l'adrénaline dans certains cas de maladie bronzée d'Addison. C. R. Soc. Biol. (Paris) **55**, 1471—1473 (1903).

BONDY, P. K., and J. R. ALTROCK: Estimation of the rate of release of adrenal 17-hydroxycorticosteroids in the human being by the venous catheter technique, with a method for determining plasma 17-hydroxycorticosteroids. J. clin. Invest. **32**, 703—710 (1953).

BONGIOVANNI, A. M., and W. J. EISENMENGER: Adrenal cortical metabolism in chronic liver disease. J. clin. Endocr. **11**, 152—172 (1951).

BONNER, C. D.: Eosinophil levels as an index of adrenal responsiveness. J. Amer. med. Ass. **148**, 634—637 (1952).

BOOKER, W. M., J. R. TUREMAN, F. M. DACOSTA, J. POULSON and S. Q. MITCHELL: Relationship between glutathione and ascorbic acid in adrenocortical function. Amer. J. Physiol. **181**, 371—374 (1955).

BORNSTEIN, A., u. C. HORNEMANN: Über Zufallserscheinungen nach Nebennieren-Extirpation III. Über die Beziehung der Nebenniere zum Zuckerstoffwechsel. Z. ges. exp. Med. **37**, 24—32 (1923).

BORST, J. G. G., S. P. TEN-HOLT, L. A. DE VRIES and J. A. MOLHUYSEN: Synergistic action of liquorice and cortisone in Addison's and Simmonds' disease. Lancet **1953 i**, 657—662.

BOUDIN, G., J.-L. FUNCK-BRENTANO et M. GAYNO: Maladie d'Addison par aplasie surrénale et syndrome para-biermérien. Bull. Soc. méd. Hôp. Paris **66**, 1736—1741 (1950a).

— — — Considérations sur une encéphalopathie addisonienne. Bull. Soc. méd. Hôp. Paris **66**, 1741—1748 (1950b).

BOURNE, G. H.: The Mammalian Adrenal Gland. Oxford: Clarendon Press 1949.

— Comparative anatomy and embryology of the adrenal glands. In: H. D. MOON, Ed., The Adrenal Cortex, Chapt. 1. New York: Paul B. Hoeber, Inc. 1961.

BOWMAN, R. H.: The influence of the adrenal medulla on muscle glycogen in normal and hypophysectomized rats. Endocrinology **58**, 158—162 (1956).

BRAMWELL, B.: A case of Addison's Disease. Brit. med. J. **1905 ii**, 1082—1083.

BRENNER, O.: Addison's disease with atrophy of the cortex of the suprarenals. Quart. J. Med. **22**, 121—144 (1928).

BRIGHT, R.: Reports of Medical Cases Selected with a View of Illustrating the Symptoms and Cure of Diseases by a Reference to Morbid Anatomy, Vol. II, 247—248 (Case CXIX). London: Longman 1831.

BRITTON, S. W.: Adrenal insufficiency and related considerations. Physiol. Rev. **10**, 617—682 (1930).

— and R. F. KLINE: Relation of the adrenal cortex to reproduction and lactation. Amer. J. Physiol. **115**, 627—631 (1936).

— and H. SILVETTE: Some observations on the cortico-adrenal hormone. Science **73**, 373—374 (1931a).

— — Further experiments on cortico-adrenal extract: its efficacy by mouth. Science **74**, 440—441 (1931b).

— — Theories of cortico-adrenal function. Science **77**, 366—368 (1933).

— — On the function of the adrenal cortex — general carbohydrate and circulatory theories. Amer. J. Physiol. **107**, 190—206 (1934).

— — A comparison of sodium, chloride and carbohydrate changes in adrenal insufficiency and other experimental conditions. Amer. J. Physiol. **118**, 594—599 (1937a).

— — The adrenal cortex and carbohydrate metabolism. Cold Spr. Harb. Symp. quant. Biol. **8**, 357—361 (1937b).

— J. C. FLIPPIN and H. SILVETTE: The oral administration of cortico-adrenal extract. Amer. J. Physiol. **99**, 44 (1931).

— R. F. KLINE and H. SILVETTE: Blood-chemical and other conditions in normal and adrenalectomized sloths. Amer. J. Physiol. **123**, 701—704 (1938).

BROLIN, S. E., and M. LINDBÄCH: The reacting capacity of the gonads after adrenalectomy as judged by the response to gonadotrophins in the rat. Acta endocr. (Kbh.) **8**, 55—71 (1951).
BROWN, F. K., and J. W. REMINGTON: Arteriolar responsiveness in adrenal crisis in the dog. Amer. J. Physiol. **182**, 279—284 (1955).
BROWNE, J. S. L.: The effect of corticotropin on the excretion of cortin-like substances and 17-ketosteroids. Trans. J. Macy Conf. Metab. Aspects Convalescence and Wound Healing **4**, 88—90 (1943).
— M. ARONOVITCH, J. C. BECK, W. LEITH and J. F. MEAKINS: The management of concomitant Addison's disease and active pulmonary tuberculosis. Trans. Ass. Amer. Phycns. **67**, 206—223 (1954).
BROWN-SÉQUARD, E.: Recherches expérimentales sur la physiologie et la pathologie des capsules surrénales. C. R. Acad. Sci. (Paris) **43**, 422—425 (1856a).
— Recherches expérimentales sur la physiologie des capsules surrénales. C. R. Acad. Sci. (Paris) **43**, 542—546 (1856b).
— Nouvelles recherches sur l'importance des fonctions des capsules surrénales. C. R. Acad. Sci. (Paris) **45**, 1036—1039 (1857).
— Influence hereuse de la transfusion de sang normal après l'extirpation des capsules surrénales chez le cobaye. C. R. Soc. Biol. (Paris) **45**, 448—449 (1893).
VAN BUCHEM, F. S. P.: Simmonds'-Sheehan's disease. Acta Endocr. (Kbh.) **19**, 165—180 (1955).
BÜCHERL, E., u. M. SCHWAB: Der Einfluß der Adrenalektomie auf das weiße Blutbild. Klin. Wschr. **30**, 113—115 (1952).
BUFFETT, R. F., and L. C. WYMAN: Changes in the width of uncalcified tibial epiphyseal cartilage of the rat following adrenalectomy and gonadectomy. Amer. J. Physiol. **180**, 16—20 (1955).
— — Changes in the region of the tibial epiphyseal cartilage of the rat following adrenalectomy and gonadectomy. Anat. Rec. **125**, 17—39 (1956).
BULLOCK, W. K., and A. E. HIRST, JR.: Metastatic carcinoma of the adrenal. Amer. J. med. Sci. **226**, 521—524 (1953).
BUNIM, J. J., R. L. BLACK, A. J. BOLLET and M. M. PECHET: Metabolic effects of metacortandralone and metacortandracin. Ann. N. Y. Acad. Sci. **61**, 358—368 (1955a).
— M. M. PECHET and A. J. BOLLET: Studies on metacortandralone and metacortandracin in rheumatoid arthritis; antirheumatic potency, metabolic effects, and hormonal properties. J. Amer. med. Ass. **157**, 311—318 (1955b).
BURNETT, C. H., D. M. SELDIN and M. WALSER: Observations on the electrolyte and water metabolism in Addison's disease during oral salt loading. Trans. Ass. Amer. Phycns. **66**, 65—71 (1953).
BURSTEIN, S., R. I. DORFMAN and E. M. NADEL: Corticosteroids in the urine of normal and scorbutic guinea pigs: isolation and quantitative determination. J. biol. Chem. **213**, 597—608 (1955).
BUSCH, F. C., T. M. LEONARD and T. WRIGHT: Further results in suprarenal transplantation. J. Amer. med. Ass. **51**, 640—642 (1908).
BUTCHER, E. O.: Hair growth in adrenalectomized and adrenalectomized thyroxin-treated rats. Amer. J. Physiol. **120**, 427—434 (1937).
— Effect of adrenalectomy on the growth of mammary glands in underfed albino rats. Proc. Soc. exp. Biol. (N. Y.) **42**, 571—573 (1939).
— Oxygen consumption of skin and hair growth after adrenalectomy in the white rat. Endocrinology **32**, 493—496 (1943).
— and R. A. RICHARDS: The relation of the adrenals to the retarded hair growth in underfed albino rats. Endocrinology **25**, 787—792 (1939).
BUTLER, L. C., and A. F. MORGAN: The content of adrenocorticotropic hormone in the pantothenic acid-deficient female rat. Endocrinology **56**, 322—326 (1955).
BUTTERFLY, J. M., L. FISHMAN, J. SECKLER and H. STEINBERG: Addison's disease secondary to metastatic carcinoma of the adrenal glands. Ann. intern. Med. **37**, 930—939 (1952).
BUTTLE, J. M., R. L. KIRK and H. WARING: The effect of complete adrenalectomy on the wallaby (Setonyx brachyurus). J. Endocr. 8, 281—290 (1952).
CALLOW, N. H., R. K. CALLOW and C. W. EMMENS: 17-Ketosteroid, androgen and oestrogen excretion in the urine of cases of gonadal or adrenal cortical deficiency. J. Endocr. **2**, 88—98 (1940).
CARR, J. L., and C. L. CONNOR: Animal experiments with adrenal cortical extracts. Ann. intern. Med. **6**, 1225—1229 (1933).
Case records of the Massachusetts General Hospital. New Engl. J. Med. **248**, 385—389 (1953).
CASTELMAN, B., and S. HERTZ: Pituitary fibrosis with myxedema. Arch. Path. **27**, 69—79 (1939).
CATTELL, J. McK.: The adrenal glands in an editor's office. Science **77**, 428—439 (1933).

CAWADIAS, A. P.: Adrenocortical cancer and undulating fever in Addison's disease. J. clin. Endocr. **6**, 507—514 (1946).
CHALMERS, T. M., and A. A. G. LEWIS: The effect of ACTH on the diuretic response to water in panhypopituitarism. Lancet **1951 ii**, 1158—1160.
CHARRIN, A., et P. LANGLOIS: Action antitoxique du tissu des capsules surrénales. C. R. Soc. Biol. (Paris) **46**, 410—411 (1894).
CHART, J. J., H. SHEPPARD, M.J. ALLEN, W. L. BENCZE and R. GAUNT: New amphenone analogs as adrenocortical inhibitors. Experientia (Basel) **14**, 151—152 (1958).
CHASE, H. B.: Growth of the hair. Physiol. Rev. **34**, 113—126 (1954).
CHASSIN, J. L., H. A. MCDOUGALL, W. STAHL, M. MACKAY and S. A. LOCALIO: Effect of adrenalectomy on wound healing in normal and in stressed rats. Proc. Soc. exp. Biol. (N. Y.) **86**, 446—448 (1954).
CHAUCHARD, P.: Hormones et système nerveux. Rev. Sci. **90**, 120—134 (1952).
CHESTER JONES, I.: The Adrenal Cortex. Cambridge: Univ. Press 1957.
CHOKAS, W. V.: Bilateral adrenal hemorrhage complicating dicoumarol therapy for myocardial infarction. Amer. J. Med. **24**, 454—460 (1958).
CHRISTIANSON, M., and I. CHESTER JONES: The interrelationships of the adrenal glands of mother and foetus in the rat. J. Endocr. **15**, 17—42 (1957).
CHRISTY, N. P.: Pituitary insufficiency. In: E. B. ASTWOOD, Ed., Clinical Endocrinology **1**, 53—67. New York and London: Grune and Stratton 1960.
— E. Z. WALLACE and J. W. JAILER: The effect of intravenously-administered ACTH on plasma 17,21-dihydroxy-20-ketosteroids in normal individuals and in patients with disorders of the adrenal cortex. J. clin. Invest. **34**, 899—906 (1955).
— — — Comparative effects of prednisone and of cortisone in suppressing the response of the adrenal cortex to exogenous adrenocorticotropin. J. clin. Endocr. **16**, 1059—1074 (1956).
CICARDO, V. H.: Physiocochemical mechanisms in experimental epilepsy. J. nerv. ment. Dis. **101**, 527—536 (1945).
— The susceptibility to convulsions of adrenalectomized, hypophysectomized animals injected with desoxycorticosterone (in Spanish). Publ. Cent. Invest. isiol. (B. Aires) **10**, 233—239 (1946).
CICCANTELLI, M. J., W. B. GALLAGHER, F. C. SKEMP and P. C. DIETZ: Fatal nephropathy and adrenal necrosis after translumbar aortography. New Engl. J. Med. **258**, 433—435 (1958).
CLARK, J. H., and L. G. ROWNTREE: Studies of the adrenal glands in health and disease. Endocrinology **18**, 256—273 (1934).
CLARK, W. G.: Effect of adrenalectomy upon intestinal absorption of sodium chloride. Proc. Soc. exp. Biol. (N. Y.) **40**, 468—470 (1934).
— and D. F. CLAUSEN: Dietary "self-selection" and appetites of untreated and treated adrenalectomized rats. Amer. J. Physiol. **139**, 70—79 (1943).
CLARKE, A. P. W., and R. A. CLEGHORN: Chemical study of tissue changes in adrenal insufficiency and traumatic shock. Endocrinology **31**, 597—606 (1942).
CLAYTON, B. E., and F. T. G. PRUNTY: Relation of adrenal cortical function to scurvy in guinea-pigs. Brit. med. J. **1951 ii**, 927—930.
— — The relation of adrenocortical function to scurvy in guinea-pigs. The effect of exogenous ACTH and adrenalectomy. J. Endocr. **9**, 370—378 (1953).
— I. H. MILLS and F. T. G. PRUNTY: Further studies on adrenocortical function in ascorbic acid deficiency. J. Endocr. **11**, vi (1954).
CLEGHORN, R. A.: Adrenal cortical insufficiency: physiological and neurological observations. Canad. med. Ass. J. **65**, 449—454 (1951).
— Endocrine influence on personality and behavior. In: Biology of Mental Health and Disease, 265—273. New York: Hoeber 1952a.
— Observations on the mechanism of the disordered psychological outlook in Addison's disease and hypopituitarism. In: Ciba Found. Colloq. Endocr., Vol. III, 141—149, 1952b.
— Psychologic changes in Addison's disease. J. clin. Endocr. **13**, 1291—1293 (1953).
— Steroid hormones in relation to neuropsychiatric disorders. In: H. HOAGLAND, Ed., Hormones, Brain Function and Behavior, 3—25. New York: Academic Press 1957.
— and C. J. PATTEE: Psychologic changes in 3 cases of Addison's disease during treatment with cortisone. J. clin. Endocr. **14**, 344—352 (1954).
— J. L. A. FOWLER and J. S. WENZEL: The assay of desoxycorticosterone acetate and its use in the treatment of Addison's disease. J. clin. Invest. **18**, 475 (1939).
— — — and A. P. W. CLARKE: The desoxycorticosterone acetate requirement of the adrenalectomized dog. Endocrinology **29**, 535—541 (1941).
— A. P. W. CLARKE and W. F. GREENWOOD: Activity of desoxycorticosterone acetate in propylene glycol by oral and intravenous routes in adrenalectomized dogs, and its effect on the cardiac arrhythmia of adrenal insufficiency. Endocrinology **32**, 170—175 (1943).

CLEGHORN, R. A., J. L. A. FOWLER, W. F. GREENWOOD and A. P. W. CLARKE: Pressor responses in healthy adrenalectomized dogs. Amer. J. Physiol. **161**, 21—28 (1950).
Clinical Pathological Conference. Hypopituitarism: clinical pathological conference at the Nationial Institutes of Health. Ann. intern. Med. **46**, 1143—1159 (1957).
Clinicopathologic Conference. Hemochromatosis versus Addison's disease. Amer. J. Med. **9**, 383—393 (1950).
COHN, C., E. SHRAGO and D. JOSEPH: Effect of food administration on weight gains and body composition of normal and adrenalectomized rats. Amer. J. Physiol. **180**, 503—507 (1955).
COMBRIDGE, C. D.: The effect of pyridoxin deficiency on certain organs of the rat. Brit. J. Nutr. **10**, 347—354 (1956).
CONN, J. W.: Electrolyte composition of sweat. Arch. intern. Med. **83**, 416—428 (1949).
— and K. P. MATTHEWS: Addison's disease in the Negro. Amer. J. med. Sci. **212**, 404—408 (1946).
— S. S. FAJANS and L. H. LOUIS: Metabolic effects in man of orally and parenterally administered corticosterone (compound B). Trans. Ass. Amer. Phycns. **64**, 269—278 (1951).
COPE, C. L., and E. G. BLACK: The behaviour of ^{14}C-cortisol and estimation of cortisol production rate in man. Clin. Sci. **17**, 147—163 (1958a).
— — The production rate of cortisol in man. Brit. med. J. **1958i**, 1020—1024.
— — The reliability of some adrenal function tests. Brit. med. J. **1959ii**, 1117—1122.
CORCORAN, W. J., and A. A. STRAUSS: Suprarenal hemorrhage in the new-born. J. Amer. med. Ass. **82**, 626—628 (1924).
COREY, E. L.: The effect of forcing fluids upon survival after bilateral epinephrectomy. Amer. J. Physiol. **79**, 633—640 (1926/27).
— and S. W. BRITTON: Carbohydrate metabolism of hypophysectomized and hypophyso-adrenalectomized rats. Amer. J. Physiol. **118**, 15—20 (1937).
CORI, C. F., and G. T. CORI: On the carbohydrate metabolism of adrenalectomized rats. Proc. Soc. exp. Biol. (N. Y.) **24**, 539—541 (1926).
COSGRIFF, S. W.: The Waterhouse-Friderichsen syndrome: observations of associated adrenal insufficiency and report of four cases. Ann. intern. Med. **21**, 187—193 (1944).
COSMETTATOS, G. F.: Hydrocéphalie et atrophie des glandes surrénales. Grèce méd. **22**, 17—19 (1920).
COWIE, A. T.: The influence of age and sex on the life span of adrenalectomized rats. J. Endocr. **6**, 94—98 (1949a).
— The relative growth of the mammary gland in normal, gonadectomized and adrenalectomized rats. J. Endocr. **6**, 145—157 (1949b).
— and S. J. FOLLEY: The role of the adrenal cortex in mammary development and its relation to the mammogenic action of the anterior pituitary. Endocrinology **40**, 274—285 (1947).
— and J. STEWART: Adrenalectomy in the goat and its effects on the chemical constituents of the blood. J. Endocr. **6**, 197—204 (1949).
CRAFTS, R. C.: The effects of the endocrines on the formed elements of the blood. I. The effects of hypophysectomy thyroidectomy and adrenalectomy on the blood of the adult female rat. Endocrinology **29**, 596—618 (1941).
— The similarity between anemia induced by hypophysectomy and that induced by a combined thyroidectomy and adrenalectomy in adult female rats. Endocrinology **53**, 465—473 (1953).
CRAIG, J. W., M. MILLER, M. S. MACKENZIE and H. WOODWARD, JR.: The influence of dietary carbohydrate deprivation on the metabolism of intravenously administered fructose and glucose in man. J. clin. Invest. **37**, 118—126 (1958).
CRAIG, M. M., L. H. SCHIFF and J. E. BOONE: Chronic moniliasis associated with Addison's disease. Amer. J. Dis. Child. **89**, 669—684 (1955).
CRAWFORD, M. D.: Suprarenal haemorrhage and necrosis in pregnancy. J. Path. Bact. **63**, 365—376 (1951).
CRISMON, M. J., and J. FIELD, II: The oxygen consumption *in vitro* of brain cortex, kidney, and skeletal muscle from adrenalectomized rats. Amer. J. Physiol. **130**, 231—238 (1940).
CRISPELL, K. R., W. PARSON and P. SPRINKLE: A cortisone-resistant abnormality in the diuretic response to ingested water in primary myxedema. J. clin. Endocr. **14**, 640—644 (1954).
— — J. HAMLIN and G. HOLLIFIELD: Addison's disease associated with histoplasmosis. Amer. J. Med. **20**, 23—29 (1956).
CROSBY, E. H.: Addison's disease associated with congenital absence of the suprarenal glands. Arch. Path. **10**, 38—50 (1930).
CROWE, S. J., and G. B. WISLOCKI: Experimental observations on the suprarenal glands with especial reference to the functions of their interrenal portions. Bull. Johns Hopk. Hosp. **25**, 287—304 (1914).
CUPPS, P. T.: The ability of certain adrenal cortical steroids to restore reproduction in adrenalectomized female rats. Endocrinology **57**, 1—4 (1955).

CURTIS, H. J.: Conduction in nerve and muscle. In: P. BARD, Ed., Medical Physiology, Ed. 10, 892—903. St. Louis: C. V. Mosby Co. 1956.
CUTLER, H. H., M. H. POWER and R. M. WILDER: Concentration of chloride, sodium and potassium in urine and blood. J. Amer. med. Ass. **111**, 117—122 (1938a).
— — — Concentrations of sodium, chloride and potassium in the blood plasma and urine of patients with Addison's disease: their diagnostic significance. Proc. Mayo Clin. **13**, 244—249 (1938b).
D'AGATI, V. C., and B. A. MARANGONI: The Waterhouse-Friderichsen syndrome. New Engl. J. Med. **232**, 1—7 (1945).
DANOWSKI, T. S., E. B. FERGUS and F. M. MATEER: The low salt syndromes. Ann. intern. Med. **43**, 643—657 (1955).
DARROW, D. C., and H. C. MILLER: The production of cardiac lesions by repeated injections of desoxycorticosterone acetate. J. clin. Invest. **21**, 601—611 (1942).
— H. E. HARRISON and M. TAFFEL: Tissue electrolytes in adrenal insufficiency. J. biol. Chem. **130**, 487—502 (1939).
DAUGHADAY, W. H., H. JAFFE and R. H. WILLIAMS: Adrenal cortical hormone excretion in endocrine and non-endocrine disease as measured by chemical assay. J. clin. Endocr. **8**, 244—256 (1948).
DA VANZO, J. P., H. C. CROSSFIELD and W. W. SWINGLE: Effect of various adrenal steroids on plasma magnesium and the electrocardiogram of adrenalectomized dogs. Endocrinology **63**, 825—830 (1958).
DAVENPORT, H. W., and V. J. CHAVRE: The lack of effect of the adrenal hormones upon gastric acid secretion. Endocrinology **47**, 193—197 (1950).
DAVENPORT, V. D.: Relation between brain and plasma electrolytes and electroshock seizure thresholds in adrenalectomized rats. Amer. J. Physiol. **156**, 322—327 (1949).
DAVIS, J. E., and A. B. HASTINGS: The relationship of the adrenal and thyroid glands to excised muscle metabolism. Amer. J. Physiol. **105**, 110—121 (1933).
DEAMER, W. C., and H. K. SILVER: Abnormalities in the secretion of the adrenal cortex during early life. J. Pediat. **37**, 490—516 (1950).
DEANE, H. W., and J. M. MCKIBBIN: The chemical cytology of the rat's adrenal cortex in pantothenic acid deficiency. Endocrinology **38**, 385—400 (1946).
DEANESLY, R., and A. S. PARKES: Further experiments on the administration of hormones by the subcutaneous implantation of tablets. Lancet **1938 ii**, 606—608.
DE MAIO, D.: Influence of adrenalectomy and hypophysectomy on cerebral serotonin. Science **129**, 1678—1679 (1959).
DEUTSCH, S., and H. MESCON: Melanin pigmentation and its endocrine control. New Engl. J. Med. **257**, 222—226; 268—272 (1957).
DEWAR, A. D.: The effect of adrenalectomy on the weight gain of pregnancy in the mouse. J. Endocr. **15**, 211—215 (1957).
DHOM, G.: Zur Morphologie und Genese der kongenitalen Nebennierenrindenhyperplasie beim männlichen Scheinzwitter. Zbl. allg. Path. path. Anat. **97**, 346—357 (1956/57).
DIAMOND, M. C.: The effect of hypophysectomy, removal of the posterior pituitary, adrenalectomy and dehydration on the antidiuretic activity of the rat hypothalamic-neurohypophyseal system. Endocrinology **58**, 461—470 (1956).
DIEKE, S. H.: The effect of removing various endocrine glands on the hair cycles of black rats. Endocrinology **42**, 315—319 (1948).
DI RAIMONDO, V. C., and P. H. FORSHAM: Pharmacophysiologic principles in the use of corticoids and adrenocorticotropin. Metabolism **7**, 5—24 (1958).
DOMM, L. V., and W. A. WELLBAND: Effect of adrenalectomy and cortisone on eruption rate of incisors in young female albino rats. Proc. Soc. exp. Biol. (N. Y.) **104**, 582—584 (1960).
DONAICH, I., and E. A. WRIGHT: Two cases of giant-cell granuloma of the pituitary gland. J. Path. Bact. **63**, 69—79 (1951).
DONÁTH, J.: Über den Einfluß der Nebennierenextirpation und des Suprarenins auf die Blutkonzentration bei Katzen. Naunyn-Schmiedeberg's Arch. exp. Path. Pharmak. **77**, 1—15 (1914).
DONE, A. K., R. S. ELY, L. R. HEISELT and V. C. KELLEY: Circulating 17-hydroxycorticosteroids in ascorbic acid-deficient guinea pigs. Proc. Soc. exp. Biol. (N. Y.) **83**, 722—724 (1953).
DONNELL, G. N., N. N. LITMAN and M. ROLDAN: Renal sodium loss, hyponatremia and hyperkalemia due to a renal tubular insensitivity to mineralocorticoids. Amer. J. Dis. Child. **97**, 813—828 (1959).
DOUGHERTY, T. F., and A. WHITE: Influence of hormones on lymphoid tissue structure and function. The role of the pituitary adrenocorticotrophic hormone in the regulation of the lymphocytes and other cellular elements of the blood. Endocrinology **35**, 1—14 (1944).

DUDLEY, H. A., J. S. ROBSON, M. SMITH and C. P. STEWART: The relationship of aldosterone excretion to the metabolic response to adrenalectomy. Clin. chim. Acta **2**, 461—466 (1957).
DUFF, G. L., and C. BERNSTEIN: Five cases of Addison's disease with so-called atrophy of the adrenal cortex. Bull. Johns Hopk. Hosp. **52**, 67—85 (1933).
DUFFIN, J. D.: Cortical necrosis of the adrenal glands associated with Addison's disease. Report of eight cases. Arch. Path. **35**, 649—666 (1943).
DUMM, M. E., H. GERSHBERG, E. M. BECK and E. P. RALLI: Effect of pantothenate deficiency on synthesis of adrenal cholesterol following stress. Proc. Soc. exp. Biol. (N. Y.) **82**, 659 to 662 (1953).
— S. H. LESLIE and E. P. RALLI: Influence of adrenalectomy and gonadectomy on capacity of rats to excrete an acid load. Proc. Soc. exp. Biol. (N. Y.) **88**, 592—594 (1955).
EDMONDS, H. W.: Pituitary, adrenal and thyroid in cyclopia. Arch. Path. **50**, 727—735 (1950).
EHRENGUT, W. VON: Morbus Addison und Wachstum. Helv. paediat. Acta **11**, 63—77 (1956).
EICHHOLTZ, F., P. HOTOVY, P. COLLISCHON u. H. KNAUER: Beeinflussung der Entgiftungszeiten von Avertin und Pentothal-Natrium an der nebennierenlosen Ratte durch Nebennieren-Rindenhormon (Pentothal-Natriumtest). Naunyn-Schmiedeberg's Arch. exp. Path. Pharmak. **207**, 576—585 (1949).
EIK-NES, K., A. A. SANDBERG, C. J. MIGEON, F. H. TYLER and L. T. SAMUELS: Changes in plasma levels of 17-hydroxycorticosteroids during the intravenous administration of ACTH. II. Responses under various clinical conditions. J. clin. Endocr. **15**, 13—21 (1955).
— — F. H. TYLER and L. T. SAMUELS: Changes in plasma level of 17-hydroxycorticosteroids induced by ACTH. III. Effects of various vehicles and modes of administration. J. clin. Endocr. **18**, 882—885 (1958).
EISENSTEIN, A. B.: Pantothenic acid and adrenocortical hormone secretion. Endocrinology **60**, 298—302 (1957).
— Relationship of vitamin B_6 to adrenocortical function in the rat. Proc. Soc. exp. Biol. (N. Y.) **100**, 111—114 (1959).
— and A. S. WENNEKER: Excretion of 17-hydroxycorticosteroid hormones by pantothenic acid-deficient guinea pigs. Proc. Soc. exp. Biol. (N. Y.) **97**, 844—847 (1958).
ELLIOTT, K. A. C., and E. FLOREY: Factor I. Inhibitory factor from brain. Assay. Conditions in brain. Stimulating and antagonizing substances. J. Neurochem. **1**, 181—191 (1956).
ELLIOTT, T. R.: Some results of excision of the adrenal glands. J. Physiol. (Lond.) **49**, 38—53 (1914/15).
— and I. TUCKETT: Cortex and medulla in the suprarenal glands. J. Physiol. (Lond.) **34**, 332—369 (1906).
ELMADJIAN, F., and G. PINCUS: The adrenal cortex and the lymphocytopenia of stress. Endocrinology **37**, 47—49 (1945).
— H. FREEMAN and G. PINCUS: The adrenal cortex and the lymphocytopenia due to glucose administration. Endocrinology **39**, 293—299 (1946).
— J. M. HOPE and G. PINCUS: Action of mono-ammonium glycyrrhizinate on adrenalectomized subjects: synergism with hydrocortisone. J. clin. Endocr. **16**, 338—349 (1956).
ELY, R. S., V. C. KELLEY and R. B. RAILE: Studies of 17-hydroxycorticosteroids in children. I. Peripheral blood levels in health and disease. J. Pediat. **42**, 38—45 (1953).
— R. B. RAILE, P. F. BRAY and V. C. KELLEY: Studies of 17-hydroxycorticosteroids. IV. Evaluation of a standard ACTH-17-hydroxycorticosteroid response test in children. Pediatrics **13**, 403—411 (1954).
EMERSON, K.: Treatment of Addison's disease. Arch. intern. Med. **98**, 100—105 (1956).
ENGEL, F. L.: In discussion of: CLEGHORN, R. A., and B. F. GRAHAM: Manifestations of altered autonomic and humoral function in psychoneurosis. Recent Progr. Hormone Res. **4**, 355—356 (1949).
— The influence of the endocrine glands on fatty acid and ketone body metabolism. Amer. J. clin. Nutr. **5**, 417—430 (1957).
— and J. L. SCOTT: The insulin-glucose tolerance test. A modified procedure for the detection of hypoglycemia unresponsiveness in pituitary and adrenal insufficiency. J. clin. Invest. **29**, 151—160 (1950).
ENGEL, G. L., and S. G. MARGOLIN: Neuropsychiatric disturbances in Addison's disease and the role of impaired carbohydrate metabolism in production of abnormal cerebral function. Arch. Neurol. Psychiat. (Chic.) **45**, 881—884 (1941).
— — Neuropsychiatric disturbances in internal disease. Metabolic factors and electroencephalographic correlations. Arch. intern. Med. **70**, 236—259 (1942).
EPPINGER, H., W. FALTA u. C. RUDINGER: Über die Wechselwirkung der Drüsen mit innerer Sekretion. Z. klin. Med. **67**, 380—398 (1909).
ERSHOFF, B. H.: Comparative effects of pantothenic acid deficiency and inanition on resistance to cold stress in the rat. J. Nutr. **49**, 373—385 (1953).
— Effects of riboflavin deficiency on insulin sensitivity in the rat. Metabolism **3**, 357—363 (1954).

ERSHOFF, B. H. and F. D. PARROTT, JR.: Effects of B vitamin deficiencies on the leukocyte response to epinephrine and corticotropin (ACTH). J. clin. Nutr. **1**, 124—133 (1953).
— R. B. A. SLATER and J. G. GAINES: Effects of pantothenic acid deficiency on pituitary-adrenal function in the rat. J. Nutr. **50**, 299—316 (1953).
ESCAMILLA, R. F., and H. LISSER: Simmonds' disease. J. clin. Endocr. **2**, 65—96 (1942).
EVERSE, J. W. R., and P. DE FREMERY: On a method of measuring fatigue in rats and its application for testing the suprarenal cortical hormone (Cortin). Acta brev. neerl. Physiol. **2**, 152 (1932).
EVERSOLE, W. J.: Studies on the effectiveness of desoxycorticosterone acetate in adrenalectomized rats given a low sodium chloride diet. Endocrinology **36**, 27—31 (1945).
— and R. GAUNT: Methods of administering desoxycorticosterone and the problem of inactivation by the liver. Endocrinilogy **32**, 51—56 (1943).
— A. EDELMANN and R. GAUNT: Effect of adrenal cortical transplants on life-maintenance and "water intoxication." Anat. Rec. **76**, 271—281 (1940).
— R. GAUNT and E. C. KENDALL: The effect of adrenal steroids in water intoxication. Amer. J. Physiol. **135**, 378—382 (1941/42).
FARBER, J. E., K. GOLDSTEIN and W. F. BESWICK: Simmonds' disease with a craniopharyngioma. J. clin. Endocr. **1**, 688—690 (1941).
FELDMAN, J. D.: Effect of estrogen on thyroid uptake of I^{131} in adrenalectomized rats. Amer. J. Physiol. **184**, 369—370 (1956).
FERREBEE, J. W., C. RAGAN, D. W. ATCHLEY and R. F. LOEB: Desoxycorticosterone esters. Certain effects in the treatment of Addison's disease. J. Amer. med. Ass. **113**, 1725—1731 (1939).
— D. PARKER, W. C. CARNES, M. K. GERITY, D. W. ATCHLEY and R. F. LOEB: Certain effects of desoxycorticosterone. Amer. J. Physiol. **135**, 230—237 (1941/42).
FIDANZA, A., and A. BONOMOLO: Pantothenic acid and the adrenals (in Italian). Acta vitamin. (Milano) **10**, 11—15 (1956).
DE FILIPPIS, V., and I. I. YOUNG: Evaluation of adrenocortical function with intramuscular injection of ACTH gel. New Engl. J. Med. **257**, 1—6 (1957).
FINZI, O.: Über Veränderungen der Magenschleimhaut bei Tieren nach Nebennierenextirpation und über experimentell erzeugte Magengeschwüre. Virchows Arch. path. Anat. **214**, 413—432 (1913).
FIROR, W. M., and A. GROLLMAN: Adrenalectomy in mammals with particular reference to the white rat. Amer. J. Physiol. **103**, 686—698 (1933).
FISHER, J. W.: The effects of adrenalectomy on cobalt polycythemia in the rat. Endocrinology **62**, 547—556 (1958).
FLANAGAN, J. B., A. K. DAVIS and R. R. OVERMAN: Mechanism of extracellular sodium and chloride depletion in the adrenalectomized dog. Amer. J. Physiol. **160**, 89—102 (1950).
FLÜCKINGER, E., and F. VERZÁR: Influence of adrenalectomy on sodium and potassium exchange in muscle. Acta endocr. (Kbh.) **17**, 80—84 (1954).
FOLLEY, S. J.: The adrenal cortex and the mammary gland. In: J. M. YOFFEY, Ed., The suprarenal Cortex, 85—90. London: Butterworths Scientific Publications 1953.
— Lactation. In: A. S. PARKES, Ed., Marshall's Physiology of Reproduction, Ed. 3, Vol. II, 525—647. London: Longmans, Green and Co. 1956.
FORBES, A. P., G. C. GRISWOLD and F. ALBRIGHT: Clinical experience with a bioassay method for the determination of urinary corticosteroids. J. clin. Endocr. **10**, 230—247 (1950).
FORKER, B. R., and A. F. MORGAN: Effect of adrenocortical hormone on the riboflavin-deficient rat. J. biol. Chem. **209**, 303—311 (1954).
FORSHAM, P. H., and G. W. THORN: The adrenals. In: R. H. WILLIAMS, Ed., Textbook of Endocrinology, Ed. 2, Chap. 4, Philadelphia and London: W. B. Saunders Co. 1955.
— — G. E. BERGNER and K. EMERSON, JR.: Metabolic changes induced by synthetic 11-dehydrocorticosterone acetate. Amer. J. Med. **1**, 105—134 (1946).
— — F. T. G. PRUNTY and A. G. HILLS: Clinical studies with pituitary adrenocorticotropin. J. clin. Endocr. **8**, 15—66 (1948).
FOWLER, J. L. A., and R. A. CLEGHORN: The response of splanchnic blood vessels and of the small intestine to vasconstrictor influences in adrenal insufficiency in the cat. Amer. J. Physiol. **137**, 371—379 (1942).
FRANK, H. A.: Present-day concepts of shock. New Engl. J. Med. **249**, 445—453 (1953).
FRANKSSON, C., C. A. GEMSELL and U. S. V. EULER: Cortical and medullary adrenal activity in surgical and allied conditions. J. clin. Endocr. **14**, 608—621 (1954).
FRASER, R. W., A. P. FORBES, F. ALBRIGHT, A. SULKOWITCH and E. C. REIFENSTEIN, JR.: Colorimetric assay of 17-ketosteroids in urine; survey of use of this test in endocrine investigation, diagnosis and therapy. J. clin. Endocr. **1**, 234—256 (1941 a).
— F. ALBRIGHT and P. H. SMITH: The value of the glucose tolerance test, the insulin tolerance test, and the glucose-insulin tolerance test in the diagnosis of endocrinologic disorders of glucose metabolism. J. clin. Endocr. **1**, 297—306 (1941 b).

FRAWLEY, T. F., and G. W. THORN: The relation of the salivary sodium-potassium ratio to adrenal cortical function. In: J. R. MOTE, Ed., Proc. Second ACTH Conference, Vol. 1, 115—122, Philadelphia: The Blakiston Company 1951.

FREDERICKSON, D. S.: Effect of massive cortisone therapy on thyroid function. J. clin. Endocr. **11**, 760 (1951).

FREEMAN, N. E.: Cortin and traumatic shock. Science **77**, 211—212 (1933).

FRENKEL, J. K.: Evaluation of infection-enhancing activity of modified corticoids. Proc. Soc. exp. Biol. (N. Y.) **103**, 552—555 (1960a).

— Pathogenesis of infections of the adrenal gland leading to Addison's disease in man: The role of corticoids in adrenal and generalized infection. Ann. N. Y. Acad. Sci. **84** (Art. 11), 391—440 (1960b).

— Infection of the adrenal cortex. In: H. D. MOON, Ed., The Adrenal Cortex. Intern. Acad. Pathol. Monog., No. 2, pp. 201—219. New York: Hoeber 1961.

FRIDERICHSEN, C.: Adrenal apoplexy in infants (in Swedish). Ugeskr. Laeg. **79**, 1817—1826 (1917).

— Nebennierenapoplexie bei kleinen Kindern. Jhb. Kinderheilk. **87**, 109—124 (1918).

— Waterhouse-Friderichsen syndrome. Acta endocr. (Kbh.) **18**, 482—492 (1955).

FRIED, J.: Biological effects of 9-α-fluorohydrocortisone and related halogenated steroids in animals. Ann. N. Y. Acad. Sci. **61** (Art. 2), 573—581 (1955).

FRIEDGOOD, H. B.: Adrenalectomy and coitus-induced ovulation in the rabbit and cat. Endocrinology **25**, 296—301 (1939).

FRIEDMAN, S. M., K. R. MACKENZIE and C. L. FRIEDMAN: Renal function in the adrenalectomized rat. Endocrinology **43**, 123—125 (1948).

FRIEDMANN, H. C.: 17-Ketosteroid excretion in Indian males. Lancet **1954 ii**, 262—266.

FRITZ, I., and R. LEVINE: Action of adrenal cortical steroids and norepinephrine on vascular responses of stress in adrenalectomized rats. Amer. J. Physiol. **165**, 456—465 (1951).

FROUIN, A.: Ablation des capsules surrénales et diabète pancréatique. C. R. Soc. Biol. (Paris) **64**, 216—217 (1908).

GAARENSTROOM, J. H., L. WATERMAN and L. LAQUEUR: A method for standardizing the cortical hormone. Acta brev. neerl. Physiol. **1**, 10—13 (1937).

GAGNON, J., et R. LEBLANC: Sclérose cérébrale diffuse avec mélanodermie et atrophie surrénale. Un. méd. Canada 88, 392—412 (1959).

GAILLARD, P. J.: Greffes à l'Homme des tissus cultivés „in vitro". J. Méd. Bordeaux **131**, 643—649 (1954).

GALIN, M. A.: Unilateral adrenal hemorrhage during ACTH therapy. New Engl. J. Med. **258**, 945—946 (1958).

GARDNER, L. I.: Newer knowledge of adrenocortical disturbances. In: W. A. REILLY, Ed., Pediat. Clinics North America 1957, 889—904. Philadelphia: W. B. Saunders Co.

GARROD, A. E., and J. H. DRYSDALE: Haemorrhage into both supra-renal capsules in an infant aged four months. Trans. path. Soc. Lond. **49**, 257—258 (1898).

GARROD, O., S. A. DAVIES and G. F. CAHILL: The action of cortisone and desoxycorticosterone acetate on glomerular filtration rate and sodium and water exchange in the adrenalectomized dog. J. clin. Invest. **34**, 761—776 (1955).

GARVIN, C. F., and H. S. REICHLE: Death presumably due to the use of the salt restriction test in the diagnosis of Addison's disease. Ann. intern. Med. **14**, 323—324 (1940).

GAUDINO, M., and M. F. LEVITT: Influence of the adrenal cortex on body water distribution and renal function. J. clin. Invest. **28**, 1487—1497 (1949).

GAUNT, R.: The adrenal cortex in salt and water metabolism. Recent Progr. Hormone Res. **6**, 247—276 (1951).

— and W. J. EVERSOLE: Notes on the history of the adrenal cortical problem. Ann. N. Y. Acad. Sci. **50**, 511—521 (1949).

— C. E. TOBIN and J. H. GAUNT: The survival of salt-treated adrenalectomized rats. Amer. J. Physiol. **111**, 321—329 (1935).

— M. LILING and C. W. MUSHETT: Disturbances of water metabolism in vitamin deficiencies and effects of adrenal cortical hormones. Endocrinology **38**, 127—132 (1946).

— J. H. BIRNIE and W. J. EVERSOLE: Adrenal cortex and water metabolism. Physiol. Rev. **29**, 281—310 (1949).

— C. HOWELL, N. ANTONCHAK and M. GLITZER: Adrenal cortical steroids and body growth. Anat. Rec. **111**, 560 (1951).

GEPPERT, L. J., W. A. SPENCER and A. M. RICHMOND: Adrenal insufficiency in infancy. J. Pediat. **37**, 1—22 (1950).

GERMUTH, F. G., JR.: The role of adrenocortical steroids in infection, immunity and hypersensitivity. Pharmacol. Rev. 8, 1—24 (1956).

GERSCHMAN, R., D. L. GILBERT, S. W. NYE, P. W. NADIG and W. O. FENN: Role of adrenalectomy and adrenal-cortical hormones in oxygen poisoning. Amer. J. Physiol. **178**, 346—350 (1954).

— — W. E. PRICE, JR. and W. O. FENN: Effect of autonomic drugs and of adrenal glands on oxygen poisoning. Proc. Soc. exp. Biol. (N. Y.) **88**, 617—621 (1955).

GERSH, I., and A. GROLLMAN: Kidney function in adrenal cortical insufficiency. Amer. J. Physiol. **125**, 66—74 (1939).

GESCHWIND, I. I.: Species variation in protein and polypeptide hormones. In: A. GORBMAN, Ed., Comparative Endocrinology, pp. 421—443. New York: Wiley 1959.

GIBELLI, C.: The relationship of adrenocortical function to fracture repair and to the etiology of gastric ulcers. Pathologica **1**, 131—133 (1908/09).

GILBERT-DREYFUS, J. SCHILLER, M. ZARA, J. C. SAVOIE, C. ALEXANDRE et J. C. RENIER: L'origine hypophysaire de la mélanodermie Addisonienne: ACTH et pigmentation. Ann. Endocr. (Paris) **14**, 46—49 (1953).

GILLMAN, J., and L. GOLDBERG: Studies in cortico-adrenal function. I. Survival of adrenalectomized rats. Endocrinology **31**, 201—208 (1942).

GILMAN, A.: Experimental sodium loss analogous to adrenal insufficiency. The resulting water shift and sensitivity to hemorrhage. Amer. J. Physiol. **108**, 662—669 (1934).

GINSBURG, M.: The secretion of ADH in response to haemorrhage and the fate of vasopressin in adrenalectomized rats. J. Endocr. **11**, 165—176 (1954).

GLOMSET, D. A.: Incidence of metastasis of malignant tumors to the adrenals. Amer. J. Cancer **32**, 57—61 (1938).

GOLD, E. M., and P. STARR: Urinary excretion of 17-ketogenic steroids in normal subjects before and after stimulation with intramuscular Zinc-ACTH. J. clin. Endocr. **19**, 907—920 (1959).

— V. C. DIRAIMONDO and P. H. FORSHAM: Quantitation of pituitary corticotropin reserve in man by use of an adrenocortical 11β- hydroxylase inhibitor (SU-4885). Metabolism **9**, 3—20 (1960a).

— B. SERENA and S. COOK: The combined estimation of cortisol and 11-desoxycortisol in plasma as Porter-Silber chromogens. J. clin. Endocr. **20**, 315—326 (1960b).

GOLDBERG, M. B., and A. F. MAXWELL: Male pseudohermaphroditism. J. clin. Endocr. **8**, 367—379 (1948).

GOLDFIEN, A., J. C. LAIDLAW, N. ABU HAYDAR, A. E. RENOLD and G. W. THORN: Fluorohydrocortisone and chlorohydrocortisone, highly potent derivatives of compound F. New Engl. J. Med. **252**, 415—421 (1955).

GOLDMAN, H., and L. G. ABOOD: Postadrenalectomy cycles of motor activity and creatine phosphate in pituitary and hypothalamus of the rat. Fed. Proc. **12**, 53 (1953).

GOLDSTEIN, M. S., E. R. RAMEY and R. LEVINE: Relation of muscular fatigue in the adrenalectomized dog to inadequate circulatory adjustment. Amer. J. Physiol. **163**, 561—565 (1950).

GOLDZIEHER, M. A.: Interrenin. Das Hormon der Nebennierenrinde. Klin. Wschr. **1928 i**, 1124—1125.

— The Adrenal Glands in Health and Disease. Philadelphia: F. A. Davis Company 1944.

GOODMAN, L. S., and A. GILMAN: The Pharmacological Basis of Therapeutics, Ed. 2. New York: Macmillan Co. 1955.

GORDAN, E. S.: The use of desoxycorticosterone and its esters in the treatment of Addison's disease. J. Amer. med. Ass. **114**, 2549—2551 (1940).

GORDAN, W. H., and M. B. SHIMKIN: Fulminating meningococcemia with purpura, meningitis and focal necrosis of pituitary. J. Amer. med. Ass. **123**, 147—148 (1943).

GORDON, A. S.: Endocrine influences upon the formed elements of blood and blood-forming organs. Recent Progr. Hormone Res. **10**, 339—394 (1954).

— and G. F. KATSH: The relation of the adrenal cortex to the structure and phagocytic activity of the macrophagic system. Ann. N. Y. Acad. Sci. **52**, 1—30 (1949).

— S. J. PILIERO and D. LANDAU: The relation of the adrenal to blood formation in the rat. Endocrinology **49**, 497—511 (1951).

GORDON, D., B. N. HORWITT and A. SEGALOFF: Adrenal response to ACTH in various clinical conditions. J. clin. Endocr. **14**, 297—308 (1954).

GORMAN, W. F., and S. B. WORTIS: Psychosis in Addison's disease. Dis. nerv. Syst. **8**, 267—271 (1947).

GOTTFRIED, S. P., M. BOGIN and N. V. LEVYCKY, JR.: The effect of corticotropin on the activity of the adrenal cortex of the male child. J. Pediat. **50**, 170—174 (1957).

GRADINESCU, A. V.: Der Einfluß der Nebennieren auf den Blutkreislauf und den Stoffwechsel. Pflügers Arch. ges. Physiol. **152**, 187—253 (1913).

GRAHAM, L. S.: Celiac accessory adrenal glands. Cancer **6**, 149—152 (1953).

GRATIOLET, P.: Note sur les effets qui suivent l'ablation des capsules surrénales. C. R. Soc. Biol. (Paris) **43**, 468—470 (1856).

GRAY, S. J., and C. G. RAMSEY: Adrenal influences upon the stomach and the gastric responses to stress. Recent Progr. Hormone Res. **13**, 583—617 (1957).
GREEN, C. H.: Clinical use of extract of the adrenal cortex. Arch. intern. Med. **59**, 759—784 (1937).
GREEN, H. D., R. N. LEWIS, N. D. NICKERSON and A. L. HELLER: Blood flow, peripheral resistance and vascular tonus, with observations on the relationship between blood flow and cutaneous temperature. Amer. J. Physiol. **141**, 518—536 (1944).
GREENBERG, R.: Variation of choline acetylase content of brain in stressed and unstressed normal and adrenalectomized rats. Fed. Proc. **8**, 61 (1949).
GREENE, R. C.: Spontaneous idiopathic bilateral adrenal apoplexy associated with hypertension. J. Amer. med. Ass. **152**, 133—134 (1953).
GREENHOW, E. H.: On Addison's Disease. Philadelphia: Lindsay and Blakiston 1875.
GREEP, R. O.: Preliminary observations on the relation of the adrenal cortex to electrolyte metabolism in the rat. In: R. C. CHRISTMAN, Ed., Pituitary-Adrenal Function, 96—101. Washington: Amer. Ass. Advanc. Sci. 1950.
— E. KNOBIL, F. G. HOFMANN and T. L. JONES: Adrenal cortical insufficiency in the rhesus monkey. Endocrinology **50**, 664—676 (1952).
GRIFFITHS, W. J., JR.: Effect of adrenalectomy on incidence of audiogenic seizures among domestic and wild rats. J. comp. physiol. Psychol. **42**, 303—312 (1949).
GROEN, J., H. PELSER, A. I. WILLEBRANDS and C. E. KAMMINGA: Extract of licorice for the treatment of Addison's disease. New Engl. J. Med. **244**, 471—475 (1951).
— — M. FRENKEL, C. E. KAMMINGA and A. F. WILLEBRANDS: Effect of glycyrrhizinic acid on the electrolyte metabolism in Addison's disease. J. clin. Invest. **31**, 87—91 (1952).
GROLLMAN, A.: The Adrenals. Baltimore: Wililams and Wilkins Co. 1936.
— Further observations on replacement therapy in experimental adrenal insufficiency with desoxycorticosterone acetate and sodium chloride. Endocrinology **29**, 862—865 (1941).
— Water and electrolyte content of tissues of the adrenalectomized and adrenalectomized-nephrectomized dog. Amer. J. Physiol. **179**, 36—38 (1954).
— and W. M. FIROR: Studies on the adrenal. II. Extraction of cortical hormone from urine. Proc. Soc. exp. Biol. (N. Y.) **30**, 669—670 (1933).
GRØNBAEK, P.: A sweat electrolyte test for adrenocortical function. Acta endocr. (Kbh.) **18**, 563—564 (1955).
GROSS, F., and P. LICHTLEN: Further evidence for a qualitative difference between aldosterone and cortexone. In: A. MULLER and C. O'CONNOR, Eds., Aldosterone, 39—55. Boston: Little, Brown and Co. 1958.
— P. LOUSTALOT u. R. MEIER: Vergleichende Untersuchungen über die hypertensive Wirkung von Aldosteron und Desoxycorticosteron. Experientia (Basel) **11**, 67—68 (1955).
GUGGENHEIM, K.: Studies on water metabolism of pyridoxine and pantothenic acid deficient rats. Endocrinology **55**, 156—162 (1954).
GUNN, F. D.: Anatomical changes following total adrenalectomy in cats, with special reference to the kidney. Bull. Buffalo General Hosp. **5**, 13—16 (1927).
GURLING, K. J., and F. RACKOW: Addison's disease complicated by pregnancy and diabetes mellitus. Lancet **1954 ii**, 316—318.
GUTTMAN, P. H.: Addison's disease. A statistical analysis of five hundred and sixty-six cases and a study of the pathology. Arch. Path. **10**, 742—785, 895—935 (1930).
HADLOW, W. J.: Adrenal cortical atrophy in the dog. Amer. J. Path. **29**, 353—361 (1953).
HAEGER, K., D. JACOBSOHN and G. KAHLSON: The levels of histaminase and histamine in the gastro-intestinal mucosa and kidney of the cat deprived of the hypophysis or the adrenal glands. Acta physiol. scand. **25**, 243—254 (1952).
HAFKENSCHIEL, J. H., C. K. FRIEDLAND and H. A. ZINTEL: The blood flow and oxygen consumption of the brain in patients with essential hypertension before and after adrenalectomy. J. clin. Invest. **33**, 57—62 (1954).
HALES, W. M., G. M. HASLERUD and D. J. INGLE: Time for development of incapacity to work in adrenalectomized rats. Amer. J. Physiol. **112**, 65—69 (1935).
HALKIER, E.: The melanophore reaction as a test for ACTH in the blood. Acta endocr. (Kbh.) **13**, 157—161 (1953).
HALL, E. M., and L. HEMKEN: The adrenal glands. A clinical and pathological study. Arch. intern. Med. **58**, 448—468 (1936).
HALL, G. E., and R. A. CLEGHORN: Cardiac lesions in adrenal insufficiency. Canad. med. Ass. J. **39**, 126—133 (1938).
HALL, T. E., B. H. MCCRACKEN and G. W. THORN: Skin pigmentation in relation to adrenal cortical function. J. clin. Endocr. **13**, 243—257 (1953).
HALLION, L., et R. GAYET: L'augmentation de la sensibilité à l'insuline à la suite de l'ablation des capsules surrénales. C. R. Soc. Biol. (Paris) **92**, 945—946 (1925).
HAMILTON, L. D., C. J. GUBLER, H. ASHENBRUCKER, G. E. CARTWRIGHT and M. M. WINTROBE: Studies on the relationship of the adrenal cortex to the experimental production of hypoferremia in rats. Endocrinology **48**, 44—45 (1951).

Hampton, H. P., and E. J. Kepler: Addison's disease: treatment and prognosis. Amer. J. med. Sci. **202**, 264—271 (1941).

Hamwi, G. J., and D. B. Brown: The effects of carbonic anhydrase inhibitor in Addison's disease. Arch. intern. Med. **97**, 778—782 (1956).

— and R. F. Goldberg: Clinical use of fluorocortisone acetate. J. Amer. med. Ass. **159**, 1598—1601 (1955).

Harlem, O. K., and E. Myhre: Congenital adrenal hypoplasia. Amer. J. Dis. Child. **94**, 696—701 (1957).

Harriman, A. E.: The effect of a preoperative preference for sugar over salt upon compensatory salt selection by adrenalectomized rats. J. Nutr. **57**, 271—276 (1955).

Harris, F. D., A. F. Hartmen, jr., D. Rolf and H. L. White: Ammonia secretion in adrenal insufficiency. Amer. J. Physiol. **168**, 20—28 (1952).

Harris, G. W., and J. W. Woods: Aetiology of Grave's disease in relation to recent experimental findings. Brit. med. J. **1956 ii**, 737—739.

Harris-Jones, J. N., and P. G. F. Nixon: Familial Addison's disease with spastic paraplegia. J. clin. Endocr. **15**, 739—744 (1955).

Harrison, H. E., and D. C. Darrow: Renal function in experimental adrenal insufficiency. Amer. J. Physiol. **125**, 631—643 (1939).

Harrop, G. A., jr., and A. Weinstein: Addison's disease treated with suprarenal cortical hormone (Swingle-Pfiffner). J. Amer. med. Ass. **98**, 1525—1531 (1932).

— J. J. Pfiffner, A. Weinstein and W. W. Swingle: Further studies on the adrenal cortical hormone. Science **73**, 683—684 (1931a).

— A. Weinstein, A. Marlow, W. W. Swingle and J. J. Pfiffner: Studies on the hormone of the suprarenal cortex. J. Amer. med. Ass. **96**, 1986 (1931b).

— L. J. Soffer, R. Ellsworth and J. H. Trescher: III. Plasma electrolytes and electrolyte excretion during suprarenal insufficiency in the dog. J. exp. Med. **58**, 17—38 (1933a).

— A. Weinstein, L. J. Soffer, and J. H. Trescher: Studies on the suprarenal cortex. II. Metabolism, circulation and blood concentration during suprarenal insufficiency in the dog. J. exp. Med. **58**, 1—16 (1933b).

— — — — The diagnosis and treatment of Addisonian disease. J. Amer. med. Ass. **100**, 1850—1855 (1933c).

— W. M. Nicholson, L. J. Soffer and M. Strauss: Extracellular and intracellular water loss during suprarenal insufficiency in the dog. Proc. Soc. exp. Biol. (N. Y.) **32**, 1312—1315 (1935).

Hart, F. D.: Treatment of adrenocortical deficiency states. Brit. med. J. **1957 i**, 417—422.

Hartenbach, W.: Über die Erkennung einer prä- und postoperativen Nebennierenrinden-insuffizienz, deren Bedeutung und Behandlung. Med. Klin. **53**, 491—494 (1958).

Hartman, F. A., and K. A. Brownell: The hormone of the adrenal cortex. Science **72**, 76 (1930).

— — Relation of adrenals to diabetes. Proc. Soc. exp. Biol. (N. Y.) **31**, 834—835 (1934).

— — The Adrenal Gland. Philadelphia: Lea & Febiger 1949.

— and H. J. Spoor: Cortin and the Na factor of the adrenal. Endocrinology **26**, 871—878 (1940).

— R. H. Waite and E. F. Powell: The relation of the adrenals to fatigue. Amer. J. Physiol. **60**, 255—269 (1922).

— C. G. MacArthur and W. E. Hartman: A substance which prolongs the life of adrenalectomized cats. Proc. Soc. exp. Biol. (N. Y.) **25**, 69—70 (1927).

— K. A. Brownell, W. E. Hartman, G. A. Dean and C. G. MacArthur: The hormone of the adrenal cortex. Amer. J. Physiol. **86**, 353—359 (1928a).

— F. R. Griffith, jr., and W. E. Hartman: Observations upon adrenalectomized cats treated with cortical hormone. Amer. J. Physiol. **86**, 360—370 (1928b).

— K. A. Brownell and W. E. Hartman: A further study of the hormone of the adrenal cortex. Amer. J. Physiol. **95**, 670—680 (1930).

— G. W. Thorn, L. M. Lockie, C. W. Greene and B. D. Bowen: Treatment of Addison's disease with an extract of suprarenal cortex (cortin). J. Amer. med. Ass. **98**, 788—793 (1932a).

— C. W. Greene, B. D. Bowen and G. W. Thorn: Further experience with cortin therapy. J. Amer. med. Ass. **99**, 1478—1483 (1932b).

— H. J. Spoor, L. A. Lewis and J. E. Gabriel: The sodium factor of the adrenal. Amer. J. Physiol. **126**, 527—528 (1939).

— L. A. Lewis and J. E. Gabriel: Further studies on the refractory state developed following repeated injections of adrenal extract. Endocrinology **26**, 879—885 (1940).

— D. E. Smith and L. A. Lewis: Adrenal functions in the opossum. Endocrinology **32**, 340—344 (1943).

Hastings, A. B., and E. L. Compere: Effect of bilateral suprarenalectomy on certain constituents of the blood of dogs. Proc. Soc. exp. Biol. (N. Y.) **28**, 376—378 (1931).

HEALEY, C. E., and C. C. GUY: Pseudohermaphroditismus masculinis externus associated with suprarenal hyperplasia and vascular hypertension. Arch. Path. **12**, 543—561 (1931).

HEARD, R. D. H., H. SOBEL and E. H. VENNING: The neutral lipid-soluble reducing substances of urine as an index of adrenal cortical function. J. biol. Chem. **165**, 699—710 (1946).

HECHTER, O.: Concerning the hypersensitivity of adrenalectomized rats to vascular stress. Endocrinology **36**, 77—87 (1945).

HENCH, P. S., E. C. KENDALL, C. H. SLOCUMB and H. F. POLLEY: Effect of hormone of adrenal cortex (17-hydroxy-11-dehydrocorticosterone: compound E) and of pituitary adreno-corticotropic hormone on rheumatoid arthritis. Proc. Mayo Clin. **24**, 181—197 (1949).

HENI, F.: Frage des Morbus Addison. Klin. Wschr. **18**, 1052—1056 (1939).

— Die Behandlung des Morbus Addison mit subcutaner Implantation von krystallinem Desoxycorticosterone. Klin. Wschr. **20**, 145—146 (1941).

— u. H. MAST: 9-α-Fluorohydrocortison in der Behandlung des Morbus Addison. Medizinische **1957** No. 27—28, 1005—1007.

HERNBERG, C. A.: The effect of adrenalectomy, cold stress and postoperative treatment with cortisone on the mean lymphocyte diameter in the rat. Acta med. scand. **147**, 149—156 (1953).

HERON, W. T., W. M. HALES and D. J. INGLE: Capacity of skeletal muscle in rats to maintain work output. Amer. J. Physiol. **110**, 357—361 (1934).

HERRICK, E. H., and O. TORSTVEIT: Some effects of adrenalectomy in fowls. Endocrinology **22**, 469—473 (1938).

HILL, A. V.: Excitation and accomodation in nerve. Proc. roy. Soc. Lond. B. **119**, 305—355 (1936).

HILL, R. T.: Grafted mouse ovaries and their adrenal cortical function. Endocrinology **42**, 339—351 (1948).

HILL, S. R., JR., R. S. REISS, P. H. FORSHAM and G. W. THORN: The effect of adrenocorticotropin and cortisone on thyroid function: thyroid-adrenocortical interrelationships. J. clin. Endocr. **10**, 1375—1400 (1950).

HILLS, A. G., T. M. CHALMERS, G. D. WEBSTER, JR., and O. ROSENTHAL: Adrenal cortical regulation of the distribution of water and electrolytes in the human body. J. clin. Invest. **32**, 1236—1247 (1953).

— E. H. VENNING, F. C. DOHAN, G. D. WEBSTER, JR., and E. M. RICHARDSON: Pregnancy and adrenocortical function: Endocrine studies of pregnancy occurring in two adrenal-deficient women. J. clin. Invest. **33**, 1466—1480 (1954).

— H. A. ZINTEL and D. W. PARSONS: Observations of human adrenal cortical deficiency. Amer. J. Med. **21**, 358—379 (1956).

HIMWICH, H. E., J. F. FAZEKAS, S. B. BARKER and M. H. HURLBURT: The metabolism of tissues excised from adrenalectomized rats. Amer. J. Physiol. **110**, 348—351 (1934).

HIRSCH, E. F., and J. A. CAPPS: Paroxysmal cyanosis associated with bilateral thrombosis of the suprarenal veins. Arch. intern. Med. **40**, 112—119 (1927).

HIRSCHOWITZ, B. I., D. H. P. STREETEN and H. M. POLLARD: Influence of adrenal cortex on gastric pepsinogen in the rat. Endocrinology **59**, 419—424 (1956).

HOAGLAND, H.: Studies of brain metabolism and electrical activity in relation to adrenocortical physiology. Recent Progr. Hormone Res. **10**, 29—63 (1954).

— Steroid hormones and events in the nervous system. In: K. A. C. ELLIOTT et al., Eds., Neurochemistry. The Chemical Dynamics of Brain and Nerve, 756—784. Springfield: Thomas 1955.

— and D. STONE: Brain and muscle potassium in relation to stressful activities and adrenal cortex function. Amer. J. Physiol. **152**, 423—435 (1948).

— J. R. BERGEN, A. G. SLOCOMBE and C. A. HUNT: Head blood flow, oxygen consumption, and electrical activity in relation to the adrenal cortex. Ann. N.Y. Acad. Sci. **56**, 659—665 (1953a).

— — — — Studies of adrenocortical physiology in relation to the nervous system. Metabolic and toxic diseases of the nervous system. Ass. Res. nerv. Dis. Proc. **32**, 40—60 (1953b).

HOEFNAGEL, D., S. VAN DEN NOORT and S. H. INGBAR: Diffuse cerebral sclerosis with endocrine abnormalities in young males. Brain **85**, 553—568 (1962).

HOELSCHER, B.: Effect of adrenalectomy on cardiac output of parabiotic and single rats. Amer. J. Physiol. **179**, 171—176 (1954).

HOFFMAN, W. C., R. A. LEWIS and G. W. THORN: The electroencephalogram in Addison's disease. Johns Hopk. Hosp. Bull. **70**, 335—361 (1942).

HOFFMANN, F., E. J. HOFFMANN and J. TAKESNIK: The influence of thyroxine and adrenal cortical extract on the oxygen consumption of adrenalectomized rats. J. Physiol. (Lond.) **107**, 251—264 (1948).

— O. CORI and A. T. DE CORI: Effect of thyroxin on the oxygen consumption and nitrogen excretion in adrenalectomized rats. Acta physiol. lat.-amer. **1**, 84—90 (1951).

HOFMANN, F. G.: Studies in adrenal cortical physiology. Thesis, Harvard University 1952.

— E. KNOBIL and R. O. GREEP: Effects of saline on the adrenalectomized rhesus monkey. Amer. J. Physiol. **178**, 361—366 (1954).

HOLUB, D. A., J. W. JAILER, J. I. KITAY and A. G. FRANTZ: Direct and indirect estimation of pituitary adrenocorticotropin reserves in man following adrenal steroid therapy. J. clin. Endocr. **19**, 1540—1552 (1959).

HOPPE, G., u. G. VOGEL: Der Einfluß der Adrenalektomie auf den Zeitwert der muskulären Erregbarkeit. Pflügers Archiv ges. Physiol. **253**, 518—532 (1951).

HOUSSAY, B. A.: Carbohydrate metabolism. New Engl. J. Med. **214**, 971—986 (1936).

— and J. T. LEWIS: The relative importance to life of the cortex and medulla of the adrenal glands. Amer. J. Physiol. **64**, 512—521 (1923).

— A. BIASOTTI and C. T. RIETTI: Diabetogenic action of anterior pituitary extract (in Spanish) Rev. Soc. argent. Biol. **8**, 469—481 (1932).

— — and R. C. DAMBROSI: Glycogen and the hypophysis (in Spanish). Rev. Soc. argent. Biol. **12**, 185—204 (1936).

HUDSON, B.: Melanocyte stimulating hormone. In: E. B. ASTWOOD, Ed., Clinical Endocrinology **1**, 649—655. New York and London: Grune and Stratton 1960.

HUDSON, P. B., A. MITTELMAN and P. MANN: Urinary steroid excretion after total adrenalectomy. I. Levels of 17-ketosteroids in cancer patients maintained on varying amounts of acetate and glycyrrhyzin. J. clin. Endocr. **13**, 1064—1069 (1953).

— — and M. PRODBEREZEC: Use of glycyrrhizin after bilateral adrenalectomy. New Engl. J. Med. **251**, 641—646 (1954).

HUGGINS, C., and W. W. SCOTT: Bilateral adrenalectomy in prostatic cancer. Ann. Surg. **122**, 1031—1041 (1945).

— D. M. BERGENSTAL and A. S. CLEVELAND. The adrenal in cancer. Recent Progr. Hormone Res. **8**, 272—282 (1953).

HUNDLEY, J. M., and R. B. ING: Effect of adrenalectomy and hypophysectomy on achromotrichia in copper deficient rats. Endocrinology **48**, 482—484 (1951).

HUNGERFORD, G. F., W. O. REINHARDT and C. H. LI: Effects of pituitary and adrenal hormones on the numbers of thoracic duct lymphocytes. Blood **7**, 193—206 (1952).

HUNT, A. B., and W. M. MCCONAHEY: Pregnancy associated with diseases of the adrenal glands. Amer. J. Obstet. Gynec. **66**, 970—987 (1953).

HUNTER, W. C., and H. P. RUSH: Amyloidosis of the adrenals as a cause of Addison's disease. Ann. intern. Med. **5**, 404—412 (1926/27).

HURLEY, L. S., and A. F. MORGAN: Carbohydrate metabolism and adrenal cortical function in the pantothenic acid-deficient rat. J. biol. Chem. **195**, 583—590 (1952).

INGALLS, T. H., and D. R. HAYES: Epiphyseal growth: the effect of removal of the adrenal and pituitary glands on the epiphyses of growing rats. Endocrinology **29**, 720—724 (1941).

INGLE, D. J.: Work capacity of the adrenalectomized rat treated with cortin. Amer. J. Physiol. **116**, 622—625 (1936).

— The work performance of adrenalectomized rats treated with corticosterone and chemically related compounds. Endocrinology **26**, 472—477 (1940).

— The quantitative assay of adrenal cortical hormones by the muscle-work test in the adrenalectomized-nephrectomized rat. Endocrinology **34**, 191—202 (1944).

— The functional interrelationship of the anterior pituitary and the adrenal cortex. Ann. intern. Med. **35**, 652—672 (1951).

— and B. L. BAKER: The inhibition of hair growth by estrogens as related to adrenal cortical function in the male rat. Endocrinology **48**, 764—771 (1951).

— and E, C. KENDALL: Atrophy of the adrenal cortex of the rat produced by the administration of large amounts of cortin. Science **86**, 245 (1937).

— and M. H. KUIZENGA: The relative potency of some adrenal cortical steroids in the muscle-work test. Endocrinology **36**, 218—226 (1945).

— and F. D. W. LUKENS: Reversal of fatigue in the adrenalectomized rat by glucose and other agents. Endocrinology **29**, 443—452 (1941).

— and J. E. NEZAMIS: The work performance of adrenalectomized rats given continuous intravenous infusions of glucose. Endocrinology **43**, 261—271 (1948).

— — The effect of adrenal cortex extract with and without epinephrine upon the work of adrenally insufficient rats. Endocrinology **44**, 559—564 (1949).

— H. W. NILSON and E. C. KENDALL: The effect of cortin on the concentrations of some constituents of the blood of adrenalectomized rats. Amer. J. Physiol. **118**, 302—308 (1937).

— G. M. HIGGINS and E. C. KENDALL: Atrophy of the adrenal cortex in the rat produced by the administration of large amounts of cortin. Anat. Rec. **71**, 363—372 (1938).

— G. B. GINTHER and J. E. NEZAMIS: Effect of diet in rats on adrenal weights and on survival following adrenalectomy. Endocrinology **32**, 410—414 (1943).

JABOULAY, M.: La greffe de corps thyroïde et des capsules surrénales dans la maladie de ces glandes. Lyon méd. **84**, 399 (1897).

JACOB, A.: Zwei Fälle von Simmonscher Krankheit (hypophysaere Kachexie) mit besonderer Berücksichtigung der Veränderungen im Zentralnervensystem. Virchows Arch. path. Anat. **246**, 151—182 (1923).

JAFFE, H. L., and A. PLAVSKA: Functioning autoplastic suprarenal transplants. Proc. Soc. exp. Biol. (N. Y.) **23**, 528—530 (1926).

JAFFE, I., and C. BYRON: Potassium tolerance test as an aid in diagnosis of adrenal insufficiency. J. clin. Endocr. **2**, 339—342 (1942).

JAUDON, J. C.: Addison's disease in an infant. J. clin. Endocr. **6**, 558—564 (1946a).

— Addison's disease in children. J. Pediat. **28**, 737—755 (1946b).

— Further observations concerning hypofunction of the adrenals during early life. J. Pediat. **32**, 641—669 (1948).

JEANNERET, O., R. BORTH et R. S. MACH: Emploi d'une préparation retard d'ACTH pour le dépistage de l'insuffisance surrénale. Acta endocr. (Kbh.) **17**, 198—205 (1954).

JEFFERSON, A.: A clinical correlation between encephalopathy and papilloedema in Addison's disease. J. Neurol. Psychiat. **19**, 21—27 (1956).

JEGHERS, H.: Pigmentation of the skin. New Engl. J. Med. **231**, 88—100; 122—136; 181—189 (1944).

JENKINS, D., P. H. FORSHAM, J. C. LAIDLAW, W. J. REDDY and G. W. THORN: Use of ACTH in the diagnosis of adrenal cortical insufficiency. Amer. J. Med. **18**, 3—14 (1955).

JENKINS, J. S., J. W. MEAKIN, D. H. NELSON and G. W. THORN: Inhibition of adrenal steroid 11-oxygenation in the dog. Science **128**, 478—480 (1958).

JOHNSSON, S., and B. HÖGBERG: Observations on the connexion between intermedin and adrenocorticotropic hormone. Nature (Lond.) **169**, 286 (1952).

— — The occurrence of the melanophore hormone in serum from human subjects in different pathological conditions and its relation to corticotrophin. Acta endocr. (Kbh.) **13**, 325—342 (1953).

— R. LUFT, B. SJÖGREN and J. WALDENSTRÖM: Two cases of chronic renal disease simulating Morbus Addison. Acta endocr. (Kbh.) **8**, 165—174 (1951).

JOHNSTON, R. F., and J. F. SMITHCORS: The effects of estrogen on mammary structure of adrenalectomized and thiouracil treated rats. Endocrinology **43**, 193—201 (1948).

JOHNSTON, W. W.: A case of Addison's disease, with marked exacerbations and remissions in the early period; treatment by suprarenal extract for five months, with gain in weight and strength but with increase of pigmentation. Trans. Ass. Amer. Phycns. **15**, 65—70 (1900).

JONES, E. S.: Cellular electrolytes and adrenal steroids. Nature (Lond.) **176**, 269 (1955).

JOSEPHSON, B.: Operated endocrine case of Cushing's syndrome changing to cachexia. Nord. med. **16**, 1396—1441 (1938).

KAISER, I. H.: Electrolyte and steroid metabolism in a pregnant Addisonian. J. clin. Endocr. **16**, 1251—1261 (1956).

KATZIN, B., and C. N. H. LONG: The effect of adrenal cortical extract on carbohydrate and protein metabolism of the rat. Amer. J. Physiol. **126**, 551 (1939).

KEKWICK, A., and G. L. S. PAWAN: Oral aldosterone. Effect in a case of Addison's disease. Lancet **1954 ii**, 162—164.

KELLOGG, R. H., and W. R. BURACK: Effect of adrenalectomy upon the diuresis produced by isotonic saline solutions in rats. Amer. J. Physiol. **177**, 38—43 (1954).

KEMPERER, C. F.: The treatment of Addison's disease by the implantation method. Ann. intern. Med. **22**, 161—169 (1945).

KENDALL, E. C.: A chemical and physiological investigation of the suprarenal cortex. Cold Spr. Harb. Symp. quant. Biol. **5**, 299—312 (1937).

— The influence of cortin, insulin and glucose on the metabolism of potassium. Proc. Mayo Clin. **13**, 519—523 (1938).

— The function of the adrenal cortex. Proc. Mayo Clin. **15**, 297—304 (1940).

— The function of the adrenal cortex. In: Glandular Physiology and Therapy, Chapter XVIII. Chicago: The Amer. Med. Ass. 1942.

— The influence of the adrenal cortex on the metabolism of water and electrolytes. Vitam. and Horm. **6**, 277—327 (1948).

— H. L. MASON, B. F. MCKENZIE, C. S. MYERS and G. A. KOELSCHE: Isolation in crystalline form of the hormone essential to life from the suprarenal cortex: its chemical nature and physiologic properties. Proc. Mayo Clin. **9**, 245—250 (1934).

— — — — and W. D. ALLERS: Recent developments in the investigation of the hormone of the suprarenal cortex. Proc. Mayo Clin. **10**, 245—246 (1935).

— — and C. S. MYERS: Concerning the chemical nature of the hormone of the adrenal cortex. Proc. Mayo Clin. **11**, 351—352 (1936).

KENIGSBERG, S., S. PEARSON and T. H. MCGAVACK: The excretion of 17-ketosteroids. 1. Normal values in relation to age and sex. J. clin. Endocr. **9**, 426—429 (1949).

KEPLER, E. J.: Diseases of the adrenal glands. A review with special reference to clinical aspects. Arch. intern. Med. **56**, 105—135 (1935).
— G. A. PETERS and H. L. MASON: Addison's disease associated with pubic and axillary alopecia and normal menses. J. clin. Endocr. **3**, 497—499 (1943).
KING, N. B., and H. L. MASON: Urinary corticosteroid values of children as determined chemically. J. clin. Endocr. **10**, 479—491 (1950).
KINNICUT, F. P.: The therapeutics of the internal secretions. Amer. J. med. Sci. **114**, 1—23 (1897).
KINSELL, L. W., D. BRYANT and F. ALBRIGHT: Rate of growth of axillary hair as a diagnostic index. J. clin. Endocr. **14**, 897—909 (1954).
KLEEMAN, C. R., M. QUINN and D. E. BASS: The control of variable influences on the electrolyte content of sweat and their application to clinical study of adrenal function. J. clin. Endocr. **14**, 1126—1133 (1954).
— M. H. MAXWELL and R. E. ROCKNEY: Mechanisms of impaired water excretion in adrenal and pituitary insufficiency. I. The role of altered glomerular filtration rate and solute excretion. J. clin. Invest. **37**, 1799—1808 (1958).
— J. KOPLOWITZ and M. H. MAXWELL: Acute effect of 6-methyl prednisolone (Medrol) on impaired water excretion of adrenal and pituitary insufficiency. Proc. Soc. exp. Biol. (N. Y.) **100**, 615—617 (1959).
KLEIN, R., J. FORTUNATO and C. PAPADATOS: Free blood corticoids in the newborn infant. J. clin. Invest. **33**, 35—37 (1954).
KLINE, E. M.: The diagnosis of Addison's disease. J. Amer. med. Ass. **108**, 1592—1593 (1937).
KLIPPEL: Encéphalopathie addisonnienne. Rev. neurol. **7**, 898—899 (1899).
KNOBIL, E., F. G. HOFMANN and R. O. GREEP: Nature of adrenal crisis in the rhesus monkey. Fed. Proc. **12**, 78 (1953).
KNOWLTON, A. I., and R. A. KRITZLER: The development of diabetes mellitus in Addison's disease. J. clin. Endocr. **9**, 36—47 (1949).
— G. H. MUDGE and J. W. JAILER: Pregnancy in Addison's disease. Report of four patients. J. clin. Endocr. **9**, 514—528 (1949).
KOELSCHE, G. A., and E. C. KENDALL: The relation of the suprarenal cortical hormone to nitrogen metabolism in experimental hyperthyroidism. Amer. J. Physiol. **113**, 335—349 (1935).
KOHN, A.: Anenzephalie und Nebenniere. Dtsch. med. Wschr. **48**, 1666 (1922).
KOJIMA, T.: The effect on the life of rabbits of the removal of the main suprarenals and the accessory suprarenal cortical tissue. Tohoku J. exp. Med. **13**, 357—378 (1929).
KOMIYA, A., and K. SHIBATA: Effect of adrenalectomy and replacement with adrenocortical steroids on barbital anesthesia in mice. J. Pharmacol. **116**, 98—106 (1956).
KOSTER, H., and L. P. KASMAN: Effect of desoxycorticosterone acetate in postoperative shock. Arch. Surg. **45**, 272—285 (1942).
KOTTKE, F. J., C. F. CODE and E. H. WOOD: Urine dilution and concentration tests in adrenalectomized dogs. Amer. J. Physiol. **136**, 229—243 (1942).
KRAMÁR, J., and M. SIMAY-KRAMÁR: The effect of adrenalectomy, surgical trauma, and ether anesthesia upon the capillary resistance of the albino rat. Endocrinology **52**, 453—562 (1953).
KUDO, M., S. MUTAI and K. URAGUCHI: The effects of cortisone, cortate, parotin and adrenalectomy on Japanese encephalitis virus infection in mice. Yokohama med. Bull. **5**, 337—344 (1954).
KÜHL, G.: Untersuchungen zur Hormonwirkung der Nebennierenrinde. Pflügers Arch. ges. Physiol. **215**, 277—290 (1927).
KUHLMANN, D., C. RAGAN, J. W. FERREBEE, D. W. ATCHLEY and R. F. LOEB: Toxic effects of desoxycorticosterone esters in dogs. Science **90**, 496—497 (1939).
KUIZENGA, M. H., J. W. NELSON and D. J. INGLE: The effect of 17-hydroxy-11-dehydrocorticosterone on the growth of young adrenalectomized rats. Amer. J. Physiol. **139**, 499—503 (1943).
— — S. C. LYSTER and D. J. INGLE: Fractionation of hog adrenal cortical extract. J. biol. Chem. **160**, 15—24 (1945).
KURIYAMA, S.: The adrenals in relation to carbohydrate metabolism. II. The influence of adrenalectomy upon the glycogenetic power of the liver. J. biol. Chem. **34**, 287—297 (1918).
KYLE, J., J. S. LOGAN, D. W. NEILL and R. B. WELBOURN: Influence of the adrenal cortex on gastric secretion in man. Lancet **1956 i**, 664—666.
KYLE, L. H., W. P. WALSH and P. D. DOOLAN: Corticotropin (ACTH) and adrenal hormones during operative stress in endocrine disease. Arch. intern. Med. **91**, 283—295 (1953).
LABHART, A.: Klinik der inneren Sekretion. Berlin-Göttingen-Heidelberg: Springer-Verlag 1957.
LACASSAGNE, A., et N. SAMSSONOW: De l'effet de la destruction totale ou partielle des capsules surrénales par le rayonnement caustique des foyers radioactifs. C. R. Soc. Biol. (Paris) **89**, 72—74 (1923).

LAIDLAW, J. C., J. F. DINGMAN, W. L. ARONS, J. T. FINKENSTAEDT and G. W. THORN: Comparison of the metabolic effects of cortisone and hydrocortisone in man. Ann. N. Y. Acad. Sci. **61**, 315—323 (1955a).

— W. J. REDDY, D. JENKINS, N. ABU HAYDAR, A. E. RENOLD and G. W. THORN: Advances in the diagnosis of altered states of adrenocortical function. New Engl. J. Med. **253**, 747—753 (1955b).

LANGLEY, F. A., K. V. LODGE and A. S. WOODCOCK: The survival of rats after adrenalectomy. Brit. J. exp. Path. **37**, 350—356 (1956).

LANGLOIS, P.: Destruction des capsules surrénales chez le chien. C. R. Soc. Biol. (Paris) **45**, 444—448 (1893a)

— Destruction des capsules surrénales. Arch. Physiol. (Paris) **5**, 488—498 (1893b).

LANMAN, J. T.: Function of adrenal cortex in premature infants. I. Infants without recognized disease. Pediatrics **11**, 120—128 (1953).

LAPIDES, J.: An operative technique for adrenalectomy. Bull. N. Y. Acad. Med. **34**, 303—310 (1958).

LEARY, O. C., JR., and J. J. MASTERS: Adrenal insufficiency produced by metastases from gastric carcinoma. Ann. intern. Med. **46**, 1161—1165 (1957).

LEATHEM, J. H.: Plasma protein concentrations in adrenalectomized rats maintained on sodium chloride. Proc. Soc. exp. Biol. (N. Y.) **60**, 260—262 (1945).

— and V. A. DRILL: The role of the adrenals and the hypophysis in the control of systolic blood pressure in the rat. Endocrinology **35**, 221 (1944).

LECOQ, R.: Appréciation chronaximétrique du retentissement nerveux des perturbations métaboliques et humorales. Rev. Path. comp. **54**, 529—542 (1954).

— P. CHAUCHARD et H. MAXOUÉ: Action de quelques préparations glandulaires et hormonales sur la régulation de l'équilibre acido-basique. Bull. Soc. Chim. biol. (Paris) **34**, 239—246 (1952).

LEDINGHAM, J. M.: The influence of the adrenal on the water and electrolyte disturbances following nephrectomy, and its relation to renoprival hypertension. Clin. Sci. **13**, 535—542 (1954).

LEIDERMAN, P. H., and R. KATZMAN: Effect of adrenalectomy, desoxycorticosterone and cortisone on brain potassium exchange. Amer. J. Physiol. **175**, 271—275 (1953).

LEITH, W., and J. C. BECK: 9-α-Fluorohydrocortisone alone and combined with hydrocortisone in the management of chronic adrenal insufficiency. J. clin. Endocr. **17**, 280—290 (1957).

LELOIR, L. F.: Surrénales et diabète pancreatique. C. R. Soc. Biol. (Paris) **117**, 459—461 (1934).

LEONARD, M. F.: Chronic idiopathic hypoparathyroidism with superimposed Addison's disease in a child. J. clin. Endocr. **6**, 493—506 (1946).

LERNER, A. B., and Y. TAKAHASHI: Hormonal control of melanin pigmentation. Recent Progr. Hormone Res. **12**, 303—320 (1956).

— K. SHIZUME and I. BUNDING: The mechanism of endocrine control of melanin pigmentation. J. clin. Endocr. **14**, 1463—1490 (1954).

LEVINSON, S. M., H. J. TAGNON, W. L. GOODPASTOR, R. W. GREEN, F. H. L. TAYLOR and C. C. LUND: Addison's disease associated with amyloidosis following thermal burns. New Engl. J. Med. **237**, 152—156 (1947).

LEVY, M., L. R. ARONSON and J. S. ROSENBLATT: Effects of prepuberal adrenalectomy on male cats maintained on DCA — a comparison with intact and prepuberally castrated males. Anat. Rec. **125**, 587 (1956).

LEVY, M. S., M. H. POWER and E. J. KEPLER: The specificity of the "water test" as a diagnostic procedure in Addison's disease. J. clin. Endocr. **6**, 607—632 (1946).

LEVY-SIMPSON, S.: The use of synthetic desoxycorticosterone acetate in Addison's disease. Lancet **1938 ii**, 557—558.

— Discussion on recent developments in the treatment of Addison's disease. Proc. roy. Soc. Med. **32**, 685—689 (1939).

LEWIN, G.: Über Morbus Addisonii. Charité-Ann. **17**, 536—739 (1890—1891).

LEWIS, J. T., et M. MAGENTA: Sur le mécanismes de récupération de la glycémie après l'action de l'insuline. C. R. Soc. Biol. (Paris) **92**, 821—822 (1925).

— and A. TORINO: The sensitivity of rats to morphine after adrenalectomy. Rev. Soc. argent. Biol. **2**, 459—473 (1926).

LEWIS, L., R. F. ROBINSON, J. YEE, L. A. HACKER and G. EISEN: Fatal adrenal cortical insufficiency precipitated by surgery during prolonged continuous cortisone treatment. Ann. intern. Med. **39**, 116—126 (1953).

LI, C. H.: Hormones of anterior pituitary gland. I. Growth and adrenocorticotropic hormones. Advanc. Protein Chem. II, 101—190 (1956).

LICHTON, I. J., M. H. GOLDBLATT and S. G. STOLPE: Extended survival of adrenalectomized rats possessing ear or spleen-grafted ovaries. Endocrinology **52**, 546—553 (1953).

LIDDLE, G. W., M. M. PECHET and F. C. BARTTER: Enhancement of biological activities of corticosteroids by substitution of halogen atoms in 9-α position. Science **120**, 496—497 (1954).

— D. ISLAND, E. M. LANCE and A. P. HARRIS: Alteration of adrenal steroid patterns in man resulting from treatment with a chemical inhibitor of 11β-hydroxylation. J. clin. Endocr. **18**, 906—912 (1958).

— H. L. ESTEP, J. W. KENDALL, JR., W. C. WILLIAMS, JR., and A. W. TOWNES: Clinical application of a new test of pituitary reserve. J. clin. Endocr. **19**, 875—894 (1959).

LIEBERMAN, A. H., and J. A. LUETSCHER: Some effects of abnormalities of pituitary, adrenal or thyroid function on excretion of aldosterone and the response to corticotropin or sodium deprivation. J. clin. Endocr. **20**, 1004—1016 (1960).

LILIENFELD, A.: The use of the low salt diet in the diagnosis of Addison's disease. J. Amer. med. Ass. **110**, 804—805 (1938).

LIPSETT, M. B., and O. H. PEARSON: Pathology and treatment of adrenal crisis. New Engl. J. Med. **254**, 511—514 (1956).

LITTLE, E. G.: Cases of purpura, ending fatally, associated with haemorrhage into the suprarenal capsules. Brit. J. Derm. **13**, 445—467 (1901).

LOCKE, W., N. B. TALBOT, H. S. JONES and J. WORCESTER: Studies on the combined use of measurements of sweat electrolyte composition and rate of sweating as an index of adrenal cortical activity. J. clin. Invest. **30**, 325—337 (1951).

LOEB, R. F.: Chemical changes in the blood in Addison's disease. Science **76**, 420—421 (1932).

— Effect of sodium chloride in treatment of a patient with Addison's disease. Proc. Soc. exp. Biol. (N. Y.) **30**, 808—812 (1933).

— The adrenal cortex. J. Amer. med. Ass. **104**, 2177—2182 (1935).

— The adrenal cortex. In: Glandular Physiology and Therapy, Chap. XX. Chicago: Amer. Med. Ass. 1935.

— Adrenal cortex insufficiency. In: Glandular Physiology and Therapy, Chap. XIX. Chicago: Amer. Med. Ass. 1942.

— D. W. ATCHLEY, E. M. BENEDICT and J. LELAND: Electrolyte balance studies in adrenalectomized dogs with particular reference to the excretion of sodium. J. exp. Med. **57**, 775—792 (1933).

— — E. B. GUTMAN and R. JILLSON: On the mechanism of sodium depletion in Addison's disease. Proc. Soc. exp. Biol. (N. Y.) **31**, 130—133 (1934).

— — and J. STAHL: The role of sodium in adrenal insufficiency. J. Amer. med. Ass. **104**, 2149—2154 (1935).

— — L. W. FERREBEE and C. RAGAN: Observations of the effect of desoxycorticosterone esters and progesterone in patients with Addison's disease. Trans. Ass. Amer. Phycns. **52**, 228—235 (1937).

LONG, C. N. H.: Studies on the "diabetogenic" action of the anterior pituitary. Cold Spr. Harb. Symp. quant. Biol. **5**, 344—356 (1937).

— and F. D. W. LUKENS: The effects of adrenalectomy and hypophysectomy upon experimental diabetes in the cat. J. exp. Med. **63**, 465—490 (1936).

— B. KATZIN and E. G. FRY: The adrenal cortex and carbohydrate metabolism. Endocrinology **26**, 309—344 (1940).

LOTSPEICH, W. D.: The effect of adrenalectomy on the renal tubular reabsorption of water in the rat. Endocrinology **44**, 314—316 (1949).

LUCAS, G. H. W.: Blood and urine findings in desuprarenalized dogs. Amer. J. Physiol. **77**, 114—125 (1926).

LUETSCHER, J. A., and B. J. AXELRAD: Sodium-retaining corticoid in the urine of normal children and adults and of patients with hypoadrenalism or hypopituitarism. J. clin. Endocr. **14**, 1086—1089 (1954).

LUFT, R., and B. SJÖGREN: The significance of renal function for the effect of desoxycorticosterone acetate (DCA) in Addison's disease. Acta endocr. (Kbh.) **2**, 365—378 (1949).

— B. SJÖGREN, D. IKKOS, H. LJUNGGREN and H. TARUKOSKI: Clinical studies in electrolyte and fluid metabolism. Recent Progr. Hormone Res. **10**, 425—470 (1954).

LUKENS, F. D. W.: Diagnosis and treatment of disorders of the adrenal glands. Med. Clin. N. Amer. **26**, 1803—1815 (1942).

— H. F. FLIPPIN and F. M. THIGPEN: Adrenal cortical adenoma with absence of the opposite adrenal. Amer. J. med. Sci. **193**, 812—820 (1937).

LÜTHY, E., and F. VERZÁR: Adenosinetriphosphatase and hexokinase in the epithelium of the small intestine in normal and adrenalectomized rats. Biochem. J. **57**, 316—320 (1954).

LYONS, W. R., C. H. LI and R. E. JOHNSON: The hormonal control of mammary growth and lactation. Recent Progr. Hormone Res. **14**, 219—254 (1958).

MAASEN, A. P.: Adrenals and bone-growth. Acta endocr. (Kbh.) **9**, 135—140 (1952).

MACH, R. S., J. FABBE, A. DUCKERT, R. BORTH et P. DUCUMMUN: Action clinique et métabolique de l'aldosterone (Electrocortine). Schweiz. med. Wschr. **84**, 407-416 (1954).

MACLEAN, J. P., M. B. LIPSETT, M. C. LI, C. D. WEST and O. H. PEARSON: Regulation of salt metabolism after hypophysectomy in man. J. clin. Endocr. **17**, 346-355 (1957).

MACLEAN, N., and J. J. IRONSIDE: Adrenal apoplexy. Brit. med. J. **1958 i**, 1162—1163.

MADDEN, R. J., and H. H. RAMSBURG: Adrenalectomy in the Shay rat. Gastroenterology **18**, 128—134 (1951a).

— — Gastric secretion in the adrenalectomized rat. Endocrinology **49**, 82—85 (1951b).

MADDOCK, W. O., R. B. LEACH, S. P. KLEIN and G. B. MYERS: Selective pituitary failure: an example characterized by deficient ACTH and gonadotrophin secretion with intact thyrotrophin secretion. Amer. J. med. Sci. **226**, 509—515 (1953).

MADSEN, P. O.: Prolonged, heavy salt loading in adrenalectomized dogs. Acta endocr. (Kbh.) **20**, 169—172 (1955).

MAES, J.: Study of the syndrome of adrenal insufficiency in the frog. Arch. int. Physiol. **45**, 135—188 (1937).

MALONEY, P. J.: Addison's disease due to chronic disseminated coccidiomycosis. Arch. intern. Med. **90**, 869—878 (1952).

MANDELSTAM, P., J.W. GOLDZIEHER, H. S. SOROFF and N. GREEN: The pituitary-adrenal axis: Acute adrenocortical insufficiency and persistent occult dysfunction following thermal injury. J. clin. Endocr. **18**, 284—299 (1958).

MANDL, A. M.: The sensitivity of the adrenalectomized rat to gonadotrophins. J. Endocr. **11**, 359—376 (1954).

MANN, F. C.: A study of gastric ulcers following removal of the adrenals. J. exp. Med. **23**, 203—209 (1916).

MÅNSSON, J., and N. OBEL: The technic of adrenalectomy in the ruminant. Cornell Vet. **48**, 197—200 (1958).

MARAÑON, G. L., P. SALA and G. ARGUELLES: Digestive symptoms in chronic suprarenal insufficiency (Addison's disease). Endocrinology **18**, 497—504 (1934).

MARENZI, A. D.: Blood potassium and suprarenal glands. Endocrinology **23**, 330—338 (1938).

MARINE, D.: Calcification of the suprarenal glands of cats. J. exp. Med. **43**, 495—499 (1926).

— and E. J. BAUMANN: Influence of glands with internal secretions on the respiratory exchange. II. Effect of suprarenal insufficiency (by removal or by freezing) in rabits. Amer. J. Physiol. **57**, 135—152 (1921).

— — Influence of glands with internal secretions on the respiratory exchange. III. Effect of suprarenal insufficiency (by removal) in thyroidectomized rabbits. Amer. J. Physiol. **59**, 353—368 (1922).

— — Duration of life after suprarenalectomy in cats and attempts to prolong it by injection of solutions containing sodium salts, glucose and glycerol. Amer. J. Physiol. **81**, 86—100 (1927).

MARINONE, G., F. CORSO and G. DORIA: Effects of adrenalectomy and the administration of cortisone on the reproductive activity of erythroblasts in the medulla of the albino rat (in Italian). Haematologica **41**, 667—680 (1956).

MARINO-ZUCCO, F., u. U. DUTTO: Chemische Untersuchungen über die Addisonische Krankheit. Untersuch. Naturlehre Menschen und Thiere **14**, 617—622 (1894).

MARSHALL, E. K., JR., and D. M. DAVIS: The influence of the adrenals on the kidneys. J. Pharmacol. exp. Ther. **8**, 525—550 (1916).

MARSHALL, P. B.: The influence of adrenal cortical deficiency on the histamine content of rat tissues. J. Physiol. (Lond.) **102**, 180—190 (1943).

MARTIN, S. J.: The effect of complete suprarenalectomy on the oestrual cycle of the white rat with reference to the suprarenal-pituitary relationship. Amer. J. Physiol. **100**, 180—191 (1932).

— and J. F. FAZEKAS: Effect of sodium chloride therapy on oestrus cycle and hypophysis of bilaterally suprarenalectomized rats. Proc. Soc. exp. Biol. (N. Y.) **37**, 369—372 (1937).

MARTINOVITCH, R. N.: Infantile rat adrenal transplanted into the anterior eye chamber of adrenalectomized hosts after cultivation in vitro. J. exp. Zool. **129**, 97—127 (1955).

MARTLAND, H. S.: Fulminating meningococcic infection with bilateral massive adrenal hemorrhage (the Waterhouse-Friderichsen syndrome). Arch. Path. **37**, 147—158 (1944).

MARZULLO, E. R., and M. B. HANDELSMAN: Pituitary necrosis and diabetes mellitus. J. clin. Endocr. **11**, 537—547 (1951).

MASON, A. S.: Metabolic response to total adrenalectomy and hypophysectomy. Lancet **1955 ii**, 632—636.

MASON, H. L.: Methods for the assessment of adrenocortical function. J. clin. Endocr. **15**, 1035—1037 (1955).

— C. S. MYERS and E. C. KENDALL: The chemistry of crystalline substances isolated from the suprarenal gland. J. biol. Chem. **114**, 613—631 (1936a).

MASON, H. L., C. S. MYERS and E. C. KENDALL: Chemical studies of the suprarenal cortex. J. biol. Chem. **116**, 267—276 (1936b).
— M. H. POWER, E. H. RYNEARSON, L. C. CIARAMELLI, C. H. LI and H. M. EVANS: Results of administration of anterior pituitary adrenocorticotropic hormone to a normal human being. J. biol. Chem. **169**, 223—224 (1947).
MATSON, V. F., and B. B. LONGWELL: The excretion of neutral lipid-soluble reducing substances by infants. J. clin. Endocr. **9**, 646—649 (1949).
MCARTHUR, J. W., R. W. RAWSON, J. H. MEANS and O. COPE: Thyrotoxic crisis. J. Amer. med. Ass. **134**, 868—874 (1947).
MCCULLAGH, E. P., L. A. LEWIS and F. L. SHIVELY, JR.: On the absorption rate of desoxycorticosterone acetate pellets. J. clin. Endocr. **3**, 493—496 (1943).
— R. W. SCHNEIDER, W. BOWMAN and M. B. SMITH: Adrenal and testicular deficiency. J. clin. Endocr. **8**, 275—294 (1948).
MCCUTCHEON, M.: The relation of Addison's disease to amyloidosis. Amer. J. med. Sci. **166**, 197—200 (1922).
MCGAVACK, T. H.: Some pitfalls in the treatment of Addison's disease. J. clin. Endocr. **1**, 68—75 (1941).
MCKAY, D. G., C. H. BURNETT and B. A. BURROWS: Panhypopituitarism: case report correlating clinical and pathologic findings. J. clin. Endocr. **10**, 540—555 (1950).
MCMILLAN, C. D., W. N. POWELL and R. D. HAINES: The diagnosis and treatment of adrenal insufficiency. Med. Clin. N. Amer. **39**, 1111—1125 (1955).
MCQUEENEY, A. J., L. L. ASBURN, F. S. DAFT and R. FAULKNER: Tissue changes and sodium balance in pantothenic acid-deficient rats. Endocrinology **41**, 441—450 (1947).
MEAKIN, J. W., D. H. NELSON and G. W. THORN: Addison's disease in two brothers. J. clin. Endocr. **19**, 726—731 (1959).
MEANS, J. H., S. HERTZ and J. LERMAN: The pituitary type of myxedema or Simmonds' disease masquerading as myxedema. Trans. Ass. Amer. Phycns. **50**, 32—53 (1940).
MEGEL, H., and A. S. GORDON: The relation of the adrenal to red blood cell fragility. Endocrinology **48**, 391—398 (1951).
MEIKLEJOHN, A. P.: The physiology and biochemistry of ascorbic acid. Vitamins and Hormones **11**, 61—96 (1953).
MELAMPY, R. M., D. W. CHENG and L. C. NORTHROP: Effect of pantothenic acid deficiency upon adrenal cortex, thymus, spleen and circulating lymphocytes in mice. Proc. Soc. exp. Biol. (N. Y.) **76**, 24—27 (1951).
MELBY, J. C., and W. W. SPINK: Comparative studies on adrenal cortical function and cortisol metabolism in healthy adults and in patients with shock due to infection. J. clin. Invest. **37**, 1791—1798 (1958).
— R. H. EGDAHL and W. W. SPINK: Investigations on the metabolism of hydrocortisone in patients with shock due to infection. J. Lab. clin. Med. **50**, 933—934 (1957).
MENDELSOHN, M. L., and O. H. PEARSON: Alterations in water and salt metabolism after bilateral adrenalectomy in man. J. clin. Endocr. **15**, 409—423 (1955).
MEYER, R.: Nebennieren bei Anenzephalie. Virchows Arch. path. Anat. **210**, 158—164 (1912).
MIGEON, C. J., F. H. TYLER, J. P. MAHONEY, A. A. FLORENTIN, H. CASTLE, E. L. BLISS and L. T. SAMUELS: The diurnal variation of plasma levels and urinary excretion of 17-hydroxycorticosteroids in normal subjects, night workers and blind subjects. J. clin. Endocr. **16**, 622—633 (1956).
MILLER, H. C., and D. C. DARROW: Relation of serum and muscle electrolyte, particularly potassium, to voluntary exercise. Amer. J. Physiol. **132**, 801—809 (1941).
MILLER, M. R.: The normal histology and experimental alteration of the adrenal of the viviparous lizard, *Xantusia vigilis*. Anat. Rec. **113**, 309—323 (1952).
— Experimental alteration of the adrenal histology of the urodele amphibian, *Triturus torosus*. Anat. Rec. **116**, 205—225 (1953).
MILTON, G. W.: The effect of adrenal ablation on the muscle action potential of rats. Aust. J. exp. Biol. **31**, 607—612 (1953).
MOEHLIG, R. C.: Addison's disease followed for nine years. J. clin. Endocr. **7**, 134—151 (1947).
MOLHUYSEN, J. A., J. GERBRANDY, L. A. DE VRIES, J. C. DE JONG, J. B. LENSTRA, K. P. TURNER and J. G. G. BORST: A liquorice extract with deoxycorticosterone-like action. Lancet **1950 ii**, 381—386.
MOOG, F.: The development of function in the adrenal cortex. In: A. GORBMAN, Ed., Comparative Endocrinology, pp. 624—638. New York: Wiley 1959.
MOOLTEN, S. E.: Symmetrical hemorrhagic necrosis of adrenal glands complicating coronary thrombosis. J. Mt. Sinai Hosp. **24**, 1042—1046 (1957).
MOORE, B., and C. O. PURINTON: On the effects of complete removal of the suprarenal glands. Amer. J. Physiol. **5**, 182—190 (1901).

MOORE, F. P., and E. G. CERMAK: Adrenal cysts and adrenal insufficiency in an infant with fatal termination. J. Pediat. **36**, 91—95 (1950).
MORGAN, A. F.: The effects of vitamin deficiencies on adrenocortical function. Vitamins and Hormones **9**, 161—212 (1951).
— and E. M. LEWIS: The modification of choline deficiency by simultaneous pantothenic acid deficiency. J. biol. Chem. **200**, 839—850 (1953).
MORRISON, A. B.: Bilateral adrenalectomy in the guinea-pig. J. Endocr. **11**, 97—101 (1954).
MORSE, W. I., W. A. COCHRANE and P. L. LANDRIGAN: Familial hypoparathyroidism with pernicious anemia, steatorrhea and adrenocortical insufficiency. New Engl. J. Med. **264**, 1021—1026 (1961).
MOSIER, H. D.: Hypoplasia of the pituitary and adrenal cortex. J. Pediat. **48**, 633—639 (1956).
DE MOWBRAY, R. R., and P. M. F. BISHOP: ACTH in diagnosis of adrenal insufficiency (Thorn test). Brit. med. J. **1953i**, 17—29.
MUDGE, G. H.: The kidney and potassium. Bull. N. Y. Acad. Med. **34**, 152—162 (1958).
MUIRHEAD, A. L.: An autograph history of a case of Addison's disease. J. Amer. med. Ass. **76**, 652 (1921).
MULLER, A. F., E. L. MANNING and A. M. RIONDEL: Diurnal variation of aldosterone related to position and activity in normal subjects and patients with pituitary insufficiency. In: A. F. MULLER and C. M. O'CONNER, Eds., Aldosterone, 141—142. Boston: Little, Brown and Company 1958.
NABARRO, J. D. N.: The adrenal cortex and renal function. In: F. R. WINTON, Ed., Modern Views on the Secretion of Urine, 148—185. Boston: Little, Brown and Co. 1956.
— and G. WALKER: Treatment of adrenal cortical insufficiency. Brit. med. J. **1957ii**, 17—19.
NADEL, E. M., and J. J. SCHNEIDER: Excretion of formaldehydogenic substances by normal and scorbutic guinea pigs. Endocrinology **51**, 5—11 (1952).
NEHER, R., and A. WETTSTEIN: Physiochemical estimation of aldosterone in urine. J. clin. Invest. **35**, 800—805 (1956).
NELSON, D. H., and L. T. SAMUELS: Method for determination of 17-hydroxycorticosteroids in blood: 17-Hydroxycorticosterone in the peripheral circulation. J. clin. Endocr. **12**, 519—526 (1952).
— — D. G. WILLARDSON and F. H. TYLER: 17-Hydroxycorticosteroids in human peripheral blood. J. clin. Endocr. **11**, 1021—1029 (1951).
NETRAVISESH, V., and H. L. WHITE: Sodium loss and gain on adrenal insufficiency and recovery in dogs. Amer. J. Physiol. **175**, 25—28 (1953).
NETTROUR, W. S., and E. H. RYNEARSON: A salt-poor diet as a provocative test for Addison's disease. Proc. Mayo Clin. **9**, 550—556 (1934).
NICHOLS, G., JR., and N. NICHOLS: The role of bone in sodium metabolism. Metabolism **4**, 438—446 (1956).
NICHOLS, J., O. L. LESCURE and C. J. MIGEON: Levels of 17-ketosteroids in maternal and cord plasma in term anencephaly. J. clin. Endocr. **18**, 444—452 (1958).
NICHOLSON, W. M., and L. J. SOFFER: Cardiac arrhythmia in experimental suprarenal insufficiency in dogs. Bull. Johns Hopk. Hosp. **56**, 236—243 (1935).
NILSON, H. W.: Corticoadrenal insufficiency: metabolism studies on potassium, sodium and chloride. Amer. J. Physiol. **118**, 620—631 (1937).
NITSCHKE, A., u. B. KRATSCHELL: Blutveränderungen bei der toxischen Diphtherie und ihre Beziehung zum Nebennieren-Ausfall. Klin. Wschr. **17**, 374—376 (1938).
NOBLE, R. L.: Physiology of the adrenal cortex. In: G. PINCUS and K. V. THIMANN, Eds., The Hormones, Vol. II, 65—180. New York: Academic Press 1950.
— Physiology of the adrenal cortex. In: G. PINCUS and K. V. THIMANN, Eds., The Hormones, Vol. III, 685—819. New York: Academic Press 1955.
— Hormonal regulation of tumor growth. Pharmacol. Rev. **9**, 367—426 (1957).
NOVELLI, G. D.: Metabolic function of pantothenic acid. Physiol. Rev. **33**, 525—543 (1953).
O'DONNELL, W. M.: Changing pathogenesis of Addison's disease. Arch. intern. Med. **86**, 266—279 (1950).
OLDBERG, E.: An attempt to auto-transplant the adrenal. Amer. J. Physiol. **91**, 275—283 (1929—1930).
OLIVER, G., and E. A. SCHÄFER: The physiological effects of extracts of the suprarenal capsules. J. Physiol. (Lond.) **18**, 230—276 (1895).
OPPENHEIM, R., et M. LOEPER: Insuffisance surrénale chronique expérimentale par injections intra-capsulaires des poisons du bacille tuberculeux humain d'Auclair. C. R. Soc. Biol. (Paris) **55**, 330—332 (1903a).
— — L'insuffisance surrénale expérimentale par lésions directs des capsules. C. R. Soc. Biol. (Paris) **55**, 332—333 (1903b).
OSLER, W.: On six cases of Addison's disease, with the report of a case greatly benefited by the use of the suprarenal extract. Int. med. Mag. **5**, 3—11 (1896—1897).

OVERBEEK, G. A.: Adrenals and fatigue. Arch. int. Pharmacodyn. **78**, 532—543 (1949a).
— Adrenals and fatigue. II. The cholinesterase activity of serum, brain and muscle in normal and adrenalectomized rats. Arch. int. Pharmacodyn. **79**, 314—322 (1949b).
OWEN, J. A., JR., F. L. ENGEL and T. B. WESTER: 9-α-Fluorohydrocortisone-induced hypertension in a male infant with adrenogenitalism, and in 6 adults with Addison's disease. J. clin. Endocr. **17**, 272—279 (1957).
PAPADATOS, C., and R. KLEIN: Addison's disease in a boy with hypoparathyroidism. J. clin. Endocr. **14**, 653—660 (1954).
PAGE, I. H., and L. A. LEWIS: Arterial pressure and serum protein patterns in adrenalectomized hypertensive and adrenalectomized normotensive dogs. Amer. J. Physiol. **164**, 61—67 (1951).
PAPPER, E. M., and G. F. CAHILL: Anesthetic problems in hormonal disorders of the adrenal glands. J. Amer. med. Ass. **148**, 174—179 (1952).
PARIS, J., M. UPSON, JR., R. G. SPRAGUE, R. M. SALASSA and A. ALBERT: Corticotropic activity of human blood. J. clin. Endocr. **14**, 597—606 (1954).
PARKES, A. S.: The adrenal-gonad relationship. Physiol. Rev. **25**, 203—254 (1945).
PAUL, F.: Die krankhafte Funktion der Nebenniere und ihr gestaltlicher Ausdruck. Virchows Arch. path. Anat. **282**, 256—326 (1931).
PAULL, A. M., and R. W. PHILLIPS: Primary myxedema with secondary adrenocortical failure. J. clin. Endocr. **14**, 554—560 (1954).
PEARL, F., and H. BRUNN: Suprarenal apoplexy, bilateral. Surg. Gynec. Obstet. **47**, 393—400 (1928).
PELLEGRINO, C., and A. TORCIGLIANI: Effect of inanition and adrenalectomy on the electrophoretic patterns of soluble lymphatic tissue proteins. Experientia (Basel) **12**, 26—27 (1956).
PENMAN, R. W. D.: Addison's disease in association with spastic paraplegia. Brit. med. J. **1960i**, 402.
PENTSCHEW, A.: Folgen der Überdosierung des Nebennierenhormons. Klin. Wschr. **18**, 1570—1572 (1939).
PERERA, G. A., D. W. BLOOD and K. H. REINHOLD: 11-Dehydrocorticosterone. Its effect on hypoadrenalism in man. Amer. J. Med. **1**, 135—143 (1946).
— K. L. PINES, H. B. HAMILTON and K. VISLOCKY: Clinical and metabolic study of 11-dehydro-17-hydroxycorticosterone acetate (Kendall compound E) in hypertension, Addison's disease and diabetes mellitus. Amer. J. Med. **7**, 56—69 (1949).
PERKINS, R., and E. H. RYNEARSON: Practical aspects of insufficiency of the anterior pituitary gland in the adult. J. clin. Endocr. **12**, 574—603 (1952).
PERKOFF, G. T., A. A. SANDBERG, D. H. NELSON and F. H. TYLER: Clinical usefulness of determination of circulating 17-hydroxycorticosteroid levels. Arch. intern. Med. **93**, 1—8 (1954).
PERLA, D., and J. MARMORSTON-GOTTESMAN: A substance in urine of normal human adults that raises the resistance of suprarenalectomized rats. Proc. Soc. exp. Biol. (N. Y.) **28**, 1024—1027 (1931).
— and M. SANDBERG: Role of the adrenal cortex in the production and utilization of vitamin C. The influence on the structure of the teeth in the albino rat. Arch. Path. **23**, 372—382 (1937).
PERLMUTTER, M.: Rapid test for adrenocortical insufficiency. J. Amer. med. Ass. **160**, 117—118 (1956).
— and M. MUFSON: The hypoglycemic and eosinopenic response to insulin: a test for pituitary-adrenal insufficiency. J. clin. Endocr. **11**, 277—288 (1951).
— R. R. ELLISON, L. NORSA and A. R. KANTROWITZ: Idiopathic hypoparathyroídism and Addison's disease. Amer. J. Med. **21**, 634—643 (1956).
PERRY, W. F., W. W. HAWKINS and G. R. CUMMING: Adrenal function in pantothenic acid deficiency. Amer. J. Physiol. **172**, 259—264 (1953).
PETERSEN, W. E.: Lactation. Physiol. Rev. **24**, 340—371 (1944).
PETERSON, R. E.: Plasma corticosterone and hydrocortisone levels in man. J. clin. Endocr. **17**, 1150—1157 (1957).
— and J. B. WYNGAARDEN: The miscible pool and turnover rate of hydrocortisone in man. J. clin. Invest. **35**, 552—561 (1956).
PFIFFNER, J. J., and W. W. SWINGLE: The preparation of an active extract of the suprarenal cortex. Anat. Rec. **44**, 225 (1929).
— — Studies on the adrenal cortex. III. The revival of cats prostrate from adrenal insufficiency with an aqueous extract of the cortex. Amer. J. Physiol. **96**, 180—190 (1931).
PHILIPEAUX: Note sur l'extirpation des capsules surrénales chez les rats albinos. C. R. Acad. Sci. (Paris) **43**, 904—906 (1856).
PHILIPS, C.: A case of Addison's disease with simple atrophy of the adrenals. J. exp. Med. **44**, 581—595 (1899).

PHILPOTT, N. W.: Addison's disease in association with amyloidosis. Ann. intern. Med. **1**, 613—620 (1927—1928).
PILIERO, S. J.: The effects of low pressure on hematopoiesis in the adrenalectomized rat. Endocrinology **57**, 676—682 (1955).
PINCUS, G.: A diurnal rhythm in the excretion of urinary ketosteroids by young men. J. clin. Endocr. **3**, 195—199 (1943).
PITTS, R. F.: Effect of adrenal cortical hormones on renal function. In: E. P. RALLI, Ed., Adrenal Cortex, Transactions of the Third Conference, 11—53. New York: Josiah Macy jr. Foundation 1952.
PLOTZ, J.: Der akute Nebennierenausfall in der Schwangerschaft. Acta endocr. (Kbh.) **14**, 61—71 (1953).
POOR, E.: Functional adrenocortical homotransplants in the golden hamster. Proc. Soc. exp. Biol. (N. Y.) **97**, 535—539 (1958).
PORAK, R., et H. CHABANIER: Altération de la sécrétion rénale après l'ablation des glandes surrénales. C. R. Soc. Biol. (Paris) **77**, 440—442 (1914).
PORGES, O.: Über Hypoglykemie bei Morbus Addison sowie bei nebennierenlosen Hunden. Z. klin. Med. **69**, 341—349 (1909—1910).
— Zur Pathologie des Morbus Addison. II. Über Glycogenschwund nach doppelseitiger Nebennierenextirpation bei Hunden. Z. klin. Med. **70**, 243—250 (1910).
PRADER, A.: Gonadendysgensie und testiculäre Feminisierung. Schweiz. med. Wschr. **87**, 278—285 (1957).
— u. H. P. GURTNER: Das Syndrom des Pseudohermaphroditismus masculinus bei kongenitaler Nebennierenrinden-Hyperplasie ohne Androgenüberproduktion (adrenaler Pseudohermaphroditismus masculinis). Helv. paediat. Acta **10**, 397—412 (1955).
— u. R. E. SIEBENMANN: Nebenniereninsuffizienz bei kongenitaler Lipoidhyperplasie der Nebennieren. Helv. paediat. Acta **12**, 569—595 (1957).
— E. EUHLINGER u. R. ILLIG: Hypercalcämie bei Morbus Addison im Kindesalter. Helv. paediat. Acta **14**, 607—617 (1959).
PRATHER, G. C.: The use of cortin following removal of a large adrenal tumor. New Engl. J. Med. **208**, 872—877 (1933).
PRICE, H. L.: General anesthesia and circulatory homeostasis. Physiol. Rev. **40**, 187—218 (1960).
PROVENZANO, R. W.: Adrenocortical hypoplasia in the newborn infant. New Engl. J. Med. **242**, 87—89 (1950).
PRUNTY, F. T. G.: Sodium and chloride depletion in acute porphyria with reference to the status of adrenal cortical function. J. clin. Invest. **27**, 690—699 (1949).
— R. R. MCSWINEY, I. H. MILLS and M. A. SMITH: The effects of aldosterone in Addison's disease and adrenal pseudohermaphrodism. Lancet **1954 ii**, 620—624.
— — — Biological effects of aldosterone with especial reference to man. Proc. roy. Soc. Med. **48**, 629—632 (1955).
PUGSLEY, L. I., and J. B. COLLIP: The effect of parathyroid hormone upon the serum calcium and calcium excretion of normal and adrenalectomized rats. Biochem. J. **30**, 1274—1279 (1936).
RABINOVICI, N.: The effect of stress upon the ketosteroid and non-carbonyl lipoid content of the adrenal cortex of the rat. Endocrinology **49**, 579—588 (1951).
RAISZ, L. G., W. F. MCNEELY, L. SAXON and J. D. ROSENBAUM: The effects of cortisone and hydrocortisone on water diuresis and renal function in man. J. clin. Invest. **36**, 767—779 (1957).
RALLI, E. P.: Relation of vitamins to adrenal cortical function. In: E. P. RALLI, Ed., Adrenal Cortex, Transactions of the First Conference, 158—189. New York: Josiah Macy Jr. Foundation 1950.
— and M. E. DUMM: Relation of pantothenic acid to adrenal cortical function. Vitamins and Hormones **11**, 133—158 (1953).
— and I. GRAEF: Stimulating effect of adrenalectomy on hair growth and melanin deposition in black rats fed diets adequate and deficient in the filtrate factors of vitamin B_1. Endocrinology **32**, 1—12 (1943).
RAMEY, E. R., and M. S. GOLDSTEIN: The adrenal cortex and the sympathetic nervous system. Physiol. Rev. **37**, 155—195 (1957).
— — and R. LEVINE: Mechanism of muscular fatigue in adrenalectomized animals. Amer. J. Physiol. **162**, 10—16 (1950).
— — — Action of nor-epinephrine and adrenal cortical steroids on blood pressure and work performance of adrenalectomized dogs. Amer. J. Physiol. **165**, 450—455 (1951).
RANDALL, H. T.: An evaluation of adrenalectomy in man. Physiological changes and the effect on advanced neoplastic disease. Bull. N. Y. Acad. Med. **30**, 278—301 (1954).

RANDOLPH, T. G.: Differentiation and enumeration of eosinophils in the counting chamber with glycol stain; a valuable technique in appraising ACTH dosage. J. Lab. clin. Med. **34**, 1696—1701 (1949).
RAWSON, R. W., and J. E. RALL: The endocrinology of neoplastic disease. Recent Progr. Hormone Res. **11**, 257—290 (1955).
READ, C. H., E. H. VENNING and M. P. RIPSTEIN: Adrenal cortical function in newly-born infants. J. clin. Endocr. **10**, 845—857 (1950).
RECANT, L., P. H. FORSHAM and G. W. THORN: Observations on the pituitary-adrenal response following epinephrine infusion in man. J. clin. Endocr. **8**, 589 (1948).
— D. M. HUME, P. H. FORSHAM and G. W. THORN: Effect of epinephrine on the pituitary adrenocortical system. J. clin. Endocr. **10**, 187—229 (1950).
REDDY, W. J., N. ABU HAYDAR, J. C. LAIDLAW, A. E. RENOLD and G. W. THORN: Determination of total 17-hydroxycorticoids in plasma. J. clin. Endocr. **16**, 380—390 (1956).
REEDER, C. F., and S. L. LEONARD: Alterations in mammary structure following adrenalectomy in the immature male rat. Proc. Soc. exp. Biol. (N. Y.) **55**, 61—62 (1944).
REFORZO-MEMBRIVES, J., M. H. POWER and E. J. KEPLER: Studies on the renal excretion of water and electrolytes in cases of Addison's disease. J. clin. Endocr. **5**, 76—85 (1945).
REHBERG, P. B.: Studies on kidney function. 1. The rate of filtration and reabsorption in the human kidney. Biochem. J. **20**, 447—460 (1926).
REICHARD, S. M., and A. S. GORDON: Adrenal influence on distribution of injected colloidal thorium. Amer. J. Physiol. **186**, 63—66 (1956).
REICHSTEIN, T.: Die wichtigsten Hormone der Nebennierenrinde. Acta endocr. (Kbh.) **17**, 375—384 (1954).
— u. J. VON EUW: Über Bestandteile der Nebennierenrinde. Isolierung der Substanzen Q (Desoxy-corticosteron) und R sowie weiterer Stoffe. Helv. chim. Acta **21**, 1197—1210 (1938).
— and C. W. SHOPPEE: The hormones of the adrenal cortex. Vitam. and Horm. **1**, 345—413 (1943).
REID, J. D.: Congenital absence of the pituitary gland. J. Pediat. **56**, 658—664 (1960).
REIDENBERG, M. M., E. A. OHLER and R. W. SEVY: Cardiovascular responses to norepinephrine in acute adrenal insufficiency. Proc. Soc. exp. Biol. (N. Y.) **97**, 889—892 (1958).
REINHARDT, W. O., and R. O. HOLMES: Thymus and lymph nodes following adrenalectomy and maintenance with NaCl in the rat. Proc. Soc. exp. Biol. (N. Y.) **45**, 267—270 (1940).
REMINGTON, J. W.: Fluid and electrolyte shifts in the normal and adrenalectomized rat after intraperitoneal injection of isotonic sugar solutions. Endocrinology **26**, 631—640 (1940).
— Circulatory factors in adrenal crisis in the dog. Amer. J. Physiol. **165**, 306—318 (1951).
— and W. F. HAMILTON: The construction of a theoretical cardiac ejection curve from the contour of the aortic pressure pulse. Amer. J. Physiol. **144**, 546—556 (1945).
— W. M. PARKINS, W. W. SWINGLE and V. A. DRILL: Efficiency of desoxycorticosterone acetate as replacement therapy in adrenalectomized dogs. Endocrinology **29**, 740—745 (1941).
RENKIN, E. M., and B. D. ZAUN: Effects of adrenal hormones on capillary permeability in perfused rat tissues. Amer. J. Physiol. **180**, 498—502 (1955).
RENOLD, A. E., P. H. FORSHAM, J. MAISTERRENA and G. W. THORN: Intravenously administered ACTH. New Engl. J. Med. **244**, 796—798 (1951).
— N. ABU HAYDAR, W. J. REDDY, A. GOLDFIEN, J. R. ST. MARC and J. C. LAIDLAW: Biological effects of fluorinated derivatives of hydrocortisone and progesterone in man. Ann. N. Y. Acad. Sci. **66**, 582—590 (1955).
RICH, A. R.: A peculiar type of adrenal cortical damage associated with acute adrenal infection, and its possible relation to circulatory collapse. Bull. Johns Hopk. Hosp. **74** 1—15 (1944.)
RICHARDS, R. K.: Toxicity of hypnotics as affected by temperature, thyroxin and adrenalectomy. Anesthesiology **2**, 37—43 (1941).
RICHARDSON, F. L.: The relative growth of the mammary glands in normal, ovariectomized and adrenalectomized female mice. Anat. Rec. **123**, 279—290 (1955).
RICHTER, C. P.: Increased salt appetite in adrenalectomized rats. Amer. J. Physiol. **115**, 155—161 (1936).
— Salt taste thresholds of normal and adrenalectomized rats. Endocrinology **24**, 367—371 (1939).
— Sodium chloride and dextrose appetite of untreated and treated adrenalectomized rats. Endocrinology **29**, 115—125 (1941).
— Total self regulatory functions in animals and human beings. Harvey Lect. **38**, 63—103 (1942—1943).
— The effect of domestication on the steroids of animals and man. Ciba Found. Colloq. Endocr. **3**, 89—107 (1952).
— P. V. ROGERS and C. E. HALL: Failure of salt replacement therapy in adrenalectomized recently captured wild Norway rats. Endocrinology **46**, 233—242 (1950).

RICKARDS, A. G., and G. M. BARRETT: Non-tuberculous Addison's disease and its relationship to "giant-cell granuloma" and multiple glandular disease. Quart. J. Med. **23**, 403—423 (1954).
— and P. W. HARVEY: "Giant-cell" granuloma and other pituitary granulomata. Quart. J. Med. **23**, 425—435 (1954).
ROBERTS, K. E.: Effects of cortisone on electrocardiographic changes in adrenal insufficiency. Proc. Soc. exp. Biol. (N. Y.) **79**, 32—34 (1952).
ROBERTSON, W. V., and F. W. DUNIHUE: Water and electrolyte distribution in cardiac muscle. Amer. J. Physiol. **177**, 292—298 (1954).
ROBILLARD, E., A. D'IORIO et J. PELLERIN: Influences endocriniennes sur la désintoxication hépatique du pentobarbital. Un. méd. Can. **83**, 853—860 (1954).
ROBINSON, F. J., M. H. POWER and E. J. KEPLER: Two new procedures to assist in the recognition and exclusion of Addison's disease: a preliminary report. Proc. Mayo Clin. **16**, 577—583 (1941).
ROBINSON, H. J., R. C. MASON and A. L. SMITH: Beneficial effects of cortisone on survival of rats infected with D. pneumoniae. Proc. Soc. exp. Biol. (N. Y.) **84**, 312—314 (1953).
ROBINSON, K. W., and W. V. MACFARLANE: The influence of environmental temperature on the level of plasma antidiuretic substances in the rat. Aust. J. biol. Sci. **9**, 130—138 (1956).
ROEMMELT, J. C., O. W. SARTORIUS and R. F. PITTS: Excretion and reabsorption of sodium and water in the adrenalectomized dog. Amer. J. Physiol. **159**, 124—136 (1949).
ROGOFF, J. M.: Experimental production of chronic and sub-acute adrenal insufficiency in dogs and cats. Proc. Soc. exp. Biol. (N. Y.) **29**, 1240—1243 (1932a).
— Addison's disease. Further report on treatment with "interrenalin" (adrenal cortical extract). J. Amer. med. Ass. **99**, 1309—1315 (1932b).
— The adrenal cortical hormone. J. Amer. med. Ass. **103**, 1764—1767 (1934).
— Experimental pathology and physiology of the adrenal cortex. Production of Addison's disease in laboratory animals. Arch. Path. **38**, 392—409 (1944).
— and R. DOMINGUEZ: Blood pressure following adrenalectomy. Amer. J. Physiol. **83**, 84—91 (1927).
— and G. N. STEWART: Studies on adrenal insufficiency in dogs. I. Control animals not subjected to any treatment. Amer. J. Physiol. **78**, 683—710 (1926a).
— — II. Blood studies in control animals not subjected to treatment. Amer. J. Physiol. **78**, 711—729 (1926b).
— — III. The influence of pregnancy upon the survival period in adrenalectomized dogs. Amer. J. Physiol. **79**, 508—535 (1926—1927).
— — The influence of adrenal extracts on the survival period of adrenalectomized dogs. Science **66**, 327—328 (1927).
— — IV. The influence of intravenous injections of Ringer's solution upon the survival period in adrenalectomized dogs. Amer. J. Physiol. **84**, 649—659 (1928a).
— — V. The influence of adrenal extracts on the survival period of adrenalectomized dogs. Amer. J. Physiol. **84**, 660—674 (1928b).
— — VII. Further blood studies (cholesterol and calcium) in control adrenalectomized dogs. Amer. J. Physiol. **86**, 25—31 (1928c).
— — Suprarenal cortical extracts in suprarenal insufficiency (Addison's disease). J. Amer. med. Ass. **92**, 1569—1571 (1929).
ROLF, D., C. M. ARMSTRONG, B. W. STEIGER, M. AUDIA and H. L. WHITE: Comparison of external loss of sodium with loss from sucrose space in adrenalectomized rats. Amer. J. Physiol. **196**, 1109—1111 (1959).
ROSE, B., and J. S. L. BROWNE: The effect of adrenalectomy on the histamine content of the tissues of the rat. Amer. J. Physiol. **131**, 589—594 (1941).
ROSENFELD, G.: Functional studies of the calf adrenal cortex following total-body exposure to a single lethal dose of γ-radiation. Amer. J. Physiol. **192**, 232—240 (1958).
ROSENOW, G., and JAGGUTIS: Der Blutzucker bei Addisonischer Krankheit und seine Beeinflussung durch Adrenalin. Klin. Wschr. **1**, 358—360 (1922).
ROWNTREE, L. G.: Subsequent course of a case of Addison's disease. J. Amer. med. Ass. **79**, 556—557 (1922).
— Studies on Addison's disease. J. Amer. med. Ass. **84**, 327—335 (1925).
— Report of three cases of "clinical" Addison's disease surviving more than fifteen years. Endocrinology **26**, 793—800 (1940).
— and C. H. GREENE: The treatment of patients with Addison's disease with the "cortical hormone" of Swingle and Pfiffner. Science **72**, 482—483 (1930).
— and A. M. SNELL: A Clinical Study of Addison's Disease. Philadelphia: Saunders 1931.
— C. H. GREENE, W. W. SWINGLE and J. J. PFIFFNER: Addison's disease. Experience in treatment with various suprarenal preparations. J. Amer. med. Ass. **96**, 231—235 (1931a).
— — R. G. BALL, W. W. SWINGLE and J. J. PFIFFNER: Treatment of Addison's disease with the cortical hormone of the suprarenal gland. J. Amer. med. Ass. **97**, 1446—1453 (1931b).

RUBIN, M. I., and E. T. KRICK: Effect of adrenalectomy on salt metabolism in rats. Proc. Soc. exp. Biol. (N. Y.) **31**, 228—229 (1933/34).
RUCKS, W. L., and J. J. HOBSON: Purpura fulminans (Waterhouse-Friderichsen syndrome). Report of a case with recovery. J. Pediat. **22**, 226—232 (1943).
RUKES, J. M., R. H. ORR and P. H. FORSHAM: Clinical uses of intravenous hydrocortisone. Metabolism **3**, 481—488 (1954).
— — — and M. GALANTE: The use of intravenous hydrocortisone in major surgery. Ann. N. Y. Acad. Sci. **61**, 448—459 (1955).
RUMMELL, P. M., L. S. BAER, L. HOLLISTER and F. KOLB: Acute and chronic adrenal insufficiency in a patient showing polyglandular deficiency and acromegaly. J. clin. Endocr. **11**, 872—877 (1951).
RUPP, J. J., and K. E. PASCHKIS: Co-existence of Addison's disease and thyrotoxicosis. J. clin. Endocr. **17**, 143—147 (1957).
RUSSELL, J. A., and L. L. BENNETT: Carbohydrate storage and maintenance in the hypophysectomized rat. Amer. J. Physiol. **118**, 196—205 (1937).
— and W. BLOOM: Hormonal control of glycogen in the heart and other tissues in rats. Endocrinology **58**, 83—94 (1956).
— and A. E. WILHELMI: Metabolism of kidney tissue in the adrenalectomized rat. J. biol. Chem. **137**, 713—725 (1941).
ST. GEORGE, S., S. C. FREED, R. H. ROSENMAN and S. WINDERMAN: Influence of potassium deprivation and adrenalectomy on potassium concentration of the myocardium. Amer. J. Physiol. **181**, 550—552 (1955).
SALASSA, R. M., W. A. BENNETT, F. R. KEATING, JR., and R. G. SPRAGUE: Postoperative adrenal cortical insufficiency. Occurrence in patients previously treated with cortisone. J. Amer. med. Ass. **152**, 1509—1515 (1953).
— H. L. MASON, V. R. MATTOX, A. L. ORVIS and M. H. POWER: Effects of aldosterone on water, electrolyte and nitrogen metabolism in Addison's disease. Proc. Mayo Clin. **32**, 201—212 (1957).
SALMOIRAGHI, G. C., and J. W. MCCUBBIN: Effect of adrenalectomy on pressor responsiveness to angiotonin and renin. Circulat. Res. **2**, 280—284 (1954).
SAMUELS, L. T., G. T. EVANS and J. L. MCKELVEY: Ovarian and placental function in Addison's disease. Endocrinology **32**, 422—428 (1943).
SANDBERG, A. A., and W. R. SLAUNWHITE, JR.: Metabolism of cortisol-C^{14} in subjects with Addison's disease. J. clin. Endocr. **18**, 662—665 (1958).
— D. H. NELSON, E. M. GLENN, F. H. TYLER and L. T. SAMUELS: 17-Hydroxycorticosteroids and 17-ketosteroids in urine of human subjects: clinical application of a method employing β-glucuronidase hydrolysis. J. clin. Endocr. **13**, 1445—1464 (1953).
— K. EIK-NES, C. J. MIGEON and G. F. KOEPF: Plasma 17-hydroxycorticosteroids in hyperfunction, suppression and deficiency of adrenal cortical function. J. Lab. clin. Med. **50**, 286—296 (1957).
SANDERSON, P. M.: Renal function in Addison's disease. Clin. Sci. **6**, 197—206 (1948).
SANDISON, A. T.: A form of lipoidosis of the adrenal cortex in an infant. Arch. Dis. Childh. **30**, 538—541 (1955).
SARTORIUS, O. W., and K. ROBERTS: Effects of pitressin and desoxycorticosterone in low dosage on excretion of sodium, potassium, and water by normal dog. Endocrinology **45**, 273—283 (1949).
— D. CALHOON and R. F. PITTS: The capacity of the adrenalectomized rat to secrete hydrogen and ammonium ions. Endocrinology **51**, 444—450 (1952).
— — — Studies on the interrelationships of the adrenal cortex and renal ammonia excretion by the rat. Endocrinology **52**, 256—265 (1953).
SASS-KORTSAK, E.: Dehydrierungsvermögen und Glykogengehalt des Hirns epinephrektomierter Ratten. Hoppe-Seylers Z. physiol. Chem. **280**, 129—135 (1944).
SATAKE, Y., T. KOJIMA and S. NASU: Notes on double suprarenalectomized dogs. Jap. J. med. Sci. **2**, 91 (1931).
SAUNDERS, F. J.: Survival of adrenalectomized mice after several types of stress and the effects of various steroids. Endocrinology **58**, 412—419 (1956).
SCHÄFER, E. A.: Present condition of our knowledge regarding the function of the suprarenal capsules. Brit. med. J. **1908 i**, 1277—1281, 1346—1351.
SCHAFER, L. E.: 9-α-Fluorohydrocortisone. Germ. med. Mth. **2**, 375—376 (1957).
SCHALLY, A. V., and R. GUILLEMIN: Purification of corticotropin releasing factor by counter current distribution and zone electrophoresis. Proc. First Internat. Congr. Endocrinology (Copenhagen), 63 (1960).
SCHIFF, M.: On the extirpation of the adrenal cortex (in Italian). L'Imparziale **3**, 234—237 (1863).
SCHMIDT, I. G.: An accessory adrenal cortex in the kidney of a rhesus monkey and the effects of pyrimethamine on this tissue. Endocrinology **59**, 454—457 (1956).

SCHOUR, I., and J. M. ROGOFF: Changes in the rat incisor following bilateral adrenalectomy. Amer. J. Physiol. **115**, 334—344 (1936).

SCHRADER, G.: Zur Frage der Entstehung der sogenannten Nebennierenapoplexie. Frankfurt Z. Path. **37**, 128—135 (1929).

SCHULTZ, R. B., R. W. WINTERS and W. A. KREHL: The adrenal cortex of the pantothenic acid deficient rat: modification of the lesion by ACTH and cortisone treatment. Endocrinology **51**, 336—343 (1952).

SCHWARTZ, H., W. S. DERRICK and E. M. PAPPER: The influence of anesthesia, operation, and obstetrical delivery upon patients with adrenal cortical insufficiency. Surg. Gynec. Obstet. **94**, 455—463 (1952).

SCHWARZ, O.: Über Stoffwechselstörungen nach der Extirpation beider Nebennieren. Pflügers Arch. ges. Physiol. **134**, 259—288 (1910).

SCHWEITZER, A.: The effect of adrenalectomy on the contractile power of skeletal muscle. J. Physiol. (Lond.) **104**, 21—31 (1945).

SECKER, J.: A note on the suprarenal cortex and the transmission of the activity of the sympathetic nerves of the cat. J. Physiol. (Lond.) **95**, 282—285 (1939).

— The role of the adrenal cortex in the maintenance of the sciatic pressor reflex. J. Physiol. (Lond.) **109**, 49—52 (1949).

SEGALOFF, A.: The effect of diet on the growth and survival of adrenalectomized rats treated with desoxycorticosterone acetate pellets. Endocrinology **38**, 26—29 (1946).

SELLERS, E. A., S. S. YOU and R. W. YOU: The influence of adrenal cortex and thyroid on the loss of nitrogen in urine after experimental burns. Endocrinology **47**, 148—155 (1950).

SEZARY, A.: Syphilis et glandes surrénales. Gaz. Hôp. (Paris) **87**, 1317—1320 (1914).

SHANNON, I. L., and J. R. PRIGMORE: Physiologic chloride levels in human whole saliva. Proc. Soc. exp. Biol. (N. Y.) **97**, 825—828 (1958).

SHARE, L., and P. W. HALL, III: Excretion of electrolytes and water by the salt-loaded adrenalectomized dog. Amer. J. Physiol. **183**, 291—298 (1955).

SHARPEY-SCHÄFER, E.: The Endocrine Organs, Ed. 2, Part I: The thyroid, the parathyroids and the suprarenal capsules. London: Longmans, Green 1924.

SHEEHAN, H. L.: Simmonds' disease due to post-partum necrosis of the anterior pituitary. Quart. J. Med. **32**, 277—309 (1932).

— Post-partum necrosis of the anterior pituitary. J. Path. Bact. **45**, 189—214 (1937).

— and R. MURDOCH: Post-partum necrosis of the anterior pituitary. Lancet **1938 ii**, 132—135.

— and V. K. SUMMERS: The syndrome of hypopituitarism. Quart. J. Med. **18**, 319—378 (1949).

SHIBATA, K.: Experimental studies on the thymus. III. Interrelation between thymus and adrenal gland. Gunma J. med. Sci. **3**, 89—96 (1954).

— and A. KOMIYA: Effect of adrenocortical steroids on duration of pentothal anesthesia in adrenalectomized mice. Proc. Soc. exp. Biol. (N. Y.) **84**, 308—310 (1953).

SHIPLEY, R. A.: Treatment of Addison's disease with pellets of desoxycorticosterone acetate. Amer. J. med. Sci. **207**, 19—29 (1944).

— The cause of abnormal retention of ingested water in adrenalectomized rats. Endocrinology **36**, 118—123 (1945).

SHIZUME, K., and A. B. LERNER: Determination of melanocyte-stimulating hormone in urine and blood. J. clin. Endocr. **14**, 1491—1510 (1954).

SHREWSBURY, M. M., JR., and W. O. REINHARDT: Relationships of adrenals, gonads, and thyroid to thymus and lymph nodes, and to blood and thoracic duct leukocytes. Blood **10**, 633—645 (1955).

SIEBENMANN, R. E.: Die kongenitale Lipoidhyperplasie der Nebennierenrinde mit Nebennierenrinden-Insuffizienz. Schweiz. Z. Path. **20**, 77—84 (1957).

SIEMERLING, E., u. H. G. CREUTZFELDT: Bronzekrankheit und sklerosierende Encephalomyelitis (diffuse Sklerose). Arch. Psychiat. Nervenkr. **68**, 217—244 (1923).

SILBER, R. H., and R. D. BUSCH: An improved procedure for the determination of hydrocortisone in human plasma. J. clin. Endocr. **16**, 1333—1336 (1956).

— and C. C. PORTER: The determination of 17,21-dihydroxy-20-ketosteroids in urine and plasma. J. biol. Chem. **210**, 923—932 (1954).

SILBERMAN, D. E., L. G. ROWNTREE and H. B. ORENSTEIN: Recovery from Addison's disease due to diffuse vascular disease with subsequent development of hypertension and glomerulonephritis. Ann. intern. Med. **32**, 760—774 (1950).

SILVETTE, H.: Chloride, carbohydrate and water metabolism in adrenal insufficiency and other conditions. Amer. J. Physiol. **108**, 535—544 (1934).

— and S. W. BRITTON: Effects of adrenalectomy and corticoadrenal extract on renal excretion and tissue fluids. Amer. J. Physiol. **104**, 399—411 (1932).

— — Carbohydrate and electrolyte changes in the opossum and marmot following adrenalectomy. Amer. J. Physiol. **115**, 618—626 (1936).

— — Renal function in normal and adrenalectomized opossums and effects of post-pituitary and corticoadrenal extracts. Amer. J. Physiol. **121**, 528—533 (1938).

SIMMONDS, M.: Über Nebennierenschrumpfung bei Morbus Addison. Virchows Arch. path. Anat. **172**, 480—492 (1903).
— Über Hypophysenschwund mit tödlichem Ausgang. Dtsch. med. Wschr. **40**, 322—323 (1914).
— Atrophie des Hypophysisvorderlappens und hypophysäre Kachexie. Dtsch. med. Wschr. **44**, 852—854 (1918).
SIMPSON, M. E., C. W. ASLING and H. M. EVANS: Some endocrine influences on skeletal growth and differentiation. Yale J. Biol. Med. **23**, 1—27 (1950/51).
SIMPSON, S. A., and J. F. TAIT: The possible role of electrocortin in normal human metabolism. Ciba Found. Colloq. Endocr. 8, 204—227 (1955).
SIMS, E. A. H., L. G. WELT, J. ORLOFF and J. W. NEEDHAM: Asymptomatic hyponatremia in pulmonary tuberculosis. J. clin. Invest. **29**, 1545—1557 (1950).
SINDRAM, I.: Der Einfluß von Cortin auf die Empfindlichkeit für Narkotika. Acta brev. neerl. Physiol. **5**, 29—30 (1935).
SLESSOR, A.: Studies concerning the mechanism of water retention in Addison's disease and in hypopituitarism. J. clin. Endocr. **11**, 700—723 (1951).
SLOCOMBE, A. G., H. HOAGLAND and J. PRAGLIN: Conduction in the central nervous system and peripheral nerve excitability in relation to adrenal cortical function. Fed. Proc. **11**, 149 (1952).
— L. S. TOZIAN and H. HOAGLAND: Influence of adrenalectomy on nerve excitability. Amer. J. Physiol. **179**, 89—92 (1954).
SLOPER, J. C.: The pathology of the adrenals, thymus and certain other endocrine glands in Addison's disease. Proc. roy. Soc. Med. **48**, 625—628 (1955).
SMART, G. A.: Discussion. In: J. M. YOFFEY, Ed., Fifth Symposium Colston Res. Soc., p. 197. London: Butterworths 1953.
SMITH, D. J.: Variations in vascular reactivity produced by season, cold stress and immaturity: role of thyroid and adrenal cortex. Amer. J. Physiol. **172**, 118—128 (1953).
SMITH, H. W.: The Kidney. New York: Oxford Univ. Press 1951.
— Principles of Renal Physiology. New York: Oxford Univ. Press 1956.
SMITH, J. M., J. S. MURPHY and G. S. MIRICK: Effect of adrenal hormones on infection of mice with peumonia virus of mice (PVM). Proc. Soc. exp. Biol. (N. Y.) **78**, 505—513 (1951).
SMITH, M. H. D.: Waterhouse-Friderichsen syndrome: its treatment with adrenal cortex extract. Amer. J. Dis. Child. **69**, 330—331 (1945).
SMITH, P. E.: The disabilities caused by hypophysectomy and their repair. J. Amer. med. Ass. 88, 158—161 (1927).
— and I. P. SMITH: The effect of intraperitoneal injection of fresh anterior lobe substance in hypophysectomized tadpoles. Anat. Rec. **23**, 38—39 (1922).
— — The function of the lobes of the hypophysis as indicated by replacement therapy with different portions of the ox gland. Endocrinology **7**, 579—591 (1923).
SMITH, W. W.: Acute KCl and histamine tolerance and adrenal weight in X-irradiated mice. Amer. J. Physiol. **167**, 321—327 (1951).
SNAPPER, I., and K. Y. CH'IN: Massive amyloidosis of both adrenal cortices. Chinese med. J. **57**, 201—215 (1940).
SNELL, A. M.: The treatment of Addison's disease. Proc. Mayo Clin. **9**, 57—61 (1934).
— The changing relations of tuberculosis and Addison's disease. Proc. Mayo Clin. **10**, 337—344 (1935).
— R. M. WILDER and R. W. CRAGG: Suprarenal atrophy following degeneration: report of a case with findings at necropsy. J. Path. Bact. **43**, 473—478 (1936).
SNELLING, C. E., and I. H. ERB: Hemorrhage and subsequent calcification of the suprarenal. J. Pediat. **6**, 22—41 (1935).
SNYDER, J. G., and L. C. WYMAN: Gonadal influences on survival after adrenalectomy in the golden hamster. Endocrinology **49**, 205—212 (1951).
SOBEL, E. H.: The influence of testosterone and cortisone on growth. In: L. I. GARDNER, Ed., Adrenal Function in Infants and Children, 137—156. New York: Grune and Stratton 1956.
— C. M. LEE, JR., V. M. ESSELBORN and L. C. CLARK, JR.: Functioning adrenal tumors in childhood. Amer. J. Dis. Child. **86**, 733—751 (1953).
SODDU, L.: On the effect of the extirpation of the suprarenal capsule of the dog (in Italian). Sperimentale (Arch. Biol.) **52**, 87—111 (1898).
SOFFER, L. G.: Diseases of the Endocrine Glands, Ed. 2. Philadelphia: Lea and Febiger 1956.
— and F. L. ENGEL: A short history of the treatment of Addison's disease. J. Mt. Sinai Hosp. **9**, 11—16 (1942).
SOLANDT, D. Y.: Measurement of "accomodation" in nerve. Proc. roy. Soc. Lond. B **119**, 355—379 (1936).
SOLOMON, A. K.: Equations for tracer experiments. J. clin. Invest. **28**, 1297—1307 (1949).
SOLOMON, N., R. H. TRAVIS and G. SAYERS: Corticosteroids and the functional capacity of the rat heart-lung preparation. Endocrinology **64**, 535—541 (1959).

SOMERVILLE, W., H. D. LEVINE and G. W. THORN: The electrocardiogram in Addison's disease. Medicine **30**, 43—79 (1951).
SORKIN, S. Z.: Addison's disease. Medicine **28**, 371—425 (1949).
— The centenary of Addison's disease. Bull. N. Y. Acad. Med. **32**, 819—843 (1956).
— The adrenals before Addison. J. Mt. Sinai Hosp. **24**, 1238—1249 (1957).
SOUTHAM, C. M., and V. I. BABCOCK: Effect of cortisone, related hormones and adrenalectomy on susceptibility of mice to virus infections. Proc. Soc. exp. Biol. (N. Y.) **78**, 105—109 (1951).
SPECTOR, W. G.: Substances which affect capillary permeability. Pharmacol. Rev. **10**, 475—505 (1958).
SPINK, W. W.: The pathogenesis and management of shock due to infection. Arch. intern. Med. **106**, 433—442 (1960).
SPIRO, H. M., R. W. REIFENSTEIN and S. J. GRAY: The effect of adrenocorticotropic hormone upon uropepsin secretion. J. Lab. clin. Med. **35**, 899—910 (1950).
SPOOR, H. J., and E. P. RALLI: Chemical studies of melanogenesis in normal and adrenalectomized rats. Endocrinology **35**, 325—337 (1944).
SPRAGUE, R. G., C. F. GASTINEAU, H. L. MASON and M. H. POWER: Effects of synthetic 11-dehydrocorticosterone (compound A) in a subject with Addison's disease. Amer. J. Med. **4**, 175—185 (1948).
STADIE, W. C.: Aspects of carbohydrate and phosphate metabolism in diabetes. Bull. N. Y. Acad. Med. **34**, 5—20 (1958).
STANTON, E. R., H. W. JONES, JR., and A. MARBLE: Coexisting diabetes mellitus and Addison's disease. Arch. intern. Med. **93**, 911—920 (1954).
STATLAND, H., and J. LERMAN: Function of the adrenal cortex in myxedema, with some observations on pituitary function. J. clin. Endocr. **10**, 1401—1416 (1950).
STEBBINS, R. B.: Impaired water metabolism in pyridoxine deficiency and effects of pyridoxine and adrenal cortical hormone. Amer. J. Physiol. **166**, 538—540 (1951a).
— Cytochemical changes in the adrenal cortex of the rat in pyridoxine deficiency. Endocrinology **49**, 23—25 (1951b).
STEELMAN, S. L., and R. GUILLEMIN: Adrenocorticotropic activity of alpha melanocyte stimulating hormone (α-MSH). Proc. Soc. exp. Biol. (N. Y.) **101**, 600—601 (1959).
STEIN, K. F., and C. MARTIN: Effects of adrenalectomy on blood cells of mice with special reference to eosinophiles. Proc. Soc. exp. Biol. (N. Y.) **78**, 513—517 (1951).
STEIN, L., and E. WERTHEIMER: Hair-loss as deficiency test of medullectomy in rats. J. Endocr. **2**, 418—427 (1941).
STEINBACH, M. M.: The effect of restricted diet and suprarenalectomy on experimental tuberculosis in the white rat. Proc. Soc. exp. Biol. (N. Y.) **27**, 142—144 (1929).
STEINBERG, A., F. R. SHECHTER and H. I. SEGAL: True pituitary Addison's disease — a pituitary unitropic deficiency. J. clin. Endocr. **14**, 1519—1529 (1954).
STEMPIEN, S. J., and A. DAGRADI: The histamine response of the gastric mucosa in a patient with adrenal insufficiency: effect of cortisone administration. Gastroenterology **27**, 358—362 (1954).
STENGER, K.: Blutveränderungen bei schweren Infektionskrankheiten und bei Verbrennungen. Ihre Beziehung zur Nebennierenschädigung. Klin. Wschr. **18**, 576 (1939).
STEPHENS, D. J.: Chloride excretion in hypopituitarism with reference to adrenocortical function. Amer. J. med. Sci. **199**, 67—75 (1940).
STERN, T. N., V. V. COLE, A. C. BASS and R. R. OVERMAN: Dynamic aspects of sodium metabolism in experimental adrenal insufficiency using radioactive sodium. Amer. J. Physiol. **164**, 437—449 (1951).
STEVENS, R. C., and A. J. TOMSKYKOSKI: Bilateral adrenal hemorrhage and calcification. Amer. J. Dis. Child. **87**, 475—477 (1954).
STEWART, C. T., R. J. SALMON and C. D. MAY: Factors determining effect of insulin on metabolism of glucose in ascorbic acid deficiency and scurvy in the monkey. Amer. J. Dis. Child. **84**, 677—691 (1952).
STEWART, G. N.: Adrenalectomy and the relation of the adrenal bodies to metabolism. Physiol. Rev. **4**, 163—190 (1924).
— and J. M. ROGOFF: Studies on adrenal insufficiency. Proc. Soc. exp. Biol. (N. Y.) **22**, 394—397 (1925).
— — IX. The influence of extracts of adrenal cortex (sheep and cattle) on the survival period of adrenalectomized dogs and cats. Amer. J. Physiol. **91**, 254—264 (1929—1930).
STILL, G. F.: Haemorrhage into the supra-renal capsules in infants. Trans. path. Soc. Lond. **49**, 252—257 (1898).
STILLING, H.: Über die compensatorische Hypertrophie der Nebennieren. Virchows Arch. path. Anat. **118**, 569—575 (1889).

STOERK, H. C.: Thymus weight in relation to body weight in castrated and in adrenalectomized rats. Endocrinology **34**, 329—334 (1944).
— H. M. JOHN and H. N. EISEN: Turnover of serum protein in adrenalectomized rats. Proc. Soc. exp. Biol. (N. Y.) **66**, 25—28 (1947).
STONER, H. B., and D. DEXTER: Glomerular filtration rate after adrenal medullectomy in the rat. Endocrinology **54**, 708—709 (1954).
STORRIE, V. M.: Convulsions in Addison's disease. Bull. Tufts-New Engl. med. Cent. **1**, 229 to 237 (1953).
STREHL, H., u. O. WEISS: Beiträge zur Physiologie der Nebenniere. Pflügers Arch. ges. Physiol. **86**, 107—121 (1901).
SULMAN, F. G.: Hypersecretion of ACTH chromatotropic factor in Addison's disease and its bearing on skin melanosis. Acta endocr. (Kbh.) **14**, 108—112 (1953).
— Chromatophorotropic activity of human blood: review of 1,200 cases. J. clin. Endocr. **16**, 755—774 (1956).
SUNDBERG, C. G.: Sur l'action de l'insuline après l'extirpation des capsules surrénales. C. R. Soc. Biol. (Paris) **89**, 807—810 (1923).
SWANN, H. G.: The effect of diet on the survival of adrenalectomized rats. Amer. J. Physiol. **118**, 798—805 (1937).
— The pituitary-adrenocortical relationship. Physiol. Rev. **20**, 493—521 (1940).
— and J. W. FITZGERALD: Insulin shock in relation to the components of the adrenals and the hypophysis. Endocrinology **22**, 687—692 (1938).
SWEAT, M. C.: Adrenocorticosteroids in peripheral and adrenal venous blood of man. J. clin. Endocr. **15**, 1043—1056 (1955).
SWINGLE, W. W.: Studies on the functional significance of the adrenal cortex. I. Blood changes following bilateral epinephrectomy in the cat. Amer. J. Physiol. **79**, 666—678 (1926—1927).
— Experimental studies on the function of the adrenal cortex. Cold Spr. Harb. Symp. quant. Biol. **5**, 327—343 (1937).
— and A. J. EISENMAN: Studies on the functional significance of the suprarenal cortex. II. The acid-base equilibrium of epinephrectomized cats. Amer. J. Physiol. **79**, 679—688 (1926/27).
— and J. J. PFIFFNER: An aqueous extract of the suprarenal cortex which maintains the life of bilaterally adrenalectomized cats. Science **71**, 321—322 (1930a).
— — Further observations on adrenalectomized cats treated with aqueous extract of the suprarenal cortex. Science **71**, 489—490 (1930b).
— — The revival of comatose adrenalectomized cats with an extract of the suprarenal cortex. Science **72**, 75—76 (1930c).
— — Studies on the adrenal cortex. I. The effect of a lipid fraction upon the life-span of adrenalectomized cats. Amer. J. Physiol. **96**, 153—163 (1931a).
— — II. An aqueous extract of the adrenal cortex which maintains the life of bilaterally adrenalectomized cats. Amer. J. Physiol. **96**, 164—179 (1931b).
— and J. W. REMINGTON: The role of the adrenal cortex in physiological processes. Physiol. Rev. **24**, 89—127 (1944).
— and W. F. WENNER: Sulfate retention in dogs following bilateral adrenal extirpation. Proc. Soc. exp. Biol. (N. Y.) **25**, 169—171 (1927).
— J. J. PFIFFNER and B. WEBSTER: Effect of adrenal cortical hormone upon respiratory metabolism of adrenalectomized cats. Proc. Soc. exp. Biol. (N. Y.) **28**, 728—730 (1931).
— — H. M. VARS, P. A. BOTT and W. M. PARKINS: The function of the adrenal cortical hormone and the cause of death from adrenal insufficiency. Science **77**, 58—64 (1933).
— — — and W. M. PARKINS: The effect of fluid deprivation and fluid intake upon the revival of dogs from adrenal insufficiency. Amer. J. Physiol. **108**, 144—150 (1934a).
— — — — The effect of sodium chloride administration upon adrenalectomized dogs not receiving extract. Amer. J. Physiol. **108**, 159—167 (1934b).
— H. M. VARS and W. M. PARKINS: A study of the blood volume of adrenalectomized dogs. Amer. J. Physiol. **109**, 488—501 (1934c).
— W. M. PARKINS and A. R. TAYLOR: Revival from insufficiency and maintenance of adrenalectomized dogs with low serum sodium and chloride levels. Proc. Soc. exp. Biol. (N. Y.) **34**, 75—77 (1936a).
— — — and H. W. HAYS: Relation of serum sodium and chloride levels to alterations of body water in the intact and adrenalectomized dog, and the influence of adrenal cortical hormone upon fluid distribution. Amer. J. Physiol. **116**, 438—445 (1936b).
— G. BARLOW, E. COLLINS, E. J. FEDOR, W. J. WELCH and J. M. RAMPONA: Serum potassium and electrocardiographic changes in adrenalectomized dogs maintained on cortisone acetate. Endocrinology **51**, 353—361 (1952).

SWINGLE, W. W., L. J. BRANNICK and A. F. PARLOW: Effect of 2-methyl-9α-fluorohydrocortisone on internal distribution of fluid and electrolytes of fasted adrenalectomized dogs. Proc. Soc. exp. Biol. (N. Y.) **94**, 156—159 (1957).
— J. P. DAVANZO, D. GLENISTER, H. C. CROSSFIELD and G. WAGLE: Role of gluco- and mineralocorticoids in salt and water metabolism of adrenalectomized dogs. Amer. J. Physiol. **196**, 283—286 (1959).
SWINYARD, E. A., D. O. SCHIFFMANN and L. S. GOODMAN: Effect of variations in extracellular sodium concentration on the susceptibility of mice to pentylenetetrazol (Metrazol)-induced seizures. J. Pharmacol. **114**, 160—166 (1955).
SYDNOR, K. L., G. SAYERS, H. BROWN and F. H. TYLER: Studies on blood ACTH in man. J. clin. Endocr. **13**, 891—897 (1953).
SZENT-GYÖRGYI, A.: On the chemistry of the adrenal cortex. Amer. J. Physiol. **90**, 536 (1929).
TAGGART, J. V.: Mechanisms of renal tubular transport. Amer. J. Med. **24**, 774—784 (1958).
TAIT, J. F., S. A. SIMPSON and H. M. GRUNDY: The effect of adrenal extract on mineral metabolism. Lancet **1952 I**, 122—124.
TALBOT, E.: Cases of haemorrhage into the supra-renal capsules. St. Barth. Hosp. Rep. **36**, 207—211 (1900).
TALBOT, N. B., and A. M. BUTLER: Urinary 17-ketosteroid assays in clinical medicine. J. clin. Endocr. **2**, 724—729 (1942).
— and E. H. SOBEL: Certain factors which influence the rate of growth and duration of growth of children. Rec. Progr. Hormone Res. **1**, 355—369 (1947a).
— — Endocrine and other factors determining the growth of children. In: S. Z. LEVINE, Ed., Advances in Pediatrics II, 238—297. New York: Interscience Publishers 1947b.
— A. M. BUTLER and E. A. MACLACHLAN: Alpha and beta neutral ketosteroids (androgens): preliminary observations on their normal urinary excretion and clinical usefulness of their assay in differential diagnosis. New Engl. J. Med. **223**, 369—373 (1940).
— A. H. SALTZMAN, R. L. WIXOM and J. K. WOLFE: The colorimetric assay of urinary corticosteroid-like substances. J. biol. Chem. **160**, 535—546 (1945).
— F. ALBRIGHT, A. H. SALTZMAN, A. S. ZYGMUNTOWICZ and R. L. WIXOM: The excretion of 11-oxycorticosteroid-like substances by normal and abnormal subjects. J. clin. Endocr. **7**, 331—350 (1947).
— M. S. WOOD, J. WORCESTER, E. CHRISTO, A. M. CAMPBELL and A. S. ZYGMUNTOWICZ: Further observations on urinary excretion of water-soluble corticosteroids by normal and abnormal subjects. J. clin. Endocr. **11**, 1224—1236 (1951).
— E. H. SOBEL, J. W. MCARTHUR and J. D. CRAWFORD: Functional Endocrinology from Birth through Adolescence. Cambridge: Harvard University Press 1952.
TALBOTT, J. H., W. F. LEVER and W. V. CONSOLAZIO: Metabolic studies on patients with pemphigus. J. invest. Derm. **3**, 31—68 (1940).
— L. J. PECORA, R. S. MELVILLE and W. V. CONSOLAZIO: Renal function in patients with Addison's disease and in patients with adrenal insufficiency secondary to pituitary panhypofunction. J. clin. Invest. **21**, 107—119 (1942).
TAUSSKY, H., R. SWAN and E. SHORR: An inquiry into the specificity of the uric acid-creatinine ratio as a measure of adrenal cortical responsiveness. In: J. R. MOTE, Ed., Proc. Second ACTH Conference, Vol. I, 273—281. Philadelphia: Blakiston 1951.
TAYLOR, A. B., A. ALBERT and R. G. SPRAGUE: Adrenotrophic activity of human blood. Endocrinology **45**, 335—343 (1949).
TAYLOR, D. W.: Effects of adrenalectomy on oxygen poisoning in the rat. J. Physiol. (Lond.) **140**, 23—26 (1958).
THER, L., u. M. MUELLER: Kinetik der Glukogenphosphorylierung bei Nebenniereninsuffizienz und Dinitrophenolvergiftung. Naunyn-Schmiedeberg's Arch. exp. Path. Pharmak. **209**, 194—203 (1950).
THOMPSON, D. D., and R. F. PITTS: Effects of alterations of renal arterial pressure on sodium and water excretion. Amer. J. Physiol. **168**, 490—499 (1952).
THORN, G. W.: Adrenal cortical hormone therapy. Amer. J. med. Sci. **197**, 718—729 (1939).
— Clinical use of extracts from adrenal cortex. J. Amer. med. Ass. **125**, 10—15 (1944).
— The Diagnosis and Treatment of Adrenal Insufficiency, Ed. 2. Springfield: Charles C. Thomas 1951.
— and M. CLINTON, JR.: Metabolic changes in a patient with Addison's disease following the onset of diabetes mellitus. J. clin. Endocr. **3**, 335—344 (1943).
— H. R. GARBUTT, F. A. HITCHCOCK and F. A. HARTMAN: The effect of cortin on the sodium, chloride, inorganic phosphorus and total nitrogen balance in normal subjects and in patients with Addison's disease. Endocrinology **21**, 202—212 (1937a).
— — — — The effect of cortin upon the renal excretion of sodium, potassium, chloride, inorganic phosphorus and total nitrogen in normal subjects and in patients with Addison's disease. Endocrinology **21**, 213—219 (1937b).

THORN, G. W., R. P. HOWARD, K. EMERSON, JR., and W. M. FIROR: Treatment of Addison's disease with pellets of crystalline adrenal cortical hormone (synthetic desoxy-corticosterone acetate) implanted subcutaneously. Bull. Johns Hopk. Hosp. **64**, 339—365 (1939a).

— — — Treatment of Addison's disease with desoxy-corticosterone acetate, a synthetic adrenal cortical hormone (preliminary report). J. clin. Invest. **18**, 449—467 (1939b).

— — and H. DAYMAN: Electrolyte changes in pulmonary tuberculosis with special reference to adrenal cortical function. Bull. Johns Hopk. Hosp. **67**, 345—364 (1940a).

— G. F. KOEPF, R. A. LEWIS and E. F. OLSEN: Carbohydrate metabolism in Addison's disease. J. clin. Invest. **19**, 813—832 (1940b).

— R. L. GREIF, S. O. CUTINO and H. EISENBERG: Relative effectiveness of several methods of administering desoxycorticosterone acetate. J. clin. Endocr. **1**, 967—976 (1941).

— S. S. DORRANCE and E. DAY: Addison's disease: evaluation of synthetic desoxycorticosterone acetate therapy in 158 patients. Ann. intern. Med. **16**, 1053—1059 (1942).

— G. F. KOEPF and M. CLINTON, JR.: Renal failure simulating adrenocortical insufficiency. New Engl. J. Med. **231**, 76—85 (1944).

— P. H. FORSHAM, F. T. G. PRUNTY and A. G. HILLS: A test for adrenal cortical insufficiency. J. Amer. med. Ass. **137**, 1005—1009 (1948).

— — L. L. BENNETT, M. ROCHE, R. S. REISS, A. SLESSOR, E. B. FLINK and W. SOMERVILLE: Clinical and metabolic changes in Addison's disease following administration of compound E acetate. Trans. Ass. Amer. Phycns **62**, 233—244 (1949).

— — T. F. FRAWLEY, D. L. WILSON, A. E. RENOLD, D. S. FREDRICKSON and D. JENKINS: Advances in the diagnosis and treatment of adrenal insufficiency. Amer. J. Med. **10**, 595—611 (1951).

— D. JENKINS, J. C. LAIDLAW, F. C. GOETZ, J. F. DINGMAN, W. L. ARONS, D. H. P. STREETEN and B. H. McCRACKEN: Pharmacologic aspects of adrenocortical steroids and ACTH in man. New Engl. J. Med. **248**, 232—245, 284—294, 323—337, 369—378, 414—423, 588—601, 632—646 (1953a).

— — W. L. ARONS and T. F. FRAWLEY: Use of desoxycorticosterone trimethylacetate in the treatment of Addison's disease. J. clin. Endocr. **13**, 957—973 (1953b).

— R. H. SHEPPARD, W. L. MORSE, W. J. REDDY, P. M. BIEGELMAN and A. E. RENOLD: Comparative action of aldosterone and 9-α-fluorohydrocortisone in man. Ann. N. Y. Acad. Sci. **66**, 609—619 (1955).

— D. H. NELSON and A. E. RENOLD: Current status of the treatment of adrenal disorders. J. Amer. med. Ass. **168**, 2130—2137 (1958).

TIMIRAS, P. S., D. M. WOODBURY and L. S. GOODMAN: Effect of adrenalectomy, hydrocortisone acetate and desoxycorticosterone acetate on brain excitability and electrolyte distribution in mice. J. Pharmacol. **112**, 80—93 (1954).

TIPTON, S. R.: The effect of extracts of the adrenal glands on the respiration of isolated brain and liver slices. Amer. J. Physiol. **127**, 710—715 (1939).

— The effects of adrenalectomy on the activity of cytochrome oxidase and the concentration of cytochrome c in rats. Endocrinology **34**, 181—186 (1944).

— M. J. LEATH, I. H. TIPTON and W. L. NIXON: The effects of feeding thyroid substance and of adrenalectomy on the activities of succinoxidase and cytochrome oxidase in the liver tissue of rats. Amer. J. Physiol. **145**, 693—698 (1946).

TOBIN, C. E.: The effect of adrenalectomy on pregnancy and survival of untreated and sesame oil treated rats. Endocrinology **28**, 419—425 (1941).

TOLKSDORF, S., M. L. BATTIN, J. W. CASSIDY, R. M. MACLEOD, F. H. WARREN and P. L. PERLMAN: Adrenocortical properties of $\Delta^{1,4}$-pregnadiene-17 α, 21-diol-3,11,20-trione (Meticorten) and $\Delta^{1,4}$-pregnadiene-11β,17 α, 21-triol-3,20-dione (Meticortelone). Proc. Soc. exp. Biol. (N. Y.) **92**, 207—214 (1956).

TOOKE, T. B., JR., M. H. POWER and E. J. KEPLER: The tolerance of patients suffering from Addison's disease to potassium while such patients are being treated with desoxycorticosterone acetate. Proc. Mayo Clin. **15**, 365—368 (1940).

TORDA, C., and H. G. WOLFF: Effect of pituitary hormones, cortisone and adrenalectomy on some aspects of neuromuscular function and acetylcholine synthesis. Amer. J. Physiol. **169**, 140—149 (1952a).

— — Effects of various concentrations of adrenocorticotrophic hormone on electrical activity of brain and on sensitivity to convulsion-inducing agents. Amer. J. Physiol. **168**, 406—413 (1952b).

TOWNE, J. C., and J. O. SHERMAN: Failure of acute bilateral adrenalectomy to influence brain serotonin levels in the rat. Proc. Soc. exp. Biol. (N. Y.) **103**, 721—722 (1960).

TREAGER, H. S., G. J. GABUZDA, N. ZAMCHECK and C. S. DAVIDSON: Response to adrenocorticotrophic hormone in clinical scurvy. Proc. Soc. exp. Biol. (N. Y.) **75**, 517—520 (1950).

TRENDELENBURG, P.: Die Hormone, Ihre Physiologie und Pharmakologie, Vol. 1. Berlin: Springer 1929.

TRENTIN, J. J., and C. W. TURNER: Effect of adrenalectomy on the mammary gland of the castrated and estrogen treated castrated male rat. Endocrinology **41**, 127—134 (1947).
TROUSSEAU: Capsules surrénales (Maladie bronzée d'Addison). Arch. gén. Méd. 8, 478—481 (1856).
TUERKISCHER, E., and E. WERTHEIMER: Adrenalectomy and gastric secretion. J. Endocr. **4**, 143—151 (1945).
TURCATTI, E. S.: The suprarenals and pancreatic diabetes (in Spanish). Rev. méd. Rosario **19**, 269—279 (1929).
TUREMAN, J. R., W. M. BOOKER and C. FROIX: Response of adrenalectomized rats to pentobarbital sodium. Amer. J. Physiol. **167**, 833 (1951).
— A. H. MALONEY, W. M. BOOKER, C. FROIX and W. JONES: Further studies on the depressant action of pentobarbital sodium on adrenalectomized rats. J. Pharmacol. exp. Ther. **106**, 420 (1952).
UNGAR, F., G. ROSENFELD, R. I. DORFMAN and G. PINCUS: Irradiation and adrenal steroidogenesis. Endocrinology **56**, 30—36 (1955).
UNVERRICHT: Insulinempfindlichkeit und Nebenniere. Dtsch. med. Wschr. **52**, 1298—1299 (1926).
VACCARI, F., and M. ROSSANDA: Upon the modifications induced by cortisone and desoxycorticosterone in the glycogen and the total carbohydrates in the brains of normal and adrenalectomized animals (in Italian). Boll. Soc. ital. Biol. sper. **27**, 734—736 (1951).
VALETTE, G., et H. HUIDOBRO: Effets de la surrénalectomie sur la teneur en histaminase du sang et de l'intestin chez le rat. J. Physiol. (Paris) **44**, 339—341 (1952).
VANDER, A. J., W. S. WILDE and R. L. MALVIN: Stop flow analysis of aldosterone and steroidal antogonist SC-8109 on renal tubular sodium transport kinetics. Proc. Soc. exp. Biol. (N. Y.) **103**, 525—527 (1960).
VENNING, E. H., and J. S. L. BROWNE: Excretion of glycogenic corticoids and of 17-ketosteroids in various endocrine and other disorders. J. clin. Endocr. **7**, 79—101 (1947).
— J. P. RANDALL and P. GYORGY: Excretion of glucocorticoids in the newborn. Endocrinology **45**, 430—434 (1949).
— C. J. PATTEE and J. C. BECK: The adrenal cortex. In: Glandular Physiology and Therapy, Chap. 4. Philadelphia: Lippincott 1954.
VERZÁR, F., u. E. SAILER: Glukoseresorption und alkalische Phosphatase des Dünndarmes nach Adrenalektomie, bei mit NaCl behandelten Tieren. Helv. physiol. pharmacol. Acta **10**, 247—258 (1952).
VICTOR, M.: Über plötzliche Todesfälle im Säuglingsalter als Folge von akuter Nebenniereninsuffizienz. Z. Kinderheilk. **30**, 44—54 (1921).
VIDOVIC, V. L., and V. POPOVIC: Studies on the adrenal and thyroid glands of the ground squirrel during hibernation. J. Endocr. **11**, 125—133 (1954).
VINCENT, SWALE: Internal Secretion and the Ductless Glands. Ed. 2. London: Arnold 1922.
VOELCKER, A. F.: Pathological reports, 1894. Middlesex Hosp. Rep. Reg., 244—406 (1895).
VOGT, M.: Chronic suprarenal deficiency and its effects on the responses of the isolated intestine in the rabbit. J. Physiol. (Lond.) **102**, 239—257 (1943a).
— The output of cortical hormone by the mammalian suprarenal. J. Physiol. (Lond.) **102**, 341—356 (1943b).
WADE, N. J., and L. A. HAZELWOOD: Effect of removal of ovaries and adrenals on opening of vagina in the albino rat. Endocrinology **28**, 624—628 (1941).
WALKER, S. M.: The response of the triceps surae of normal, adrenalectomized, desoxycorticosterone acetate-treated and KCl-treated rats to direct and indirect, single and repetitive stimulation. Amer. J. Physiol. **149**, 7—23 (1947).
— Action potentials in rat muscle with twitch tension potentiated by KCl treatment, adrenalectomy, tetanus and treppe. Amer. J. Physiol. **154**, 63—72 (1948).
WALLACE, E. Z., N. P. CHRISTY and J. W. JAILER: Clinical application of the simplified Silber-Porter method for determining plasma 17-hydroxycorticosteroids. J. clin. Endocr. **15**, 1073—1088 (1955).
WALLACH, J. B., and W. B. SCHARFMAN: Addison's disease due to metastatic bronchiogenic carcinoma. J. Amer. med. Ass. **148**, 729—731 (1952).
WALSH, F. B.: Papilledema associated with increased intracranial pressure in Addison's disease. Arch. Ophthal. **47**, 86 (1952).
WALTERS, W., R. M. WILDER and E. J. KEPLER: The suprarenal cortical syndrome with presentation of ten cases. Ann. Surg. **100**, 670—688 (1934).
WARMING-LARSEN, A., and M. SPRECHLER: Clinical and metabolic studies in a case of Addison's disease treated with desoxycorticosterone acetate. Acta endocr. (Kbh.) **12**, 35—40 (1953).
WATERHOUSE, C., and E. H. KEUTMANN: Kidney function in adrenal insufficiency. J. clin. Invest. **27**, 372—379 (1948).
WATERHOUSE, R.: A case of suprarenal apoplexy. Lancet **1911 i**, 577—578.

WATERMAN, L., I. E. UYLDERT, J. THOMASSEN, and F. OESTREICHER: The examination of the blood of normal and adrenalectomized dogs in relation to cortin treatment. Endocrinology **25**, 885—887 (1939).

WEBSTER, B., J. J. PFIFFNER and W. W. SWINGLE: Effect of adrenal cortical hormone upon respiratory metabolism of normal, thyroidectomized and thyroidectomized-adrenalectomized cats. Proc. Soc. exp. Biol. (N. Y.) **28**, 1021—1022 (1931).

WEINER, H. A.: So-called atrophy of the adrenal cortex with intranuclear inclusions. Amer. J. Path. **12**, 411—422 (1936).

WEISKOTTEN, H. G.: Fatal superficial burns and the suprarenals. J. Amer. med. Ass. **69**, 776 (1917).

WELBOURN, R. B., and C. F. CODE: Effects of cortisone and of adrenalectomy on secretion of gastric acid and on occurrence of gastric ulceration in the pylorus-ligated rat. Gastroenterology **23**, 356—362 (1953).

WELCH, R. G.: Addison's disease in a nine-year-old girl. Brit. med. J. **1957 i**, 980—982.

WELLS, B. B.: The influence of crystalline compounds separated from the adrenal cortex on gluconeogenesis. Proc. Mayo Clin. **15**, 294—297 (1940).

— and E. C. KENDALL: A quantitative difference in the effect of compounds separated from the adrenal cortex on distribution of electrolytes and on atrophy of the adrenal and thymus glands of rats. Proc. Mayo Clin. **15**, 133—139 (1940).

WELLS, H. G.: Addison's disease with selective destruction of the suprarenal cortex ("suprarenal cortex atrophy"). Arch. Path. **10**, 499—523 (1930).

WELSH, J. B., and G. B. MEHLIN: Congenital adrenal aplasia. Amer. J. Dis. Child. **87**, 319—320 (1954).

WEST, H. F.: Discussion. In: J. M. YOFFEY, Ed., Proc. Fifth Symp. Colston Res. Soc., p. 226. London: Butterworths 1953.

WEST, K. M.: Adrenal cortical insufficiency associated with eosinopenia; report of a case. J. clin. Endocr. **13**, 791—794 (1953).

WHEELER, T. D., and S. VINCENT: The question as to the relative importance to life of cortex and medulla of the adrenal bodies. Trans. roy. Soc. Can. **11**, 125—127 (1917).

WHITAKER, J. A., B. H. LANDING, V. M. ESSELBORN and R. WILLIAMS: The syndrome of familial juvenile hypoadrenocorticism, hypoparathyroidism and superficial moniliasis. J. clin. Endocr. **16**, 1374—1387 (1956).

WHITE, A., P. HANDLER, E. L. SMITH and DE W. STETTEN, JR.: Principles of Biochemistry, Ed. 2. New York: McGraw-Hill Book Co. 1959.

WHITE, A. G., P. S. ENTMACHER, G. RUBIN and L. LEITER: Physiological and pharmacological regulation of human salivary electrolyte concentrations; with a discussion of electrolyte concentrations of some other exocrine secretions. J. clin. Invest. **34**, 246—255 (1955).

WHITE, H. L., and D. ROLF: Whole tissue electrolyte analyses in normal and adrenalectomized rats. Amer. J. Physiol. **180**, 287—295 (1955).

— P. HEINBECKER and D. ROLF: Further observations on the depression of renal function following hypophysectomy. Amer. J. Physiol. **156**, 67—78 (1949).

WILBUR, D. M. J., and A. R. RICH: A study of the role of adrenocorticotropic hormone (ACTH) in the pathogenesis of tubular degeneration of the adrenals. Bull. Johns Hopk. Hosp. **93**, 321—347 (1953).

WILDER, R. M.: Progress in treatment of Addison's disease. Proc. Mayo Clin. **15**, 273—277 (1940).

— A. M. SNELL, E. J. KEPLER, E. H. RYNEARSON, M. ADAMS and E. C. KENDALL: Control of Addison's disease with a diet restricted in potassium: a clinical study. Proc. Mayo Clin. **11**, 273—283 (1936).

— E. C. KENDALL, A. M. SNELL, E. J. KEPLER, E. H. RYNEARSON and M. ADAMS: Intake of potassium, an important consideration in Addison's disease. Arch. intern. Med. **59**, 367 to 393 (1937).

WILKINS, L.: The Diagnosis and Treatment of Endocrine Disorders in Childhood and Adolescence, Ed. 2. Springfield: Charles C Thomas 1957.

— and C. P. RICHTER: A great craving for salt by a child with cortico-adrenal insufficiency. J. Amer. med. Ass. **114**, 866—868 (1940).

WILKS, S.: Morbus Addisonii, melanaemia, anaemia idiopathica, leucocythemia splenica et lymphatica. Guy's Hosp. Rep. **5**, 89—118 (1859).

— On disease of the suprarenal capsules; or Morbus Addisonii. Guy's Hosp. Rep. 8, 1—63 (1862).

— Historical notes on Bright's disease, Addison's disease and Hodgkin's disease. Guy's Hosp. Rep. **22**, 259—274 (1877).

WILLIAMS, A., and M. J. ROBINSON: Addison's disease in infancy. Arch. Dis. Childh. **31**, 265—269 (1956).

WILLIAMS, G. A., K. R. CRISPELL and W. PARSON: Adrenocortical function in myxedema. J. clin. Endocr. **17**, 1347—1353 (1957).
WILLIAMS, H.: Meningococcal infections in infancy and childhood — Part II. Med. J. Aust. **2**, 557—561 (1942).
WILLIAMS, R. H., J. L. WHITTENBERGER, G. W. BISSELL and A. R. WEINGLASS: Treatment of adrenal insufficiency. J. clin. Endocr. **5**, 163—180 (1945).
WILLSON, D. M., and F. W. SUNDERMAN: Studies in serum electrolytes. XII. The effect of water restriction in a patient with Addison's disease receiving sodium chloride. J. clin. Invest. **18**, 35—43 (1939).
— F. J. ROBINSON, M. H. POWER and R. M. WILDER: Diagnosis of Addison's disease; further experience with the Cutler-Power-Wilder sodium restriction test. Arch. intern. Med. **69**, 460—469 (1942).
WILSON, D. A. W., and D. ROTH: Adrenal apoplexy occurring during corticotropin therapy of ulcerative colitis. J. Amer. med. Ass. **152**, 230—231 (1953).
WINKLER, A. W., and O. F. CRANKSHAW: Chloride depletion in conditions other than Addison's disease. J. clin. Invest. **17**, 1—6 (1938).
WINTERS, R. W., R. B. SCHULTZ and W. A. KREHL: The adrenal cortex of the pantothenic acid-deficient rat: eosinophile and lymphocyte responses. Endocrinology **50**, 377—384 (1952a).
— — — The adrenal cortex of the pantothenic acid-deficient rat: carbohydrate metabolism. Endocrinology **50**, 388—398 (1952b).
— — — Studies on adrenal cortex of pantothenic acid-deficient rats. IV. Adrenal and serum cholesterol levels. Proc. Soc. exp. Biol. (N. Y.) **79**, 695—696 (1952c).
WOLFSON, W. Q., H. S. GUTERMAN, R. LEVINE, C. COHEN, H. D. HUNT and E. F. ROSENBERG: An endocrine finding apparently characteristic of gout: Very low 17-ketosteroid excretion with clinically normal androgenic function. J. clin. Endocr. **9**, 497—513 (1949).
WOODBURY, D. M.: Extrarenal effects of desoxycorticosterone, adrenocortical extract and adrenocorticotrophic hormone on plasma and tissue electrolytes in fed and fasted rats. Amer. J. Physiol. **174**, 1—19 (1953).
— Effects of hormones on brain excitability and electrolytes. Recent Progr. Hormone Res. **10**, 65—107 (1954).
— Relation between the adrenal cortex and the central nervous system. Pharmacol. Rev. **10**, 275—357 (1958).
WRIGHT, D. A., and L. B. REPPERT: Fulminating meningococcemia with vascular collapse (Waterhouse-Friderichsen syndrome); report on four adult patients who recovered. Arch. intern. Med. **77**, 143—150 (1946).
WRIGHT, E., and E. J. LESTER: Effect of adrenalectomy and cortisone on peripheral nerve function. Amer. J. Physiol. **196**, 1057—1062 (1959).
— and W. ADELMAN: Accomodation in three single motor axons of the crayfish claw. J. cell. comp. Physiol. **43**, 119—132 (1954).
— and P. COLEMAN: Excitation and conduction in crustacean single motor axons. J. cell. comp. Physiol. **43**, 133—164 (1954).
WYMAN, L. C.: Studies on suprarenal insufficiency. III. The effect of thyroidectomy and of forcing fluid on suprarenal insufficiency in rats. Amer. J. Physiol. **87**, 42—50 (1928).
— The effect of adrenalectomy on propulsive motility of the small intestine. Proc. Soc. exp. Biol. (N. Y.) **84**, 303—304 (1953).
— and C. TUM-SUDEN: The effect of adrenalectomy on the epiphyseal cartilage of the rat. Endocrinology **36**, 340—346 (1945).
— and B. S. WALKER: Studies on suprarenal insufficiency. IV. The blood sugar in suprarenalectomized rats. Amer. J. Physiol. **89**, 215—222 (1929).
— G. P. FULTON, F. N. SUDAK and G. N. PATTERSON: Electrocardiograms and serum electrolyte levels in hamsters with adrenal insufficiency and hypercorticalism. Proc. Soc. exp. Biol. (N. Y.) **84**, 280—283 (1953a).
— — and M. H. SHULMAN: Direct observations on the circulation in the hamster cheek pouch in adrenal insufficiency and experimental hypercorticalism. Ann. N. Y. Acad. Sci. **56**, 643—658 (1953b).
WYNN, V., and C. G. ROB: Water intoxication. Differential diagnosis of the hypotonic syndromes. Lancet **1954 i**, 587—594.
YONKMAN, F. F.: Acid intoxication of adrenal insufficiency in dogs. Proc. Soc. exp. Biol. (N. Y.) **24**, 786—789 (1926).
— The acid intoxication of adrenal insufficiency in dogs. Amer. J. Physiol. **86**, 471—482 (1928).
YOUNG, I. I., V. DE FILIPPIS, F. L. MEYER and W. G. WOLFSON: Maintenance of adrenal cortical responsiveness during prolonged corticoid therapy. Arch. intern. Med. **100**, 1—10 (1957).

ZAHLER, H., u. G. LITZKA: Über klinische Versuche mit synthetischen Corticosteron (Cortenil). Münch. med. Wschr. **86**, 446 (1939).
ZAK, S. J., R. H. GOOD and R. A. GOOD: A technique for one-stage bilateral adrenalectomy in the rabbit. Nature (Lond.) **179**, 100—102 (1957).
ZECKWER, I. T.: The accelerating effect of adrenalectomy on regrowth of hair in thyroidectomized rat. Endocrinology **52**, 39—48 (1953a).
— The accelerating effect of adrenalectomy on regrowth of hair in the rat: regrowth of hair in parabiotic rats, one partner being adrenalectomized. Endocrinology **53**, 326—331 (1953b).
ZOELCKLER, S. J., and J. A. ORBISON: Panhypopituitarism following epidemic hemorrhagic fever. I. Clinical features: report of a case. Ann. intern. Med. **43**, 1316—1319 (1955).
ZOLOTAREVA, V. S., and M. M. GISSINA: A case of chronic suppurative inflammation of the adrenals (in Russian). Klin. Mad. (Moskva) **2**, 74—76 (1956); Abstracts Soviet Med. **1**, 350—351 (1957).
ZUCKERMAN, S.: The adreno-genital relationship. In: J. M. YOFFE, Ed., Proc. Fifth Symposium Colston Research Society, 69—83. London: Butterworths 1953.
ZWEIFACH, B. W., E. SHORR and M. M. BLACK: The influence of the adrenal cortex on behavior of terminal vascular bed. Ann. N. Y. Acad. Sci. **56**, 626—633 (1953).
ZWEMER, R. L.: An experimental study of the adrenal cortex. I. The survival value of the adrenal cortex. Amer. J. Physiol. **79**, 641—657 (1926/27a).
— An experimental study of the adrenal cortex. II. Prolongation of life after complete epinephrectomy. Amer. J. Physiol. **79**, 658—665 (1926/27b).
— The adrenal cortex and electrolyte metabolism. Endocrinology **18**, 161—169 (1934).
— and C. LYONS: Leucocyte changes after adrenal removal. Amer. J. Physiol. **86**, 545—551 (1928).
— and R. C. SULLIVAN: Corticoadrenal influence on blood sugar mobilization. Endocrinology **18**, 730—738 (1934).
— and R. TRUSZKOWSKI: Potassium: a basal factor in the syndrome of corticoadrenal insufficiency. Science **83**, 558—560 (1936).
— — The importance of corticoadrenal regulation of potassium metabolism. Endocrinology **21**, 40—49 (1937).

Author Index

Page numbers in *italics* refer to the bibliography

Subject Index

MIX
Papier aus verantwortungsvollen Quellen
Paper from responsible sources
FSC® C105338

If you have any concerns about our products,
you can contact us on
ProductSafety@springernature.com

In case Publisher is established outside the EU,
the EU authorized representative is:
Springer Nature Customer Service Center GmbH
Europaplatz 3, 69115 Heidelberg, Germany

Printed by Libri Plureos GmbH
in Hamburg, Germany